ELECTROPHORESIS

Edited by **Kiumars Ghowsi**

Electrophoresis
http://dx.doi.org/10.5772/2905
Edited by Kiumars Ghowsi

Contributors

Kiumars Ghowsi, Hosein Ghowsi, Constantina P. Kapnissi-Christodoulou, Tomasz Piasecki, Aymen Ben Azouz, Brett Paull, Mirek Macka, Dermot Brabazon, Elsa Lamy, Ana R. Costa, Célia M. Antunes, Rui Vitorino, Francisco Amado, Ying Zhou, Vibeke Simonsen, Dario G. Pighin, Lucía González-Arenzana, Rosa López, Pilar Santamaría, Isabel López-Alfaro, Malgorzata K. Sulkowska, Marcelo G.M. Chacur, Theopoline Omagano Itenge, Flavia Simone Munin, Maria Regina de Aquino-Silva, Maria Luiza Barcellos Schwantes, Vera Maria Fonseca de Almeida-Val, Arno Rudi Schwantes, Yoshimi Sato

Published by InTech

Janeza Trdine 9, 51000 Rijeka, Croatia

Publishing Process Manager Marina Jozipovic
Technical Editor InTech Prepress, Novi Sad
Cover InTech Design team

Additional hard copies can be obtained from orders@intechopen.com

Electrophoresis, Edited by Kiumars Ghowsi
p. cm.
ISBN 978-953-51-0846-7

For EU product safety concerns:
IN TECH d.o.o., Prolaz Marije Krucifikse Kozulić 3, 51000 Rijeka, Croatia,
info@intechopen.com or visit our website at intechopen.com.

Meet the editor

Professor Kiumars Ghowsi obtained a B.S. degree in Electrical Engineering Science in 1981, and a M.S. degree in Electrical Engineering Science in 1984 from Louisiana State University, USA. He changed his field to chemistry and obtained a Ph.D with a major in analytical chemistry and a minor in physical chemistry from Louisiana State University, USA in 1990. Currently Prof. Ghowsi is a member of IEEE, and a faculty member at Texas Tech Chemistry and Biochemistry Department USA, and Isfahan Technological University, Iran. In addition, Prof. Ghowsi is a faculty member in Chemistry Department, Majlesi branch Islamic Azad University, Iran. His research interest are lab on chip, Insulator Electrochemistry liquid state, interfacial phenomena, electronic properties of biological molecules and HPCE (High Performance Capillary Electrophoresis). Prof. Ghowsi is cited frequently for the discovery of novel phenomena in the capillary called field effect electroosmosis and invention of the first liquid state transistor Flow FET (Field Effect Transistor).

Contents

Preface

Electrophoresis experiment was first carried out by Tiselius in 1930. In his thesis titled, "The Moving Boundary Method to Study the Electrophoresis of Proteins", Tiselius utilized the electric charge carried by the macromolecules to achieve some pioneering separation of blood plasma proteins in free solution on a photographic film.

Electrophoresis is defined as the transport of electrically charged particles in a direct current electric field. Electrophoretic separation is based on differential rates of migration in the bulk of the liquid phase and is not concerned with reactions occurring at the electrodes. In the early days, electrophoresis was carried out either in free solution or in the supporting media such as paper, cellulose acetate, starch, agarose, and polyacry lamide gel. In between 1950 to 1970, an enumeration of techniques and instrumentation for Electrophoresis were developed.

Gel electrophoresis has been rarely used for the separation and identification of small charged molecules of molecular weight less than about 1000 Dalton. In addition, the major draw back of gel electrophoresis is lack of complete automation.

To overcome the low efficiency and reduce thermal effects, Hjerten carried out electrophoresis in narrow diameter tubes of 300 μm internal diameter the first time in 1967. This was the birth of open tubular capillary electrophoresis. However, in the following decade, capillary electrophoresis did not draw enough attention from researchers. Only until 1981 when Jorgenson and Lukacs demonstrated the use of narrow capillaries to produce high efficiency for the separation of dansy and fluorescamine derivatives of amino acids, dipeptides and simple amines, high performance capillary electrophoresis was born and a new era of capillary electrophoresis has begun. After the introduction of commercial CE instrument in late 1988 that allowed the full automation of CE analysis to be possible, more and more research publications and industrial applications have made capillary electrophoresis one of the dominant technologies in the separation field. In 1985 Terabe et. al added a new dimension to capillary electrophoresis. They added micelles to the aqueous electrolyte and were able to separate neutral molecules such as benzene and phenol.

Using this technique it is possible to separate various drugs which are neutral or even charged. One that can separate enantiomers using this technique is called micellar electrokinetic capillary chromatography (MECC).

The present book contains few fundamentals on capillary electrophoresis and diverse application of electrophoresis in general. We hope this collection will entertain the readers who are interested in fundamental as well as applications of electrophoresis in general.

Dr. Kiumars Ghowsi
Majlessi Branch, Islamic Azad University,
Iran

Chapter 1

New Looks at Capillary Zone Electrophoresis (CZE) and Micellar Electrokinetic Capillary Chromatography (MECC) and Optimization of MECC

Kiumars Ghowsi and Hosein Ghowsi

Additional information is available at the end of the chapter

http://dx.doi.org/10.5772/45760

1. Introduction

The origin of theoretical plates for electrophoresis in general sense has been obtained from the work of Giddings[1]. Giddings has derived the following equation for the number of theoretical plates.

$$N = \frac{-\Delta\mu^{ext}}{2\Theta RT} \tag{1}$$

where coefficient Θ=unity when molecular diffusion acts alone and exceed unity when other process contributes. R=gas constant and T=temperature and μ^{ext}=chemical potential conventionally substution of $\Delta\mu^{ext} = ZFV$ into eqn.(1) gives for the number of theoretical plates.

$$N = \frac{ZFV}{2RT} \tag{2}$$

For an ideal process, in which Θ=1 and T=298K,this equation reduces

$$\mathrm{N} = 20\mathrm{ZV} \tag{3}$$

where we have used the Faraday constant F=96500 coulombs/mol. This voltage drops V in the range of 100-10000 v with charge number Z=1 capable of yielding 2000-200000 theoretical plates, a range comparable to that found in chromatographic system. Jorgenson and Lukacs[2], also in their land mark paper provided a theory for capillary zone electrophoresis (CZE) in which they proposed two fundamental equations for resolution and migration time. Resolution and number of theoretical plates are the focus of this work. This is how number of theoretical plates are derived for capillary zone electrophoresis and used in resolution equation by Jorgenson and Lukacs[2] by starting with Gidding number of theoretical plates.

$$N = \frac{ZFV}{2RT} \tag{4}$$

V=potential across capillary is substituted by EX where E=electric field and X=length of the capillary. Z=charge number for ion is assumed to be 1, hence

$$N = \frac{FFX}{2RT} \tag{5}$$

$$X = t[v_{eo} + (v_{ep})_{AB}] \tag{6}$$

Eqn.6 where $(v_{ep})_{AB} = \frac{1}{2}[(v_{ep})_A + (v_{ep})_B]$ is substituted by X in eq.(5). t is the time it takes the analyte travels across the capillary and v_{eo} is the electroosmotic velocity $(v_{ep})_A$ and $(v_{ep})_B$ are the electrophoretic velocity of two analytes A and B with close electrophoretic velocities. By substitution of equation(6) into equation (5) one obtains

$$N = [(\frac{E}{2RT})Et[v_{eo} + (v_{ep})_{AB}]] \tag{7}$$

There is mistake occurs in Jorgenson and Lukacs work[2], t in eqn. (7) is replaced by eqn.(8), this is the case when electroomosis is absent.

$$t = \frac{X}{v_{ep}} \tag{8}$$

By making this mistake substitution eqn (8) , what has been obtained by Jorgenson and Lukacs for N , eqn.(2) is

$$N = \frac{F}{2RT}E[(\frac{X}{v_{ep}})[(v_{eo}) + (v_{ep})_{AB}]] \tag{9}$$

E, X, is the V, voltage across the capillary and v_{ep} the electrophoretic velocity of the analyte and v_{eo} the electroosmotic flow in the capillary are substituted in eqn.(9) by $\mu_{ep}E$ and $\mu_{eo}E$ where μ_{ep} and μ_{eo} are the electrophoretic and electroosmotic mobilities and E is the electric field. By this substitution the following eqn.(10) is obtained for N

$$N = (\frac{FV}{2RT})[\frac{\mu_{eo} + (\mu_{ep})_{AB}}{(\mu_{ep})_{AB}}] \tag{10}$$

By using Einstain relation eqns.(11) and (12) we substitute for μ_{ep} from eqn.(12)

$$\frac{D}{\mu_{ep}} = \frac{RT}{F} \tag{11}$$

$$\mu_{ep} = \frac{FD}{RT} \tag{12}$$

Where D is the Diffusion constant of the analyte by this substitution and simplification eqn.13 is obtained

$$N = \frac{V}{2D}[\mu_{eo} + (\mu_{ep})_{AB}] \tag{13}$$

This is the equation for the number of theoretical plates the Jorgenson and Lukacs use to obtain the resolution equation. N is driven by mistake[2]. The resolution equation is given by them

$$R_s = \frac{(\mu_{ep})_A - (\mu_{ep})_B}{\mu_{eo} + (\mu_{ep})_{AB}} . \frac{\sqrt{N}}{4} \tag{14}$$

substituting eqn.(13) for N into eqn.(14), the resolution can be expressed as

$$R_s = \frac{((\mu_{ep})_A - (\mu_{ep})_B)}{(\mu_{eo} + (\mu_{ep})_{AB})^{1/2}} . \frac{\sqrt{V}}{4\sqrt{2D}} \tag{15}$$

The above eqn.(15) is the same equation that was obtained by Jorgenson and Lukacs[2].

This equation is incorrect. In the present work equation for resolution which is cosidered correct equation is proposed.

2. Theory

According to Gidding's theory[1] the evaluation of the ultimate capabilities of zone electrophoresis is possible. To calculate the number of theoretical plates and separable zone achieveable in ideal zone electrophoresis, the electrostatic force exerted on a mole of charged particles on electric field of strength E is

$$\text{Force} = \text{ZFE} = \text{f} \tag{16}$$

where z=net charge of a single particle in proton units and F=Faraday constant the negative chemical potential drop across the separation path.

$$-\Delta\mu^{ext} = fX' \tag{17}$$

In which X′ is the distance where f, force applied in capillary eletrophoresis. For conventional mode of capillary zone electrophoresis, electroosmotic and electrophoretic velocities are in opposite direction. This conventional mode is similar to tread mill and electric stairs where the moveing object has two movements one walking and the other one movement of the stairs.

There X′ in eqn. (17) is not the length of capillary, it is effective distance where the electric force which is applied is greater than the length of capillary is called

$$-\Delta\mu^{ext} = fX' \tag{18}$$

Then by substituting eqn.(18) into eqn.(1) , the eqn.(19) for the number of theoretical plates results.

$$N = \frac{fX'}{2RT} \tag{19}$$

X′=effective length which looks like a treadmill, the solute is like a person which can run for miles on the mill but actually he has stayed stationary, f=FE force is Faraday constant times electric field. By substituting force in eqn. (19) one gets

$$N = \frac{F}{2RT}EX' \tag{20}$$

By definition, the effective distance the analyte travels under the force of electric field, X′ divided by the relation times, t, is equivalent to electrophoretic $(v_{ep})_{AB}$ velocity , $(v_{ep})_{AB}$, as the following

$$(v_{ep})_{AB} = \frac{X'}{t} \tag{21}$$

Where $(v_{ep})_{AB} = \frac{1}{2}[(v_{ep})_A + (v_{ep})_B]$

By the help of eqn.(21) X′ can be substituted into eqn.(20). Now eqn.(22) can be rewritten to include electrophoretic velocity.

$$N = (\frac{F}{2RT})(v_{ep})_{AB}Et \tag{22}$$

Replacing the electrophoretic velocity variable with the product of electrophoretic mobility and electric field $v_{ep} = \mu_{ep}E$ yields the following expression:

$$N = [(\frac{F}{2RT})][(\mu_{ep})_{AB}E^2t] \tag{23}$$

In order to observe the dependence, efficiency has on capillary length and electroosmosis, a substitution for the time variable in eqn.(21) is made. Net displaccment of the analyte or capillary length, X, is related to the retention time, t as shown

$$N = t[v_{eo}) + (v_{ep})_{AB}] \tag{24}$$

Rearragement of eqn.(24) given eqn.(25)

$$t = \frac{X}{v_{eo} + (v_{ep})_{AB}} \tag{25}$$

substituting eqn.(25) into eqn.(20) yields the expression,

$$N = \frac{F}{2RT}[\frac{(v_{ep})_{AB}}{v_{eo} + (v_{ep})_{AB}}]XE \tag{26}$$

By making additional substitution for electrophoretic and electroosmotic velocity produced, $v_{AB} = (\mu_{ep})_{AB}E$ and $v_{eo} = \mu_{eo}E$, a final equation for efficiency is

$$N = (\frac{F}{2RT})(\frac{(\mu_{ep})_{AB}}{\mu_{eo} + (\mu_{ep})_{AB}})XE \tag{27}$$

plate height is the ratio of effective length X′ to efficiency N is

$$H = \frac{X'}{N} \tag{28}$$

Substituting equation into eqn.(28) yields an expression for plate height.

$$H = \frac{2RT}{FE} \tag{29}$$

This interesting result shows that the theoretical plate height is independent of electroosmotic flow when it is based on the effective distance the analyte travels rather than the capillary length. Instead, plate height has a simple inverse relation with the electric field strength.

Fig .1 Shows the invers relation between theoretical plate height and electric field strength. Electric field strength is the voltage across two ends of capillary divided by the length of capillary.

Equation for resolution: Based width resolution is the quantitative measure of ability to separate two analytes. For two adjacent peaks with similar elution times, peak showed be nearly identical:

$$W_A \approx W_B \approx W_{AB} \tag{30}$$

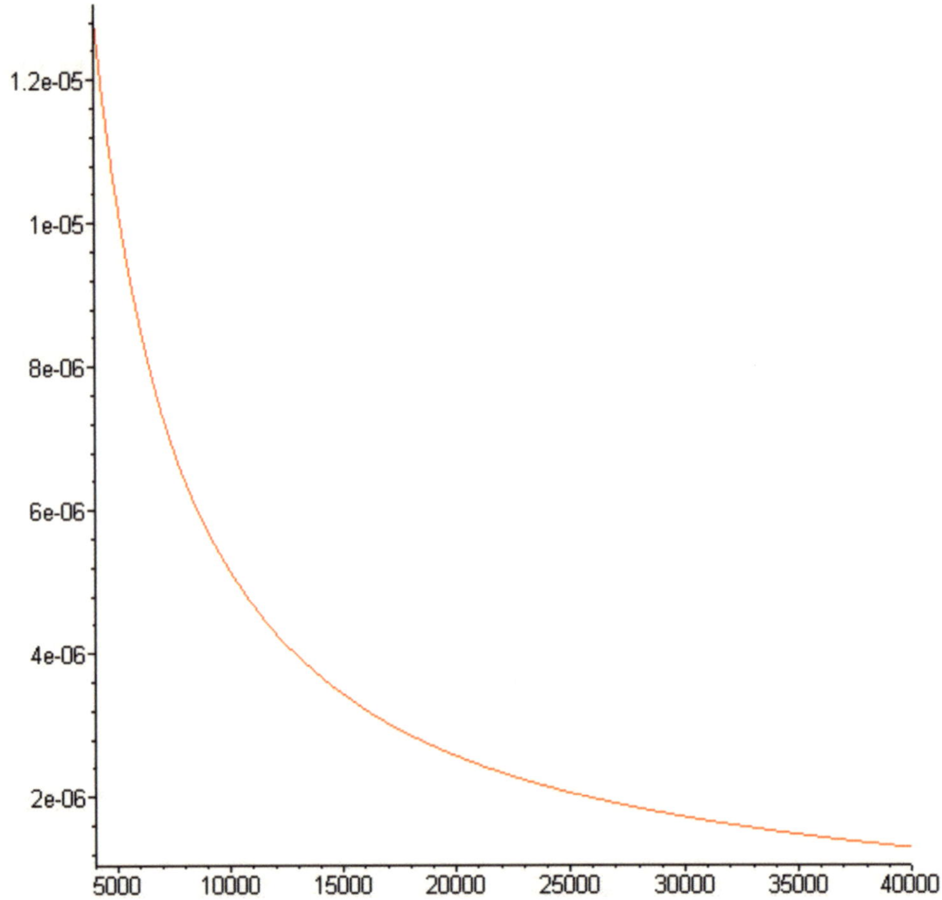

Figure 1. Theoretical plate height versus electric field for CZE.

where $W_{AB} = \frac{1}{2}(W_A + W_B)$.

Assuming eqn.(30) is true, resolution for species A and B expressed in terms of their retention times and the peak base width for either species[2].

$$R_s = \frac{(t_R)_B - (t_R)_A}{W_{AB}} \tag{31}$$

The conventional expression for separation can be written with parameters related to either species A or B, shown here using the retention time and peak base width for species B[2].

$$N = 16[\frac{(t_R)_B}{W_{AB}}]^2 \tag{32}$$

By combining eqns. (31) and (32) an equation for chromatography is produced that expresses resolution in terms of efficiency and retention time[4].

$$R_s = \frac{\sqrt{N}}{4}[\frac{(t_R)_B - (t_R)_A}{(t_R)_{AB}}] \quad (33)$$

The resolution time variable t_R and the efficiency N are eliminated by inserting eqns.(25) and (26) into eqn.(33) for the analyte B:

$$R = [(\frac{F}{32RT})(\frac{(v_{ep})_{AB}}{v_{eo} + (v_{ep})_{AB}})XE]^{1/2}[\frac{(v_{ep})_A - (v_{ep})_B}{v_{eo} + (v_{ep})_{AB}}] \quad (34)$$

when electrophoretic velocity, v_{ep}, is replaced with $\mu_{ep}E$ and applied patential, V is substituted for XE one obtains the following equation for the resolution:

$$R_s = [\frac{FV(\mu_{ep})_{AB}}{32RT}]^{1/2}[\frac{(\mu_{ep})_A - (\mu_{ep})_B}{(\mu_{eo} + (\mu_{ep})_{AB})^{3/2}}] \quad (35)$$

In order to make a comparison between new resolution eqn.(35) obtained in present work and Jorgenson Lukacs equation for resolution eqn.(15) could be transformed to the following equation by using Einstein relation relating diffustion constant to mobility $\frac{D}{\mu} = \frac{RT}{F}$ and

$$R_s = [\frac{FV}{32RT(\mu_{ep})_{AB}}]^{1/2}[\frac{(\mu_{ep})_A - (\mu_{ep})_B}{(\mu_{eo} + (\mu_{ep})_{AB})^{1/2}}] \quad (36)$$

Jorgenson and Lukacs equation (eqn.36) and new derived Ghowsi equation(eqn.38) for resolution are given as

$$R_s = [\frac{FV}{32RT(\mu_{ep})_{AB}}]^{1/2}[\frac{(\mu_{ep})_A - (\mu_{ep})_B}{(\mu_{eo} + (\mu_{ep})_{AB})^{3/2}}] \quad (37)$$

This interesting observation that for the absence of electrosmotic flow in Jorgenson and Lukacs equation the resolution eqn.36 is converted to

$$R_s = [\frac{FV}{32RT}]^{1/2}[\frac{(\mu_{ep})_A - (\mu_{ep})_B}{(\mu_{ep})}] \quad (38)$$

The other resolution equation obtained in present work Ghowsi's eqn.(35) with presence of electroosmotic flow R_s is converted to

$$R_s = [\frac{FV}{32RT}]^{1/2}[\frac{(\mu_{ep})_A - (\mu_{ep})_B}{(\mu_{ep})_{AB}}] \quad (39)$$

It is interesting that only for this case when electroosmosis is absent the resolution equation of Jorgenson and Lukacs eqn.(15) with the help of Einstein relation is equal to Ghowsi's derived equation eqn.(38) at present for capillary zone electrophoresis.

3. Micellar electrokinetic capillary chromatography (MECC)

By converting figure of Merits in MECC to electrochemical parameters[5] and pursuing similar procedure we applied to capillary electrophoresis and using effective length[3] solute travels rather than length of capillary and then converting the resolution equation in terms of chromatography parameters new equation for resolution could be found which is published in another paper[7].

4. Optimization of micellar electrokinetic capillary chromatography (MECC) as a nano separation technique using three dimensional and two dimensional plottings of characteristic equations

Feyman with the lecture of plenty of room at the bottom at an American Physical Society at Caltech on December 29, 1959 considered the possibility of direct manipulation of individual atoms as a more powerful forms of synthetic chemistry than those used at the time . In conventional chromatography there are two phases involved one is the stationary phase and one is the mobile phase[6]. Terabe et al proposed Micellar Electrokinetic Capillary Chromatography[8], MECC, which has the smallest pseudo stationary phase within nano range called micelle.

The very high strength of separation comes from these nano sized materials. That is why we call this technique MECC.

5. Nano separation technique

There are several work which have been done to final the optimum conditions of this Nano Separation Technique[5,8-10].

In all optimization characteristic equation is the focus.

What is the characteristic equation?

In column chromatography the resolution equation is given as[6]

$$R_s = \frac{N^{1/2}}{4} . \frac{\alpha - 1}{\alpha} . \frac{k'}{1 + k'} \tag{40}$$

Where k′ is the capacity factor, α is the selectivity and N is the number of theoretical plates. Terabe[8] et al proposal MECC for the first time and they proposed the resolution equation for MECC:

$$R_s = \frac{N^{1/2}}{4} . \frac{\alpha - 1}{\alpha} . \frac{k'}{1 + k'} . \frac{1 - t_o / t_{mc}}{1 + t_o / t_{mc}} \tag{41}$$

Where N,α and k′ were already defined , a new term is appearing in equation (41) for resolution is $\frac{1 - t_o / t_{mc}}{1 + t_o / t_{mc}}$. In this term k′ is the capacity factor and t_o and t_{mc} are retention times of the aqueous and micellar phases respectively. The characteristic equation for MECC to optimize is from equation(41):

$$f(k',t_o/t_{mc}) = \frac{k'}{1+k'} \cdot \frac{1-t_o/t_{mc}}{1+(t_o/t_{mc})k'} \tag{42}$$

This characteristic equation has two variables k′ and the ratio t_o/t_{mc}. In a recent work a new model for the MECC using a model based on effective length solute migrated as a similar to tread mill case were proposed[7]. Based an this model

$$R_s = \frac{1}{4}[(D_{ep})_B RT/4]^{1/2}(N_{pseudo}{}^{1/2})[\frac{\alpha-1}{\alpha}][\frac{k'}{k'+1}]^{3/2} \cdot \frac{1-(t_o/t_{mc})}{1+(t_o/t_{mc})k'} \tag{43}$$

Variables of this equations were defined in the previous work[7]. The characteristic equation of R_s for equation (43) is the last two terms.

$$f(k',t_o/t_{mc}) = (\frac{k'}{1+k'})^{3/2} \cdot (\frac{1-t_o/t_{mc}}{1+(t_o/t_{mc})k'}) \tag{44}$$

In a comparison between the characteristic equation (42) obtained by Terabe[8] and the characteristic equation obtained based on the new model[7],equation(44) , the difference is the power of $(\frac{k'}{k'+1})$ term where in equation (42) the power of this term is 1 and in equation (44) the power of this term is 3/2. The rest of these two equations (42) and (44) are the same. There are two other characteristic equations need to be obtained. For the first time J. P. Foley obtained an equation for a compromise between resolution and the migration time. He introduced $R_s/t_R . t_R$ is given in Terabe's work[8] as

$$t_R = [\frac{1+k'}{1+(t_o/t_{mc})k'}]t_o \tag{45}$$

Two characteristic equations for R_s/t_R are obtained one is for Terabe R_s equation (41) and the other one is for R_s in equation (43). The two characteristic equations for R_s/t_R are given as below correspondingly.

$$f(k',t_o/t_{mc}) = \frac{k'}{(1+k')^2} \cdot (1-\frac{t_o}{t_{mc}}) \tag{46}$$

$$f(k',t_o/t_{mc}) = \frac{k'}{(1+k')^{5/2}} \cdot (1-\frac{t_o}{t_{mc}}) \tag{47}$$

6. Two dimensional and tree dimensional plots of characteristic equations

In present work by the help of modern technology of computer and three dimensional software of Dplot direct access to the plot of characteristic equations (42) and (44) are possible.

Figure 2 shows the three dimensional plot of characteristic equation (42) obtained by Terabe[8] et al.

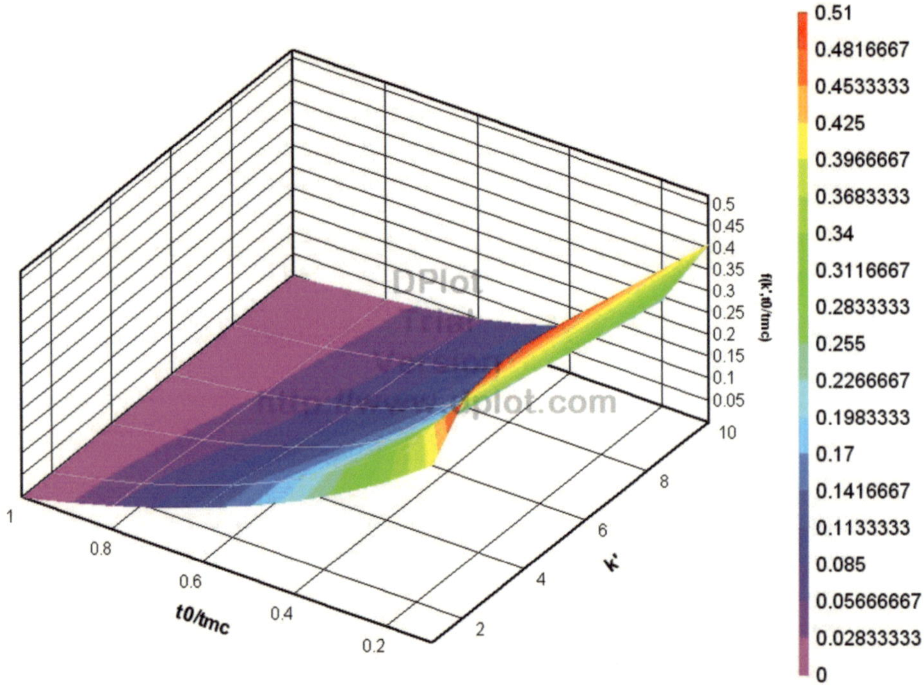

Figure 2. Three dimensional plot of Terabe's characteristic equation for resolution of MECC.

In this work $f(k',t_o/t_{mc})$ is a function of two independent variables k′ and t_o/t_{mc} .k′ is changing between 1 to 10 and t_o/t_{mc} changes between 0.1 to 1. In figure 2 the surface of the three dimensional plot shows maximums. The greatest maximum occurs at $f(k'_{\max},t_o/t_{mc\max})=0.51$.

These maximums are local maximums.

In figure 3 the characteristic equation (44) based an proposal thread mill model[7] is plotted. This plot shows no local maximums but there is a single maximum occurs at $t_o/t_{mc}=0$,k′=10 where $f(k',t_o/t_{mc})=2.25$. Similar information could be obtained for equation (42) from level curves plotted by Dplot software presented at this work fig.4. This two dimensional plot is the image of three dimensional plot on the surface of the plan $t_o/t_{mc}=0.1$, k′=1.

Similar information could be obtained for equation (44) from level curves plotted by Dplot software presented at this work fig.5. This two dimensional plot is the image of three dimensional plot on the surface of the plane $t_o/t_{mc}=0.1$,k′=1.

Foley's characteristic equation $t_o/t_{mc}=0.1$ (46) has also been plotted three dimensionally figure 6 and two dimensionaly level curves figure 7. The information obtained from either these two plots is as following. Maximum $f(k',t_o/t_{mc})=0.225$ occurs at $t_o/t_{mc}=0.1$,k′=1 and minimum $f(k',t_o/t_{mc})=0$ occurs at $t_o/t_{mc}=1$,k′=1.

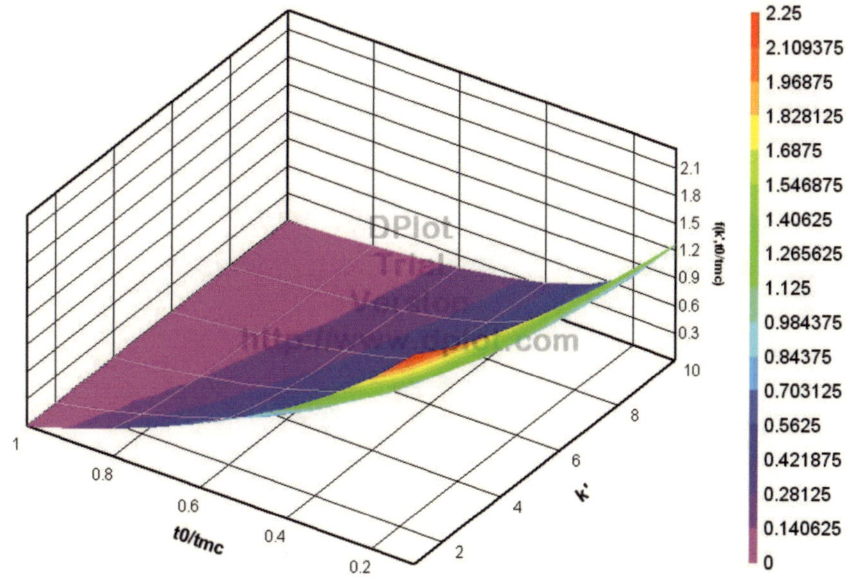

Figure 3. Three dimensional plot of Ghowsi's characteristic equation for resolution of MECC.

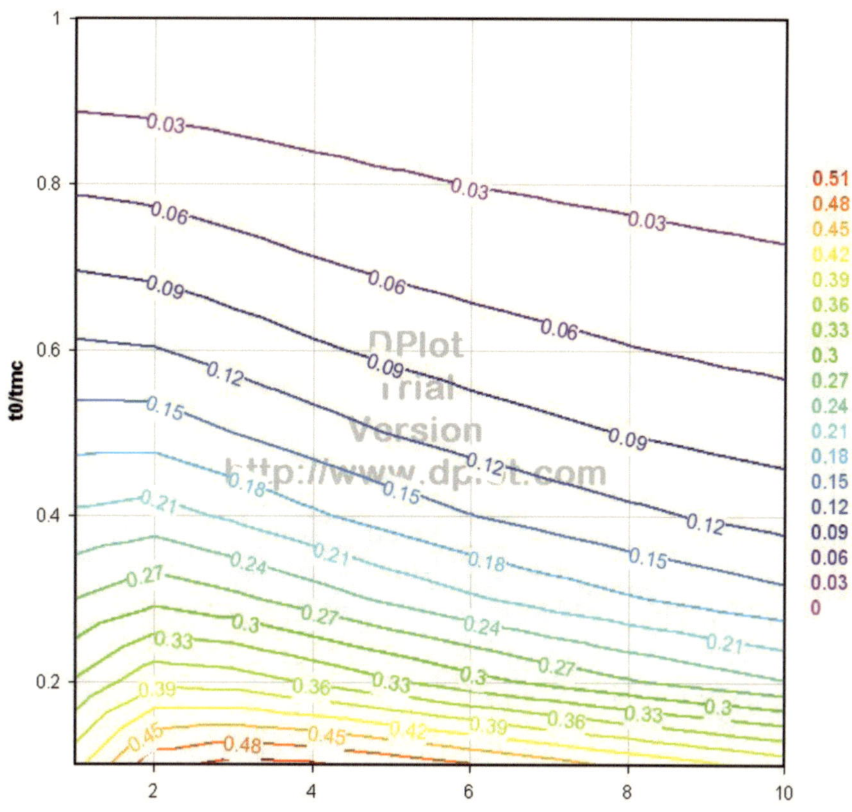

Figure 4. Two dimensional plot, level curves, of Terabe's characteristic equation for resolution of MECC.

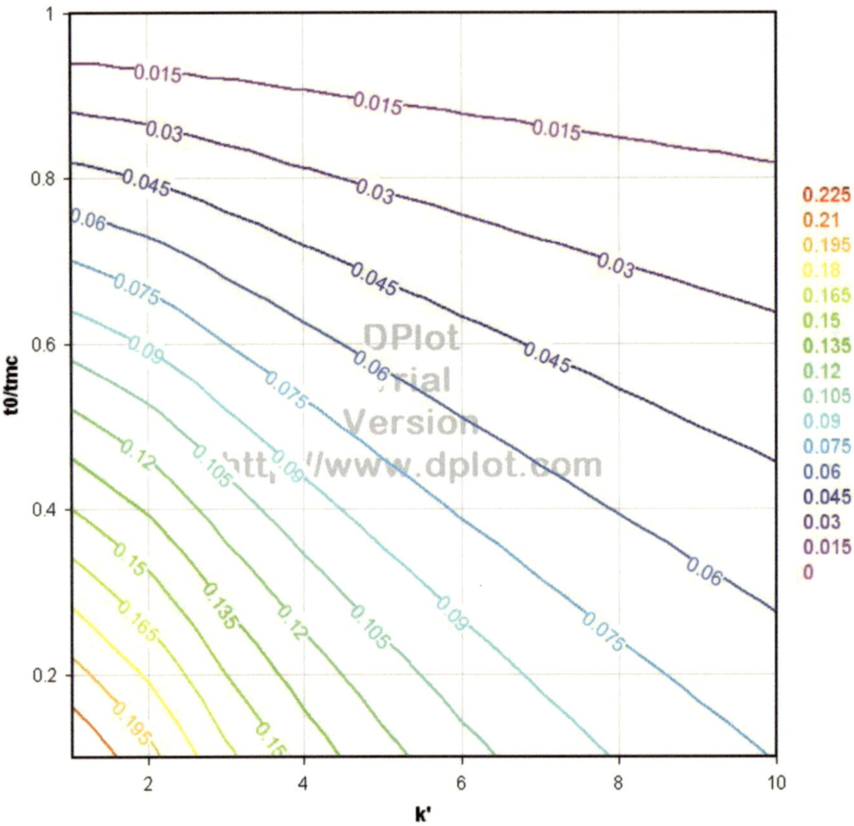

Figure 5. Two dimensional plot , level curves , of Ghowsi's characteristic equation for resolution of MECC.

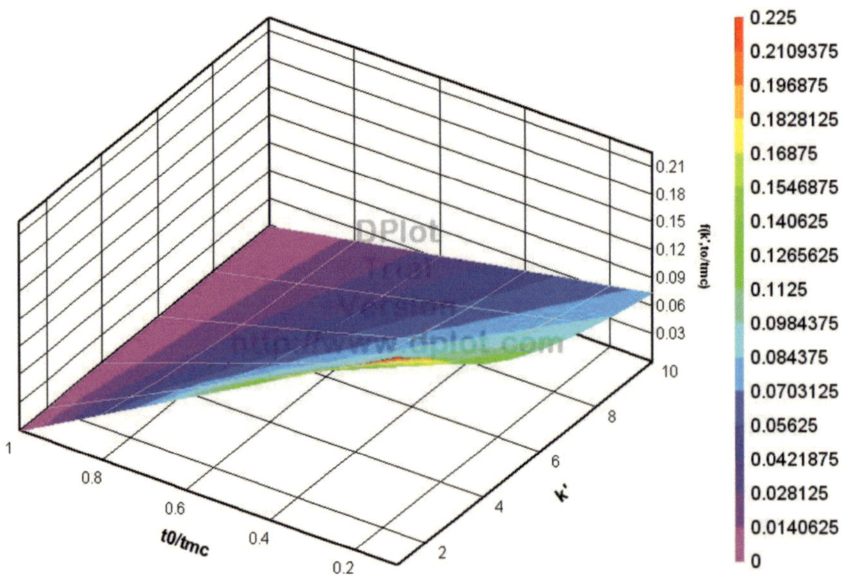

Figure 6. Three dimensional plot, Foley's characteristic equation for resolution per unit time R_s / t_R.

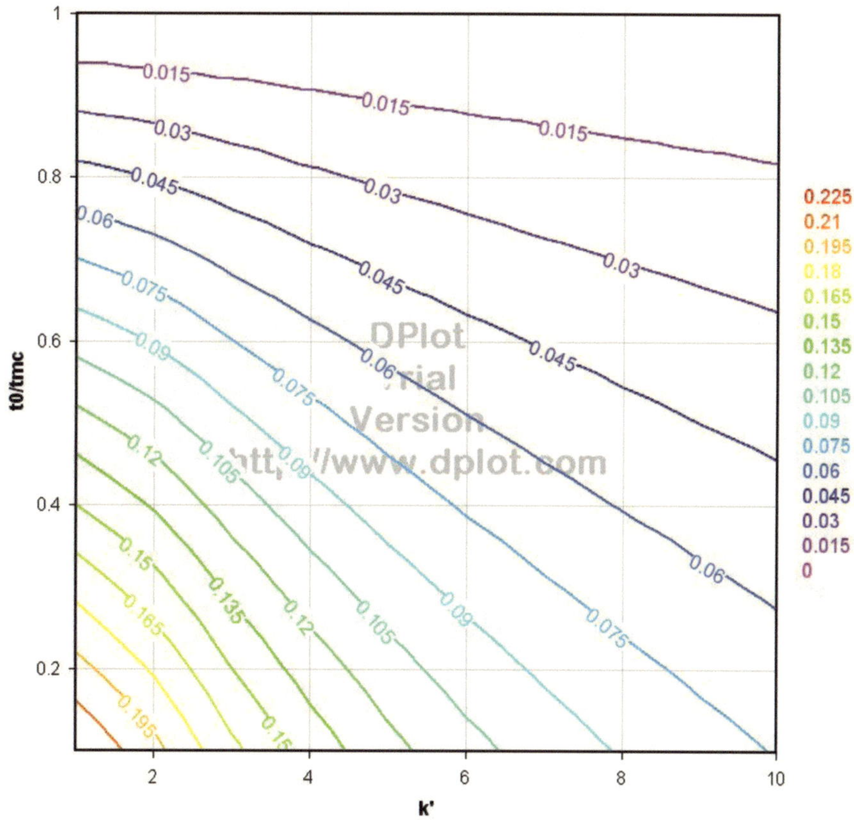

Figure 7. Two dimensional plot, level curves, of Foley's characteristic equation for resolution per unit time R_s / t_R.

Ghowsi's characteristic equation (47) has also been plotted three dimensionally figure 8 and two dimensional level curves figure 9. The information obtained from either these two plots is the following: Maximum $f(k',t_o / t_{mc}) = 0.16$ occurs at $t_o / t_{mc} = 0.1$, k′=2 and minimum $f(k',t_o / t_{mc}) = 0$ occurs at $t_o / t_{mc} = 1$,k′=1.

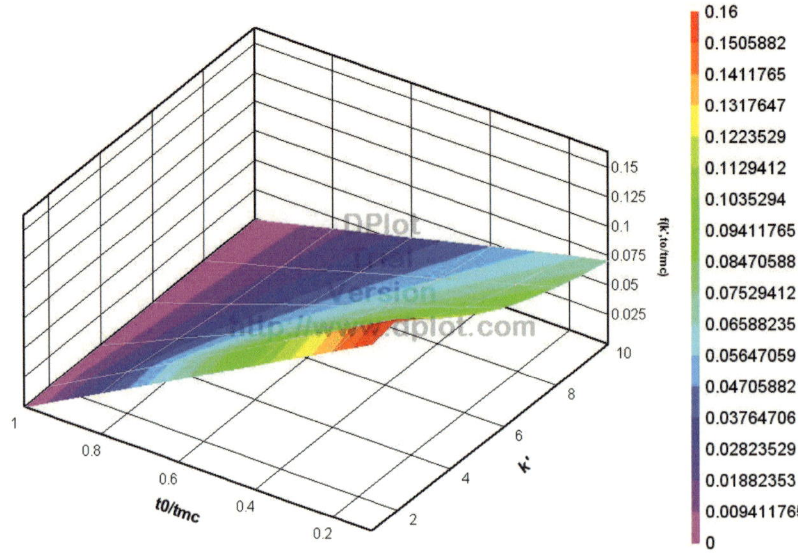

Figure 8. Tree dimensional plot of Ghowsi's characteristic equation for resolution per unit time R_s / t_R.

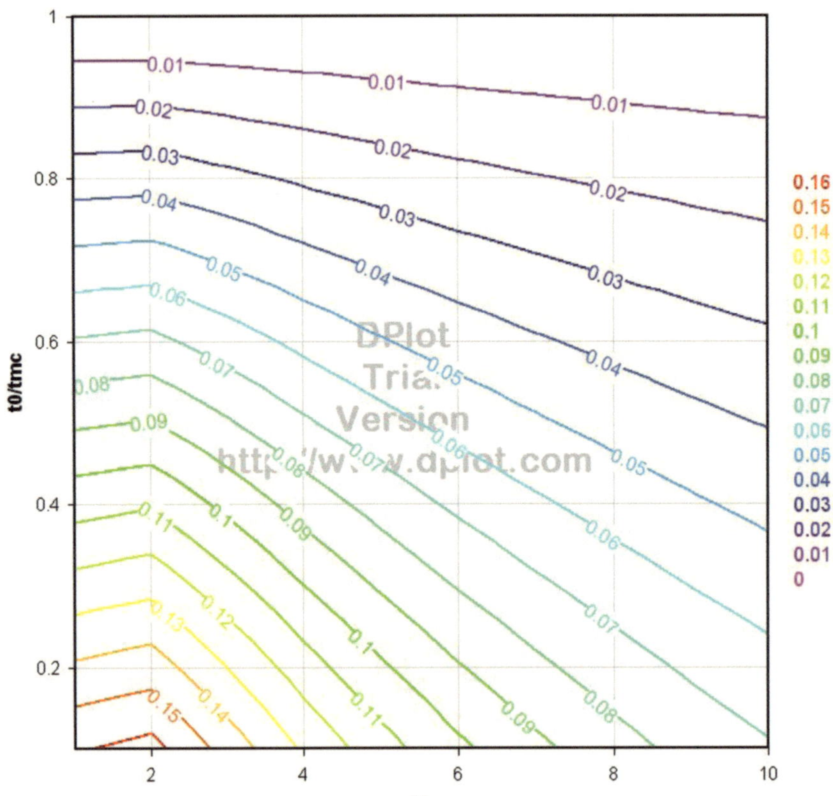

Figure 9. Two dimensional plot level curves of Ghowsi's characteristic equation for resolution per unit time R_s / t_R.

7. Conclusion

The fundamental eqn.12 for the number of theoretical plates has been used to obtain resolution equation by Jorgenson eqn.14. The number of theoretical plates equation in previous work by Jorgenson is wrong, automatically makes the resolution equation wrong.

With the consideration that length X in eqn. 17 is the distance where force is applied in capillary electrophoresis, number of theoretical plates which depends on eqn.17 is discussed.

Another equation is the result obtained in eqn.28 which indicates the new result based on effective length. In comparison with eqn.12 Jorgenson equation different eqn.26 has been used to obtain eqn.34 which is the new resolution equation. Similar procedure could be applied to obtain the number of theoretical plates for effective length for MECC and the new resolution equation could be obtained too. This work for MECC was shown in reference 7.

In present work eight graph are presented as characteristic equations of MECC. Four of them are three dimensional and four of them are two dimensionals. It is very interesting observation that the Terabe's et al characteristic equation shows local maximums in three dimensional plots but the three dimensional plot of characteristic equation obtained by us based on thead mill model shows no local maximums on the surface of three dimensional plot.

Author details

Kiumars Ghowsi
Department of Chemistry Majlesi Branch, Islamic Azad University, Isfahan, I.R. Iran

Hosein Ghowsi
Department of Mathematics, Payame Noor University, Tehran, I.R. Iran

8. References

[1] J.C. Gidding, Sep.Sci.,4,181 (1969).
[2] J. Jorgenson and K.D. Lukacs , Anal. Chem., 53, 1298 (1981).
[3] C. D. Dunn, M. G. Hankins and K.Ghowsi , Sep.Sci. Tecnol. , 29,2419 (1994).
[4] W. R. Jones and J. Jandik, J. Chromatogr., 546, 445 (1991).
[5] K. Ghowsi, J. P. Foley and R. G. Gale , Anal. Chem.,62, 2714 (1990).
[6] D. A. Skoog, Principles of Instrumental Analysis, Saunder college Publishing Philadelphia, edn.5 (1992).
[7] K. Ghowsi and H. Ghowsi, Asian J. Chem., 23,3084 (2011).
[8] S. Terabe and T. Ando, Anal. Chem., 57,834 (1985).

[9] J. P. Foley, Anal. Chem. 62,1302 (1990).
[10] K. Ghowsi and H. Ghowsi submitted to Asian. J. Chem. Accepted.

Chapter 2

Method Development by Use of Capillary Electrophoresis and Applications in Pharmaceutical, Biological and Natural Samples

Constantina P. Kapnissi-Christodoulou

Additional information is available at the end of the chapter

http://dx.doi.org/10.5772/45834

1. Introduction

Electrophoretic methods compose a family of related techniques that use narrow-bore fused-silica capillaries to perform high efficiency separations of both small and large molecules. These methods are commonly known as capillary electrophoretic methods. Capillary electrophoresis (CE) has, over the years, demonstrated its powerful separation ability in the area of chiral and achiral analysis. This is contributed to the advantages it offers when compared to chromatographic techniques: (1) low consumption of samples and solvents; (2) high separation efficiency and resolution; (3) versatility [1,2].

Two of the most important modes of CE, which will be discussed in this chapter, are capillary zone electrophoresis (CZE) and micellar electrokinetic chromatography (MEKC). CZE is the simplest and the most widely used mode of CE [1]. The separation mechanism of the analytes is based on their difference in charge-to-size ratios and their difference in electrophoretic mobilities, which, in turn, result in different velocities. However, due to the fact that neutral species do not possess an electrophoretic mobility, they cannot be separated by use of this mode.

In order to circumvent this problem, new modes of CE have been suggested as alternatives. MEKC, which combines the best features of both electrophoresis and chromatography, is considered an alternative mode because it can be used for the separation of charged as well as neutral compounds. It involves the introduction of a surfactant at a concentration above the critical micellar concentration (CMC), at which micelles are formed. MEKC was first introduced by Terabe *et al.* in 1984 [3]. Although it is a form of CE, its separation principle is more similar to HPLC than to CE. In this mode, analytes are separated according to their partitioning between the mobile and stationary phase and, when charged, their electrophoretic mobility. The driving force for the partitioning of analytes is hydrophobicity.

In addition, hydrogen bonding, dipole-dipole, and dispersive interactions can contribute to the solute partitioning between the two phases [4-6].

CE was originally considered as a powerful analytical tool for the analysis of biological macromolecules. It has though, over the years, been extensively used for the separation of other compounds, such as chiral drugs, food additives, pesticides, inorganic ions, organic acids, and others. In this chapter, the ability of CE, and particularly CZE and MEKC, to be used for the qualitative and quantitative determination of compounds in pharmaceutical, biological and natural samples is investigated. In each approach, a number of studies are reported and discussed. These studies involve establishment of optimum separation conditions, method validation, optimization of sample-preparation procedure and application for the determination of the analytes under study in real samples. The first part of this chapter involves the determination of polyphenolic compounds using CZE with UV-Vis detector in red and white wines, while the second part involves the determination of pharmaceutical compounds in biological samples, such as blood and urine, using the hyphenated technique CE-MS (mass spectrometry). The third and final part emphasizes the importance of MEKC in chiral analysis since it has been known that usually only one enantiomer is active, while the other may be less active, inactive or has adverse effects.

2. Determination of polyphenolic compounds in natural samples

Polyphenolic compounds exist in a variety of natural products, such as fruits, vegetables, beverages (tea, wine and juices), honey, cacao and herbs. They attract a lot of interest due to their beneficial implication in human health. They have been widely studied due to their antioxidant capacity and their association with several pathological conditions, such as hypertension, cardiovascular disease, dementia, and even cancer [7-9]. Therefore, due to their health significance, numerous analytical methods have, over the last decades, been developed for their separation, identification and quantitation in natural products [10-13].

According to literature, the simplest CE method, CZE, proved to be the best method for the determination of polyphenolic compounds in wine samples [14-17]. In such studies, and in each case, when the optimum CZE method was applied to different red and white wines, it was established that red wines have higher levels of polyphenolic compounds than white wines and that the polyphenolic composition varies among different wines.

2.1. Method development and validation

In this part of the chapter, a representative study performed recently in Cypriot wines is briefly described [17]. The influence of several experimental parameters is initially illustrated in order to obtain improved selectivity and resolution for the separation of seven flavonoids, which constitute the most important group of polyphenols, and trans-rasveratrol that are usually present in wine. This is accomplished by use of CZE and by examining different sample preparation procedures. Due to the low concentrations of flavonoids in wine and the high complexity of wine matrices, preconcentration methods are required, which can simplify the electropherograms. The optimized CZE and pre-treatment methods proved to be effective in characterizing flavonoids in red and white wine samples.

The effect of column temperature, and concentration and pH of background electrolyte (BGE) were investigated. These parameters, along with the applied voltage, are the most common parameters that are required to be examined in order to optimize a separation in CZE. Figure 1 illustrates the influence of the pH on the resolution and the analysis time.

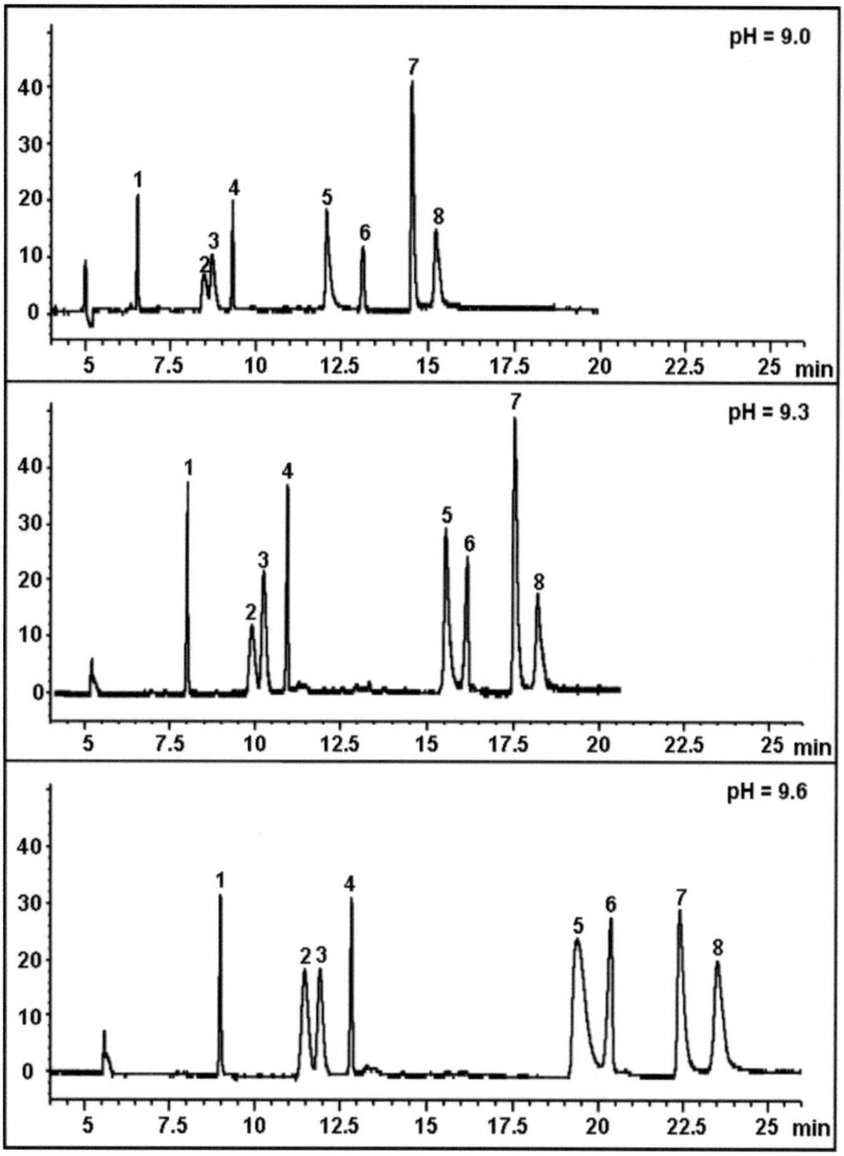

Figure 1. Effect of pH value on the separation of the eight polyphenols. (A) pH 9; (B) pH 9.3; (C) pH 9.6. Conditions: BGE 50 mM borate, 10 mM phosphate and 20 mM SDS; pressure injection, 30 mbar for 3 sec; applied voltage, 25 kV; temperature, 25 ºC; fused-silica capillary, 64 cm (55.5 cm effective length) x 50 μm i.d.; detection, 205 nm. Peak identification: trans-resveratrol (1), epicatechin (2), catechin (3), naringenin (4), kaempferol (5), apigenin (6), myricetin (7), quercetin (8) [17].

The last two increased with increasing the pH, possibly due to an increase in the negative charge, which resulted in a greater affinity and a higher complexation between borate and phenols. Taking into consideration the migration times, the peak efficiency and the sufficient resolution, the following parameters provided a baseline separation of all polyphenolic compounds: BGE containing 50 mM borate and 10 mM phosphate at pH 9.6 and column temperature of 25 ºC (Figure 1C). The use of alkaline borate-based BGEs, in CZE, resulted in a sufficient separation of polyphenols due to the complex-formation ability of borate. In addition, an increase in the borate concentration from 25 to 50 mM and an increase in the pH value from 9 to 10 resulted in an increase in the migration times of all analytes, while the resolution was significantly improved. At pH 10 though, the analysis time was very long (~ 50 min) and joule heating effects, such as high current generation and peak broadering, were observed. An increase in pH increased the negative charge of the analytes, which, in turn, favored a greater affinity for the buffer and a higher complexation between borate and phenols [18].

The method was then validated by the terms of linearity, precision and LOD. Linearities for the eight analytes were very good, and precision, which was based on the relative standard deviation, was below 1%, indicating an excellent reproducibility. In addition, LODs, which were calculated as three times the standard deviation via the slope of the calibration curve, were between 0.03 and 5.05 μg/mL for all eight polyphenolic compounds.

2.2. Application

The qualitative and quantitative analysis of analytes in real samples is often difficult due to interruptions caused by different interfering substances found in the sample matrix. Therefore, a sample-preparation procedure is a necessary step prior to the electrophoretic analysis, in order to isolate the analytes under study from real samples. Different preconcentration methods have been used over the years, including solid-phase extraction (SPE) with C-18, silica, or other cartridges [14,16,19] and liquid-liquid extraction (LLE) with different organic solvents [10,20,21].

In the study performed in Cypriot wines, the sample preparation procedure was optimized in order to determine the one that was simple, fast and reliable [17]. Therefore, three LLE-procedures (C,D,E), a SPE-procedure (F), a procedure that involve evaporation and reconstitution of wine sample (B) and a direct injection of wine sample after dilution and filtration (A) were compared and the most effective method was applied to Cypriot wines. The electropherograms obtained by use of each sample preparation procedure are illustrated in Figure 2. When no extraction was performed, the electropherograms were complex, while SPE was found to be ineffective for the isolation of polyphenolic compounds from wine samples. LLE with diethyl ether, followed by evaporation of organic layer by nitrogen stream and reconstitution in ethanol proved to be the optimum sample pre-treatment method. When the optimum method was applied to Cypriot wine samples, the quantification of polyphenolic compounds was successfully achieved. It was observed that epicatechin and catechin exist in all wine samples in comparable concentrations, whereas myricetin and quercetin exist only in two of the three wine samples. Polyphenolic composition varies among different wines, because it depends on several factors, such as the type of grapes used, the vivification process used, the type of yeast that participates in the fermentation, weather variations and other biological effects [22].

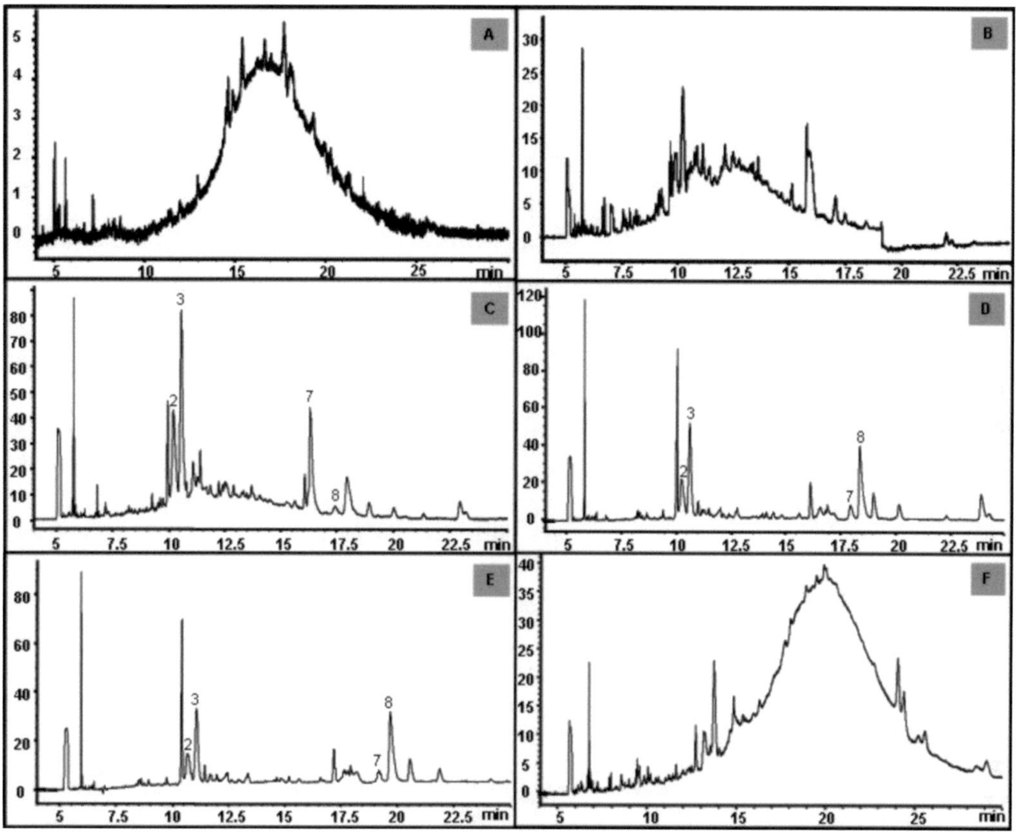

Figure 2. Electropherograms of the wine samples obtained using six different sample preparation procedures under optimum conditions. Conditions: BGE 50 mM borate, 10 mM phosphate and 20 mM SDS (pH 9.6); pressure injection, 30 mbar for 3 sec; applied voltage, 25 kV; temperature, 25 ºC; fused-silica capillary, 64 cm (55.5 cm effective length) x 50 μm i.d.; detection, 205 nm. Peak identification: epicatechin (2), catechin (3), myricetin (7), quercetin (8) [17].

Another important observation was that in white wine, the only flavonoid that was detected was catechin at a concentration of 7.3 μg/mL. This was not surprising since the majority of flavonoids in wine come from the extraction derived from grape's solids. White wine is made by pressing the juice away from the grape's solids, and then, by allowing it to ferment. So, red wines have higher levels of polyphenolic compounds [23].

3. Determination of pharmaceutical compounds in biological samples

Quantification of drugs in biological fluids, like plasma, has an important role in drug discovery and development. There are two main aspects that are taken into account in order to make the identification of drugs in biological fluids possible. The first aspect is the development of an accurate analytical method, with high sensitivity, capable to identify desirable compounds in concentrations comparable to that in biological fluids. The second one is the exploration of the optimum extraction method that can effectively extract the drug from the biological matrix.

Over the years, CE coupled to electrospray ionization-mass spectrometry (ESI-MS) has been utilized as a bioanalytical tool for the analysis of drug compounds in biological samples [24-29]. Even though the most common detector in CE is the UV detector due to its easy manageability and low cost, it has the drawback of low sensitivity due to the short optical path length. An alternative to this is the use of MS. The coupling of CE with MS is a well-established technique, which combines the high efficiency and resolution that are provided by CE and the detection sensitivity and selectivity and the identification potential that are provided by MS [25,30].

In recent years, a large number of publications have been provided on the general developments and biological applications of CE-ESI-MS [24-29]. Zheng *et al.* developed a CZE-ESI-MS method for monitoring the antiepileptic drug lamotrigine in human plasma [27]. The optimum conditions were obtained by varying a big number of BGE, sheath liquid and MS spray chamber parameters. In each case, both the CZE separation, as well as the MS detection sensitivity, were evaluated, and the parameter that provided a reasonable compromise between resolution and detection sensitivity was chosen as the optimum. The developed method was then applied to assay blank samples spiked with lamotrigine in order to set up the calibration curve and estimate the limit of detection (LOD). Both linearity of calibration curve and LOD (0.05 μg/mL) were good, and the optimum method was applied to 14 human plasma samples collected from a lamotrigine-treated subject over a period of 96 h after oral administration of 50 mg lamotrigine.

In a 2011 study, Elhamili *et al.* analyzed the anticancer drug Imatinib by use of CE coupled to ESI time-of-flight MS in human plasma [29]. The CE separation and ESI parameters were initially investigated and optimized in regard to peak efficiency, peak intensity and electrospray stability. The LOD and limit of quantitation (LOQ) were evaluated by injections of standard solutions of the drug compound, and they were determined to 5 and 20 ng/mL, respectively. In addition, the extraction recovery of Imatinib from human plasma using a common liquid-liquid extraction (LLE) method and a new strong cation exchange (SCX) solid-phase extraction (SPE) column was investigated and compared. The highest extraction recoveries were obtained by using the latter method. The SCX-SPE extraction followed by CE-ESI-TOF-MS analysis in patient plasma samples demonstrated good repeatability, linearity and sensitivity for possible therapeutic monitoring of Imatinib level. The authors, in this manuscript, also conclude that this method could be applied for the analysis, quantification, and clinical assessment of other drug compounds and their metabolites.

3.1. Method development

The performance and usefulness of CE-MS is also demonstrated here by providing a more in-depth analysis of a research work that was performed in a blood sample obtained from a patient with Alzheimer's Disease (AD) [24]. In this study, a CZE-ESI-MS method was developed for the analysis of the acetylcholinesterase inhibitor rivastigmine, using neostigmine bromide as an internal standard, which is highly recommended in order to avoid problems that are related to sample injection [31]. Rivastigmine is a pseudo-irreversible carbamate inhibitor of acetylcholinesterase, and it is clinically used for the symptomatic treatment of mild to moderate AD [32].

In a previous paper, MEKC coupled to a diode-array detector was used for the simultaneous separation of nine acetylcholinesterase inhibitors, including rivastigmine [33]. This method was validated and successfully applied to a real blood sample that was obtained from a patient who was not under any of this medication. The sample was spiked with rivastigmine in order to establish the ability of the method to separate the drug from other components that might exist in the blood sample. In this study, the blood sample was not directly injected into the capillary, because some components that exist in the sample can be absorbed to the capillary wall and deteriorate the performance of the column [34]. The blood sample was therefore diluted ten folds with the BGE [12.5 mM Na_2HPO_4 / 12.5 mM $Na_2B_4O_7$ / 20 mM SDS (pH 10)], and it was then spiked with 25 μg/mL of rivastigmine. However, due to the low sensitivity obtained by CE with on-column UV detection, the identification of rivastigmine in biological fluids using CE remained a challenge. In order for the technique to be used for the quantitation of an acetylcholinesterase inhibitor in body fluids, the sensitivity, and consequently the LOD had to be improved. The increased interest in exploring CE-MS and its potential to serve as an alternative method allowed further investigation for the determination of rivastigmine and related drugs in complex biological matrices.

When the CZE-UV method was compared with the CZE-MS, the first demonstrated a shorter analysis time of approximately 2 min due to the shorter effective length, while the S/N for the peak of rivastigmine at the SIM mode was estimated to be eight times bigger than with UV detection. This, in turn, indicated the high specificity and selectivity of the ESI-MS detector [24]. In the CZE-ESI-MS study, several electrophoretic and ESI-MS parameters were also examined, which were classified in three categories: the BGE parameters, such as the concentration, the pH and the use of organic modifier, sheath liquid parameters, such as the composition, the methanol (MeOH) content and the flow rate, and finally some spray chamber parameters, such as the temperature and the flow rate of the drying gas and the nebulizer gas pressure. The effect of each parameter on the S/N, and consequently the LOD, was examined and the optimum one was chosen for further optimization.

In the case of BGE parameters, it was observed that ammonium acetate provided the most reproducible migration times, a concentration of 40 mM ammonium acetate resulted in the highest S/N, while a higher concentration decreased the ratio, probably due to the Joule heating effect that increases the level of noise (Figures 3a & 4a). When the pH was examined, it was concluded that at pH 9, where rivastigmine starts to have a negative charge (pKa=8.6), both the analysis time and resolution increased, and a higher S/N was obtained (Figures 3b & 4b).

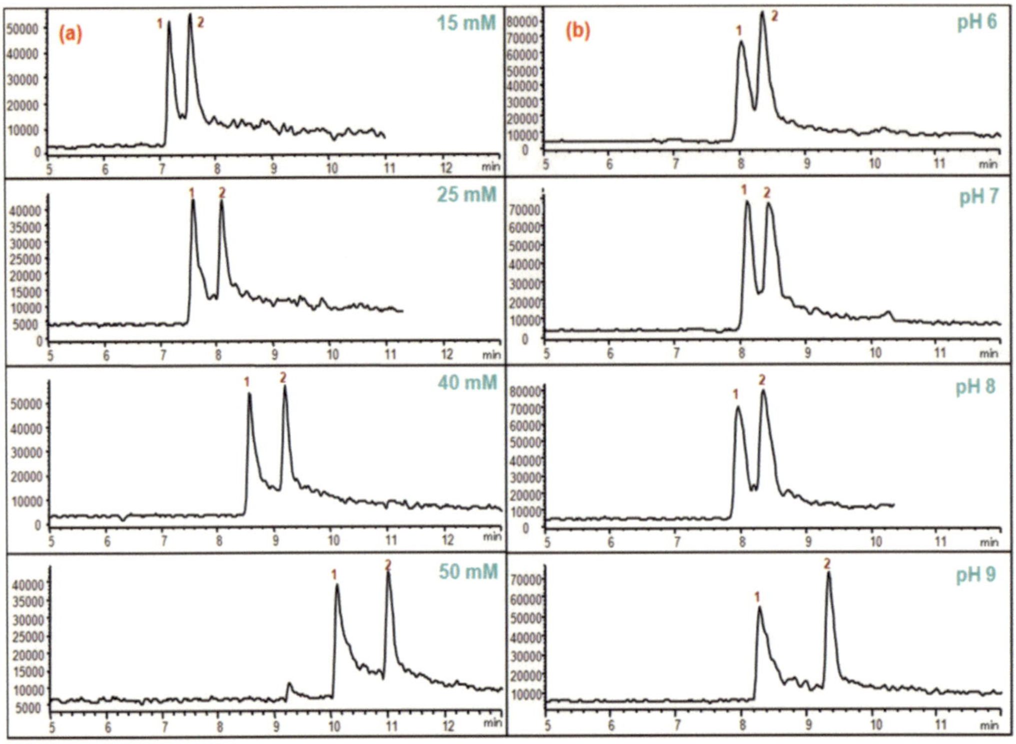

Figure 3. Effect of (a) ionic strength and (b) pH of the BGE on the separation of rivastigmine (2) and I.S (1). Conditions: BGE: ammonium acetate, sheath liquid 1 % acetic acid in water:MeOH (50:50 v/v) at a flow rate of 10 μL/min, analyte and I.S. concentrations 0.3 mg/mL. Drying gas flow rate 6 L/min and temperature 200 ºC, nebulizer gas pressure 20 psi [24].

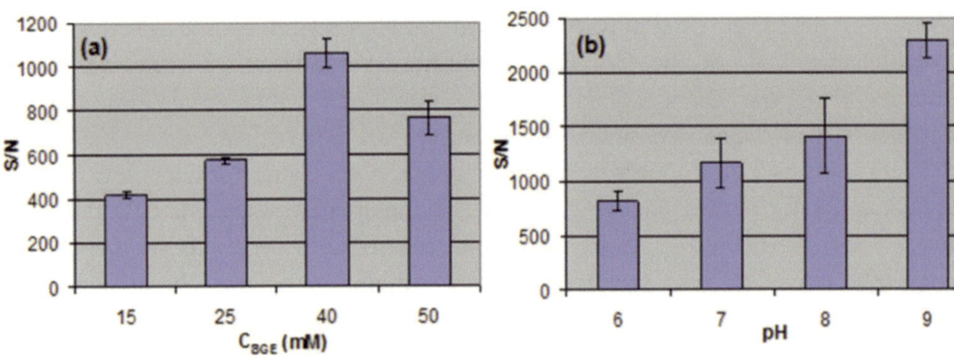

Figure 4. Effect of (a) ionic strength of the BGE and (b) pH of the BGE upon S/N ratio. Conditions: BGE: ammonium acetate, sheath liquid 1 % acetic acid in water:MeOH (50:50 v/v) at a flow rate of 10 μL/min, analyte and I.S. concentrations 0.3 mg/mL. Drying gas flow rate 6 L/min and temperature 200 ºC, nebulizer gas pressure 20 psi [24].

As far as the sheath liquid parameters are concerned, it was observed that its composition and its flow rate affected the ESI-MS sensitivity significantly. This was not a surprising observation since the sheath liquid plays an important role in the CE-MS system. The sheath liquid is used as the make-up liquid that can solve the flow-rate incompatibility problems between CE and MS [35]. These problems are encountered because the flow rate through the CE column is very low (nL/min), and it cannot support a stable electrospray, whose flow rate is typically a few μL/min. In addition, the sheath liquid is used for establishing an electrical connection at the cathode end of the CE capillary, and it provides the suitable solvent conditions for the electrospray, which does not depend on the CE BGE [36].

When different sheath liquids were evaluated, the one that was able to support the formation of positively charged ions, and consequently provide the highest S/N, was acetic acid (1%) (Figure 5a). The influence of methanol as an organic modifier in the sheath liquid was also examined, because the use of such solvents allows an easier protonation of the analytes, which results in a higher signal [28]. By varying the percentage of methanol, it was concluded that 50% was the optimum since the noise level was the lowest (Figure 5b). Finally, the flow rate of the sheath liquid was set at 10 μL/min (Figure 5c). Other values were either too low to establish an electric contact that is required to achieve separation, or they affected the spray stability negatively, which, in turn, lead to higher noise levels.

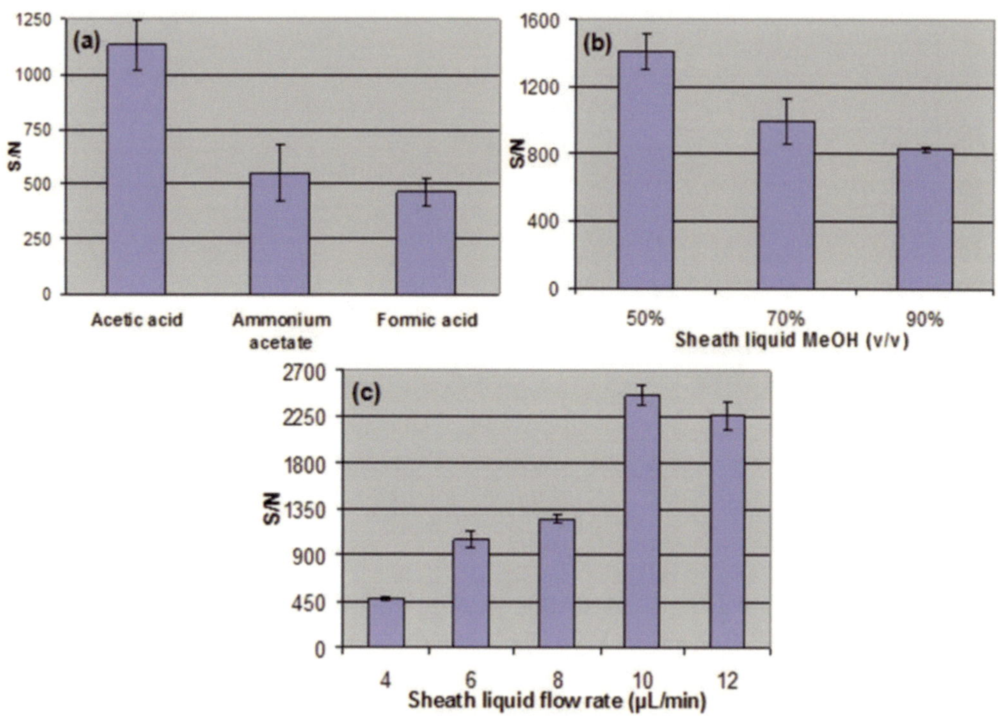

Figure 5. Effect of (a) sheath liquid composition, (b) sheath liquid organic modifier and (c) sheath liquid flow rate upon S/N ratio. Conditions: BGE: ammonium acetate 40 mM, at pH 9.0; analyte and I.S. concentrations 0.3 mg/mL. Drying gas flow rate 6 L/min and temperature 200 ºC, nebulizer gas pressure 20 psi [24].

The spray chamber parameters, which are the last parameters examined in this study, have an important effect on the response of the MS system. One of these parameters involves the drying gas, which is used for accelerating the buffer desolvation, increasing the MS sensitivity, and eliminating any undesirable ions from entering into the MS system. It was observed that the drying gas flow rate has an effect on the stability of the electrospray, and consequently, the levels of the noise. The flow rate was set at 6 L/min, because at this flow rate an increased number of ions come closer to the liquid-gas interface, and this increases the desolvation velocity [37] (Figure 6a). In addition, other flow rates that were examined in this study either caused an unstable electrospray or lowered the S/N. The drying gas temperature was varied from 150 °C to 350 °C, and the highest S/N was obtained at 200 °C, which was considered as the optimum (Figure 6b). The nebulizer gas pressure was the last parameter examined in this category, and based on the stability of the electrospray and the S/N, 20 psi was selected as the optimum. At 20 psi, the electrospray is more efficient, probably due to an improved ion evaporation process because smaller initial droplets are obtained with higher nebulizer gas pressure (Figure 6c).

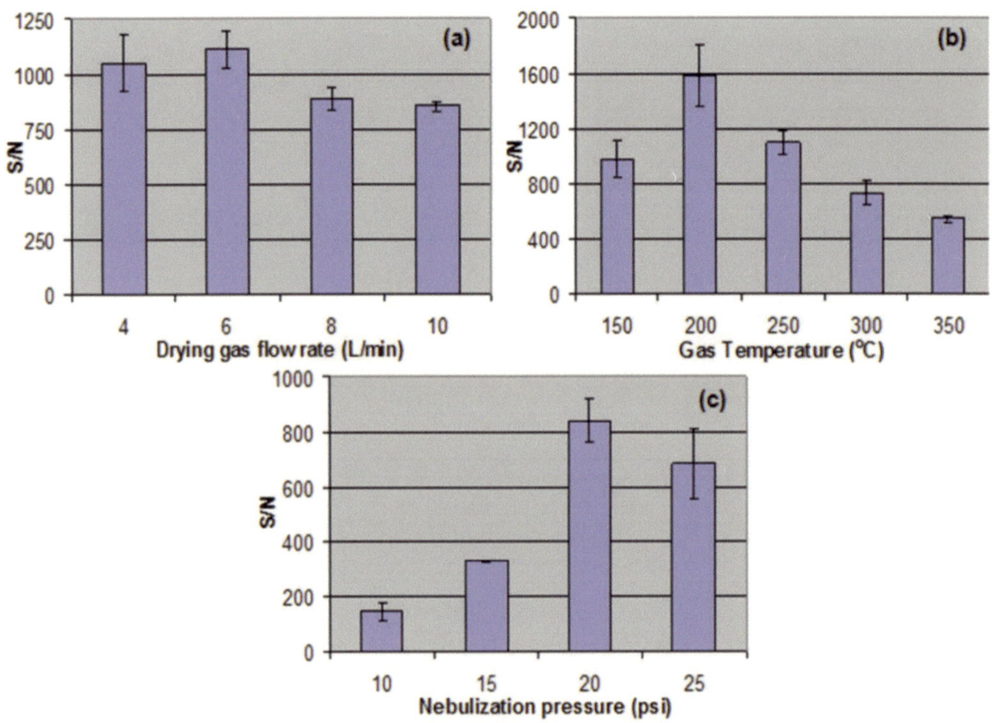

Figure 6. Effect of (a) drying gas flow rate, (b) drying gas temperature and (c) nebulizer gas pressure upon S/N ratio. Conditions: BGE: ammonium acetate 40 mM at pH 9.0, sheath liquid 1 % acetic acid in water:MeOH (50:50 v/v) at a flow rate of 10 μL/min, analyte and I.S. concentrations 0.3 mg/mL [24].

3.2. Method validation

All the parameters mentioned above are the common parameters that need to be examined in a method development process that involves a CE-MS system. These parameters affect the analysis time, resolution, response of the analyte under study, noise level, and sensitivity of the system. All these are important if the developed method is expected to be applied to biological samples for the detection and the quantification of drug and other compounds.

When the optimum conditions for the analysis of rivastigmine were determined, the method was validated in terms of linearity, precision, stability, recovery, LOD and LOQ. Two calibration curves were constructed, in human plasma and in standard solutions, and linearity was good in both cases. The precision, which was evaluated based on migration times and peak areas, was excellent, and particularly in the case where the peak area of the internal standard was also taken into consideration. The LOD and the LOQ were determined based on the standard deviation of the peak area and the slope of the calibration curve. The LOD and the LOQ were calculated as 3 and 10 times the above correlation, respectively. In the plasma sample, the LOD and the LOQ were found to be 2.8 ng/mL and 8.4 ng/mL, respectively, while in standard solutions they were 1.6 ng/mL and 5.0 ng/mL, respectively. These values are considered satisfactory for the accurate and precise quantification of rivastigmine in AD patients treated with the particular drug compound, and this is based on clinical studies that were performed in such patients [38,39].

3.3. Application

Biological matrices are among the most difficult samples to analyze because of the big number of components they contain that they may be adsorbed onto the capillary wall or interfere in the detection and/or separation process. Therefore, before plasma analysis, it is important and necessary to perform a sample preparation procedure. In addition to this, the concentration of most of the analytes in biological samples is low; so, a preconcentration step before the detection and quantitation is required. In many cases, different sample pre-treatment methods are used and compared in order to determine the most effective one, in regard to analyte recovery, difficulty, time and reproducibility. In this study, one LLE and two different SPE procedures were examined. In the case of SPE, two different SPE cartridges were used, a C18 cartridge and an Oasis HLB cartridge. LLE proved to be inefficient for rivastigmine assay, and it was time consuming because the extraction step was followed by additional steps that involved evaporation and reconstitution of the residue in an organic solvent. When the two SPE methods were compared, the C18-SPE cartridge proved to be the optimum, because the S/N was three times higher (S/N=154) than when Oasis HLB cartridge was used (S/N=52), and it provided better recoveries.

The optimum CZE-ESI-MS parameters and the optimum sample preparation procedure were finally applied for the determination of rivastigmine in a plasma sample obtained from an AD patient following rivastigmine patch administration (dose of 9.5 mg/mL rivastigmine/24-h). Figure 7 demonstrates the SIM electropherograms of C18-SPE extract of plasma sample collected 2.0 hours post-application, at m/z 223 and 251, for I.S. and rivastigmine, respectively. The mean (± S.D.) plasma concentration obtained for rivastigmine was 14.6 (± 1.7) ng/mL.

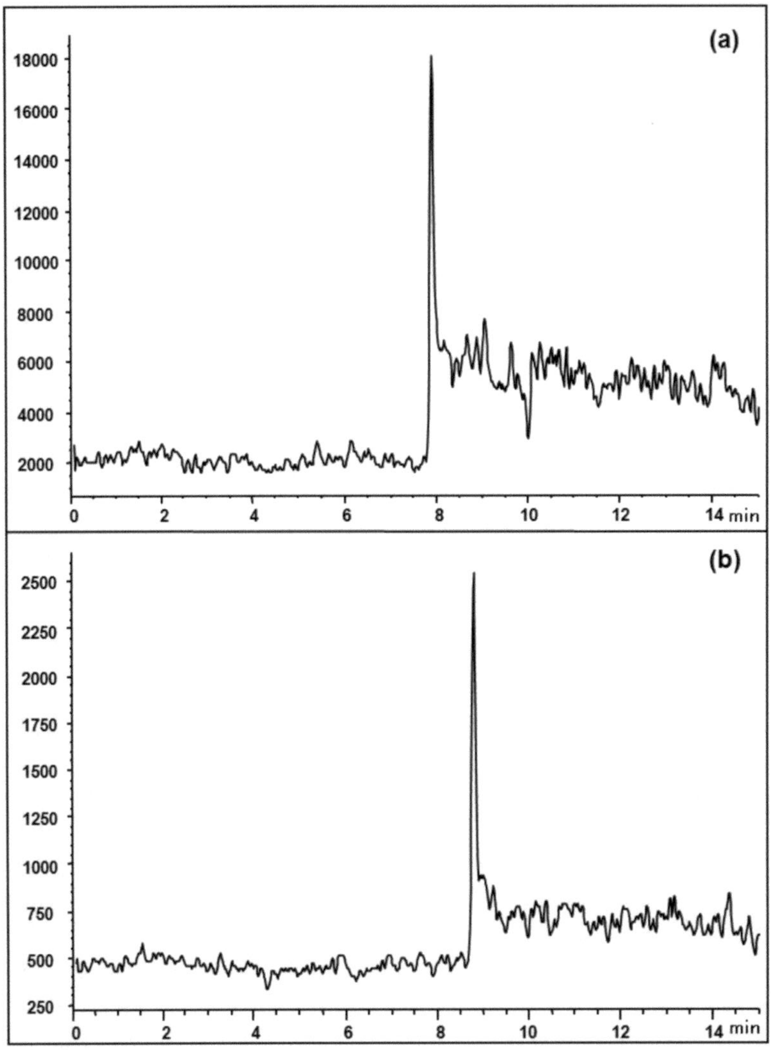

Figure 7. Electropherograms of C18-SPE extracts of plasma from an AD patient following rivastigmine patch administration in a dose of 9.5 mg/mL / 24-h in the SIM-mode at (a) m/z 223 (I.S.) and (b) m/z 251 (rivastigmine). Conditions: BGE: 40 mM ammonium acetate at pH 9, sheath liquid 1 % acetic acid in water:MeOH (50:50 v/v) at a flow rate of 10 μL/min, analyte and I.S. concentrations 0.3 mg/mL. Drying gas flow rate 6 L/min and temperature 200 ºC, nebulizer gas pressure 20 psi [24].

Based on the studies mentioned above, the CZE-ESI-MS method proved to be a promising technique in drug and pharmaceutical analysis. The development of such a method has several advantages over HPLC-MS and GC-MS. The most important ones are the reduction of the reagents cost, the low injection volume requirements, and the avoidance of disposing large volumes of organic waste. In particular, when the study described here is compared to previous studies, where HPLC-MS and GC-MS were used for the analysis of rivastigmine, the required injection volume of plasma for a single analysis is reduced from microliters [40-43] to nanoliters.

4. Determination of enantiomers in pharmaceutical samples

In the last three decades, there has been a growing interest in the separation, detection and quantification of enantiomers in pharmaceutical, clinical, environmental and food analysis. It has been known that usually only one enantiomer is active while the other may be less active, inactive or has adverse effects. Among the separation techniques, HPLC [44-48] GC [44,45,49] and CE [50-55] are most often applied in chiral analysis. Temperature and derivatization are major problems encountered in GC, and poor separation efficiency is observed in HPLC. CE has proven to be a powerful separation technique in the area of chiral analysis, since it has the major advantage of low consumption of samples and solvents.

The most common modes of chiral CE are electrokinetic chromatography (EKC) in the presence of a chiral selector, MEKC, capillary electrochromatography (CEC), where the chiral selector can be either used as a coating (OT-CEC), a packing (P-CEC) or a monolithic material (M-CEC) in the capillary, and others [55-63]. The prerequisite for separation of enantiomers in CE, as in every chromatographic system, is the formation of either stable diastereoisomers by the use of a chiral derivatization agent or reversible diastereoisomeric complexes with the addition of a chiral substance, (chiral selector). In the first case, the two enantiomers are separated based on their different physicochemical properties, while in the second case, they are separated based on their different mobilities. In general, the "three point rule," illustrated by Easson and Stedman [64], describes the interactions that are necessary for chiral discrimination. A minimum of three simultaneous interactions have to occur between the chiral selector and one of the enantiomers so that chiral separation is achieved. The other enantiomer, due to spatial restrictions, should have at least two types of interactions, which can be hydrophobic interactions between the hydrophobic core of the polymer and the analyte, electrostatic interactions between the polar head group of the polymer and the analyte, dipole-dipole forces, such as hydrogen bonding between the polar group of the chiral selector and the analyte, and secondary interactions, such as π-π interactions, ion-dipole bonds, and Van der Waals forces. This difference in the number and type of interactions between the enantiomers and the chiral selector generates a mobility difference between the enantiomer-chiral selector complexes, which is necessary for the achievement of a chiral separation.

A big number of chiral selectors have been widely used, over the years, in CE for improved chiral separations of various classes of analytes. These chiral selectors include cyclodextrins, polymeric surfactants, cyclofructans, macrocyclic antibiotics, crown ethers, and others. Cyclodextrins are molecules with large ring-like structures composed of α-(1,4)-linked D-(+)-glucopyranose units. Native cyclodextrins are cyclic oligosaccharides consisting of six (α-CD), seven (β-CD) and eight (γ-CD) glucopyranose units. The chiral recognition ability of cyclodextrins can be improved by their derivatization with different functional groups, such as methyl-, sulfate-, acetyl- and prolyl-, and with the modification of the hydroxyl groups, which are present on the rim of the CD. The mechanism of enantiomeric discrimination is the inclusion of the hydrophobic group of the analyte into the cavity and interactions of the hydroxyl groups of the C2 and C3 at the upper rim of the CD, such as hydrogen bonds and dipole-dipole interactions.

Navarro *et al.* [65] developed a CZE method for the analysis of lansoprazole enantiomers in three different pharmaceutical preparations (Davur, Alter and Cinfa). β-CD was used as a chiral selector and sodium sulphite was used as an additive. Recoveries of 91-102% of the label content were obtained, and this demonstrated the potential of the method for the routine quality control of lansoprazole enantiomers in pharmaceutical formulations.

Chai *et al.* [66] used the chiral selector hydroxypropyl-γ-cyclodextrin in order to separate the antifungal drug iodiconazole and the structurally related triadimenol analogues. This chiral selector provided the best results in regard to resolution due to its large cavity and the hydrogen bonding between the analytes and the cyclodextrin. The mechanism for the chiral discrimination of hydroxypropyl cyclodextrins possibly involves the development of secondary interactions between the chiral analyte and the hydroxypropyl groups on the cyclodextrin rim after the inclusion of the analyte into the cavity. The degree of substitution and the type of the hydroxyalkyl group on the cyclodextrin rim, which influences the depth of the cavity, can therefore change the enantiorecognition ability of the cyclodextrin [67,68].

4.1. Method development and validation

The use of CE, and particularly MEKC, in chiral analysis is demonstrated further here by providing a more in-depth analysis of a research work that was performed in a pharmaceutical formulation that contained one of the enantiomers of Huperzine A [55]. Huperzine A is considered to be a potent, highly specific and reversible inhibitor of acetylcholinesterase with high efficiency and low toxicity. The mechanism of complexation of Huperzine A with acetylcholinesterase is similar to that of other pharmaceutical drugs that are used for the treatment of AD [69]. The (-)-enantiomer of Huperzine A is three times more biologically active than the synthetically racemic mixture, and only this form behaves as a potential acetylcholinesterase inhibitor. Therefore, the development of an analytical method for the enantiomeric separation of the synthetic Huperzine A is of greatest importance.

It is important here to mention that the type of the chiral selector used in this study was the polymeric surfactant. The use of polymeric surfactants in both chiral and achiral CE has attracted considerable attention. In 1994, Wang and Warner [70] were the first to report the use of a polymeric surfactant added to the BGE in MEKC. Polymeric surfactants offer several distinct advantages over conventional micelles [63,71-73]. Firstly, polymerization of the surfactant eliminates the dynamic equilibrium due to the formation of covalent bonds between the surfactant aggregates. This, in turn, enhances stability and improves resolution. Secondly, polymeric surfactants can be used at low concentrations because they do not depend on the CMC. This usually provides higher efficiencies and rapid analysis. They have, over the years, been extensively used in a BGE [74-80], in a polyelectrolyte multilayer coating [63,74,81-83], and in a CE-MS system [84-86].

In this study, the optimal conditions, in regard to resolution, efficiency and analysis time, were initially established by varying different electrophoretic parameters. The BGE type, concentration and pH are usually the first parameters to be examined in a method development procedure. Sodium acetate at acidic and neutral pHs, where the analyte exhibits cationic behavior, was chosen as the optimum. BGEs with basic pHs did not exhibit any enantiomeric discrimination, and the analysis time was very long. The optimum pH was 5.0 because it provided slightly better peak shapes, and the optimum concentration was 50 mM because it provided higher resolution (Figure 8). The very low peak efficiency, which needs to be improved, is clearly illustrated in this figure.

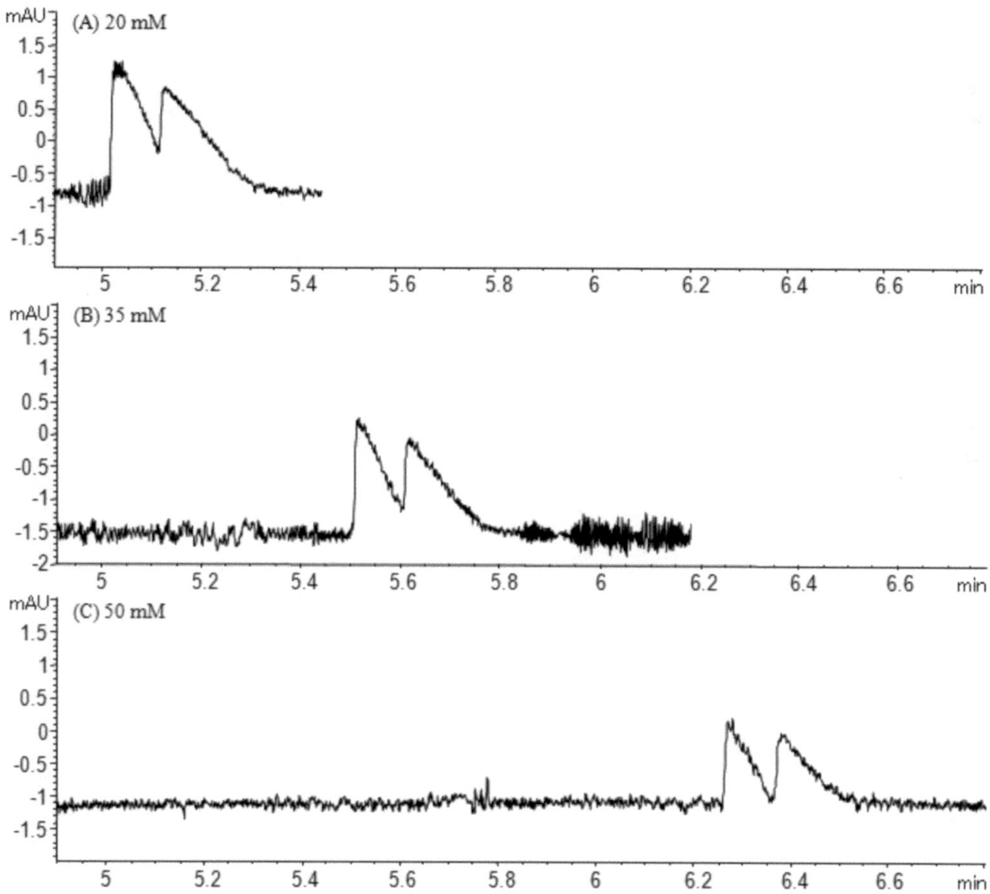

Figure 8. Effect of BGE concentration on the separation of the enantiomers of Huperzine A: (A) 20 mM, (B) 35 mM and (C) 50 mM. Separation conditions: BGE: sodium acetate (pH 5.0), 0.075% (w/v) poly-LL-SULV; pressure injection, 30 mbar for 3 s; applied voltage, 20 kV; temperature, 25 °C; fused-silica capillary, 64 cm (55.5 cm effective length) x 50 μm i.d.; detection, 230 nm [55].

As far as the chiral selector is concerned, different polymeric surfactants were examined, such as poly(sodium *N*-undecanoyl-L-leucinate) (poly-L-SUL), poly(sodium *N*-undecanoyl-LL-leucyl-leucinate) (poly-LL-SULL), poly(sodium *N*-undecanoyl-LL-leucyl-valinate) (poly-LL-SULV), poly(sodium *N*-undecanoyl-L-valinate) (poly-L-SUV), poly(sodium *N*-undecanoyl-L-valyl-glycinate) (poly-L-SUVG), poly(sodium *N*-undecanoyl-LL-alanyl-valinate) (poly-LL-SUAV), poly(sodium *N*-undecanoyl-LL-leucyl-alanate) (poly-LL-SULA), and poly(sodium *N*-undecanoyl-LL-valyl-valinate) (poly-LL-SUVV). The polymeric surfacant poly-LL-SULV, which has shown the best chiral discrimination ability for a number of pharmaceutical compounds [80], was the first to be examined in different concentrations. The concentration of 0.075% w/v was chosen as the optimum, based on analysis time, efficiency and resolution. This concentration though did not provide baseline resolution.

Another parameter examined in order to improve peak efficiency and resolution was the addition of modifiers. None of the organic solvents at different concentrations were able to improve the separation. An alternative to this was the addition of a salt, such as D- and L-alanine tert-butyl ester hydrochloride (D- and L-AlaC_4Cl). poly-LL-SULV became insoluble when the salt was added into the BGE. Therefore, the other polymeric surfactants mentioned above were examined at different concentrations. In each case, D- and L-AlaC_4Cl were used individually as additives, the electropherograms were obtained, and resolution and efficiency were estimated. Based on this, the combination of poly-LL-SUAV at a concentration of 0.20% (w/v) with L-AlaC_4Cl provided the best results.

However, the use of L-AlaC_4Cl did not provide satisfactory reproducibility of the migration time and efficiency. This is probably due to the hydrolysis of the salt in an aqueous BGE solution. An alternative involved the use of tert-butanol, one of the hydrolysis products, at different concentrations. Figure 9 clearly demonstrates the improved peak efficiency, in comparison with Figure 8. Each electropherogram was obtained at a different concentration of tert-butanol. A concentration of 10% (v/v) was the optimum, because it provided the highest resolution (1.45) and the highest peak efficiency (Figure 10).

The validation of the method demonstrated good linearities and very low relative standard deviation values, indicating excellent run-to-run and day-to-day reproducibilities. In addition, the LOD and LOQ were determined to be 4.17 μg/mL and 13.92 μg/mL, respectively.

4.2. Application

As previously shown, after method development and validation, the optimum separation conditions are applied to a real sample. In this case, the optimum parameters were applied to a pharmaceutical formulation in order to detect and quantitate the acetylcholinesterase inhibitor (-)-Huperzine A. The extraction procedure followed for extracting Huperzine A from the pharmaceutical formulation proved to be effective because the enantiomer determined in the sample was in a relatively good agreement with the amount that was stated on the bottle. Therefore, the developed MEKC-UV method is able to control the purity of (-)-Huperzine A in pharmaceutical formulations.

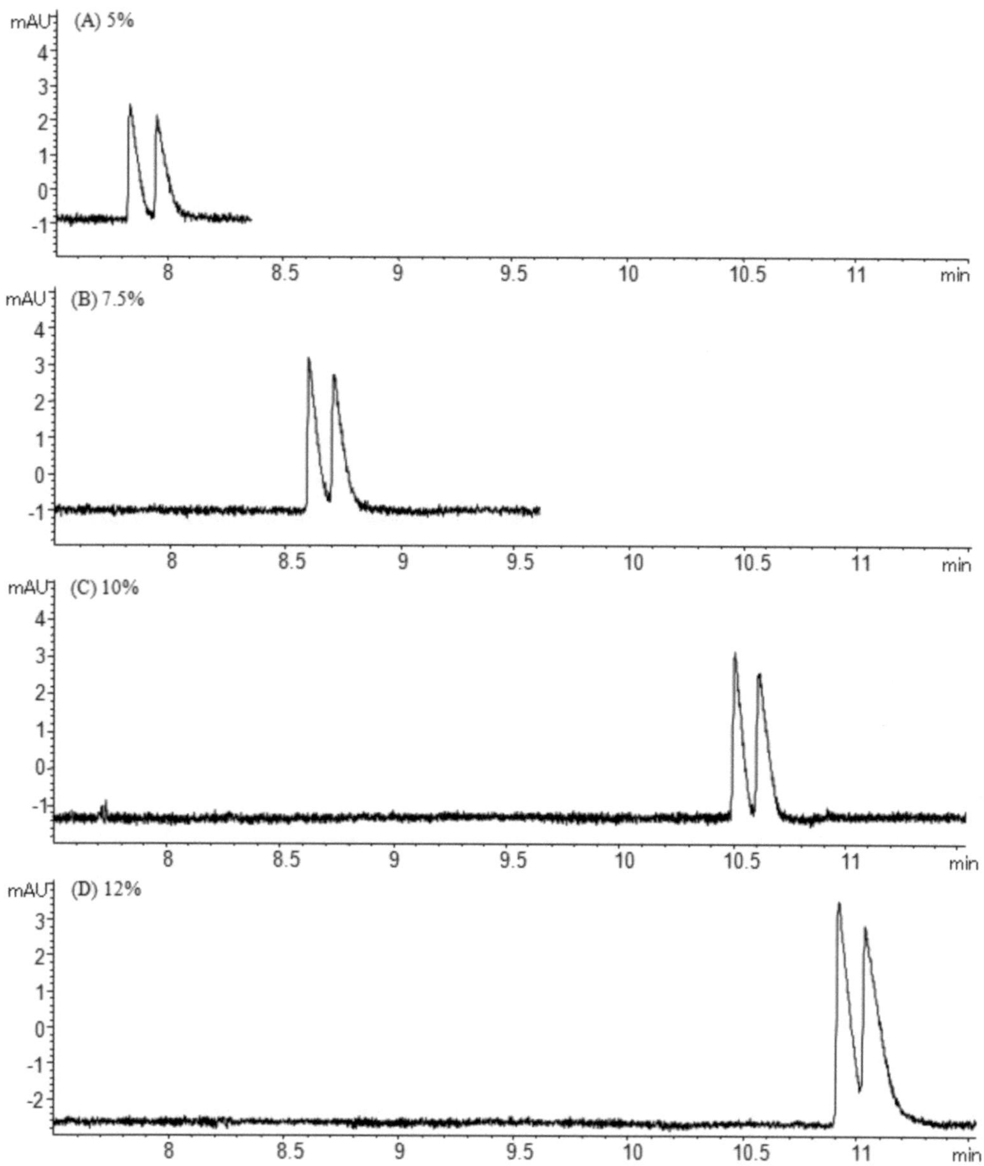

Figure 9. Effect of the concentration of tert-butanol on the separation of the enantiomers of Huperzine A: (A) 5%, (B) 7.5%, (C) 10% and (D) 12% (v/v). Separation conditions: BGE: 50 mM sodium acetate (pH 5.0), 0.2% (w/v) poly-LL-SUAV; pressure injection, 30 mbar for 3 s; applied voltage, 20 kV; temperature, 25 °C; fused-silica capillary, 64 cm (55.5 cm effective length) x 50 μm i.d.; detection, 230 nm [55].

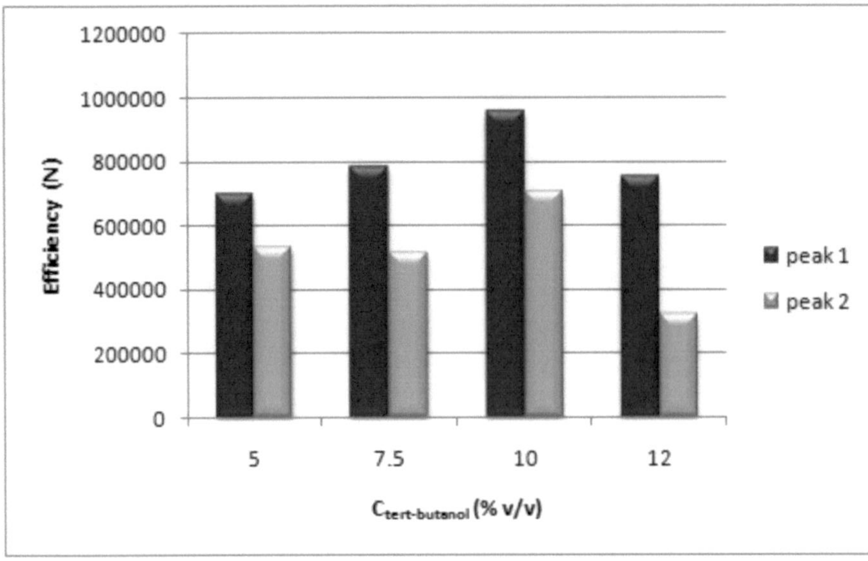

Figure 10. Effect of the concentration of tert-butanol on the efficiency. Separation conditions: Same as Fig. 4. BGE: 50 mM sodium acetate (pH 5.0), 0.2% (w/v) poly-LL-SUAV; pressure injection, 30 mbar for 3 s; applied voltage, 20 kV; temperature, 25 °C; fused-silica capillary, 64 cm (55.5 cm effective length) x 50 μm i.d.; detection, 230 nm [55].

5. Concluding remarks

Analysis of chiral and achiral analytes in natural, pharmaceutical and biological samples can be extremely difficult. Co-migration may occur, which can cause problems in detection, and the electropherograms obtained can be very complex. In addition, the analytes of interest are usually present in the matrices at very low concentrations. Therefore, all the analytical steps, including method development, detection and sample preparation, which is an essential stage in any analysis process, have to be optimized in order to obtain the desirable sensitivity, resolution, robustness and analysis time.

Among the separation techniques that have so far been used for pharmaceutical, clinical and food analysis, CE has been established as a powerful analytical tool, which has rapidly been developed and matured since its introduction. CE and its related techniques offer a number of advantages, including low consumption of sample and solvents, high separation efficiency, rapid method development, fast migration times, versatility, and simple instrumentation. Another important aspect involves its ability to separate small and large molecules, charged and neutral species, inorganic and organic molecules, synthetic and natural compounds, along with proteins and peptides.

The coupling of CE to MS provides nowadays a promising alternative to UV detection. The combination of high sensitivity, high selectivity, and high specificity provided by MS with high resolution, and high efficiency provided by CE makes it an attractive technique in different fields, such as clinical, forensic, pharmaceutical, and others. However, chiral analysis by use of CE-MS still needs some improvement, in regard to resolution and peak capacity. In addition, contamination of the ionization source induced by the chiral selector added in the BGE is still considered a main problem, even though different procedures have, in recent years, been developed in order to overcome this limitation [87].

Author details

Constantina P. Kapnissi-Christodoulou
Department of Chemistry, University of Cyprus, Nicosia, Cyprus

6. References

[1] Terabe S, Otsuka K, Nishi H (1994) Separation of Enantiomers by Capillary Electrophoretic Techniques. J. Chromatogr. A 666: 295-319.

[2] Chiari M, Nesi M, Righetti PG (1996) Capillary Electrophoresis in Analytical Biotechnology. Righetti, PG, Ed.; CRC Press, Inc.: Boca Raton, FL.

[3] Terabe S, Otsuka K, Ichikawa K, Tsuchiya A, Ando T (1984) Electrokinetic Separations with Micellar Solutions and Open-Tubular Capillaries. Anal. Chem. 56: 111-113.

[4] Terabe S, Otsuka K, Ando T (1985) Electrokinetic Chromatography with Micellar Solution and Open-Tubular Capillary. Anal. Chem. 57: 834-841.

[5] Terabe S, Ozaki K, Otsuka K, Ando T (1985) Electrokinetic Chromatography with 2-O-carboxymethyl-β-Cyclodextrin as a Moving "Stationary" Phase. J. Chromatogr. 332: 211-217.

[6] Terabe S (2008) Micellar Electrokinetic Chromatography for High-Performance Analytical Separation. The Chemical Record 8: 291-301.

[7] Ghosh D, Scheepens A (2009) Vascular Action of Polyphenols. Mol. Nutr. Food Res. 53: 322-331.

[8] Garcia-Lafuente A, Guillamon E, Villares A, Rostagno MA, Martinez JA (2009) Flavonoids as Anti-Inflammatory Agents: Implications in Cancer and Cardiovascular Disease. Inflamm. Res. 58: 537-552.

[9] Manach C, Mazur A, Scalbert A (2005) Polyphenols and Prevention of Cardiovascular Diseases. Curr. Opin. Lipidol. 16: 77-84.

[10] Rodriguez-Delgado MA, Malovana S, Perez JP, Borges T, Garcia Montelongo FJ (2001) Separation of Phenolic Compounds by High-Performance Liquid Chromatography With Absorbance and Fluorimetric Detection. J. Chromatogr. A 912: 249-257.

[11] Deng E, Zito SW (2003) Development and Validation of a Gas Chromatographic–Mass Spectrometric Method for Simultaneous Identification and Quantification of Marker Compounds Including Bilobalide, Ginkgolides and Flavonoids in Ginkgo Biloba L. Extract and Pharmaceutical Preparations. J. Chromatogr. A 986: 121-127.

[12] Ren ZY, Zhang Y, Shi YP (2009) Simultaneous Determination of Nine Flavonoids in *Anaphalis Margaritacea* by Capillary Zone Electrophoresis. Talanta 78: 959-963.

[13] Volpi N (2004) Separation of Flavonoids and Phenolic Acids from Propolis by Capillary Zone Electrophoresis. Electrophoresis 25: 1872-1878.

[14] Pazourek J, Gonzalez G, Revilla AL, Havel J (2000) Separation of Polyphenols in Canary Islands Wine by Capillary Zone Electrophoresis Without Preconcentration. J. Chromatogr. A 874: 111-119.

[15] Prasongsidh BC, Skyrray GR (1998) Capillary Electrophoresis Analysis of *trans*- and *cis*-Resveratrol, Quercetin, Catechin and Gallic Acid in Wine. Food Chem. 62: 355-358.

[16] Arce L, Tena MT, Rios A, Valcarcel M (1998) Determination of *trans*-Resveratrol and Other Polyphenols in Wines by a Continuous Flow Sample Clean-up System Followed by Capillary Electrophoresis Separation. Anal. Chim. Acta 359: 27-38.

[17] Nicolaou I, Kapnissi-Christodoulou CP (2010) Analysis of Polyphenols Using Capillary Zone Electrophoresis – Determination of the Most Effective Wine Sample Pre-Treatment Method. Electrophoresis 31: 3895-3902.

[18] Pietta P, Mauri P, Bruno A, Gardana C (1994) Influence of Structure on the Behavior of Flavonoids in Capillary Electrophoresis. Electrophoresis 15: 1326-1331.

[19] Zotou A, Frangi E (2008) Development and Validation of an SPE-LC Method for the Simultaneous Determination of *trans* -Resveratrol and Selected Flavonoids in Wine. Chromatographia 67: 789-793.

[20] Sun Y, Fang N, Chen DD, Donkor K.K (2008) Determination of Potentially Anti-Carcinogenic Flavonoids in Wines by Micellar Electrokinetic Chromatography. Food Chem. 106: 415-420.

[21] Rodriguez-Delgado MA, Perez JP, Corbella R, Gonzalez G, Garcia Montelongo FJ (2000) Optimization of the Separation of Phenolic Compounds by Micellar Electrokinetic Capillary Chromatography. J. Chromatogr. A 871: 427-438.

[22] Cordova AC, Jackson LSM, Berke-Schlessel DW, Sumpio BE (2005) The Cardiovascular Protective Effect of Red Wine. J. Am. Coll. Surg. 200: 428-439.

[23] Waterhouse AL (2002) Wine Phenolics. Ann. N.Y. Acad. Sci. 957: 21-36.

[24] Nicolaou I, Kapnissi-Christodoulou CP (2012) Development of a Capillary Electrophoresis-Mass Spectrometry Method for the Determination of Rivastigmine in Human Plasma - Optimization of the Limits of Detection and Quantitation. Electrophoresis In press.

[25] Shamsi SA (2002) Chiral Capillary Electrophoresis-Mass Spectrometry: Modes and Applications. Electrophoresis 23: 4036-4051.

[26] Peri-Okonny U, Kenndler E, Stubbs RJ, Guzman NA (2003) Characterization of Pharmaceutical Drugs by a Modified Nonaqueous Capillary Electrophoresis-Mass Spectrometry Method. Electrophoresis 24: 139-150.

[27] Zheng J, Jann MW, Hon YY, Shamsi SA (2004) Development of Capillary Zone Electrophoresis-Electrospray Ionization-Mass Spectrometry for the Determination of Lamotrigine in Human Plasma. Electrophoresis 25: 2033-2043.

[28] Varesio E, Cherkaoui S, Veuthey JL (1998) Optimization of CE-ESI-MS Parameters for the Analysis of Ecstasy and Derivatives in Urine. J. High Resol. Chromatogr. 21: 653-657.

[29] Elhamili A, Bergquist J (2011) A Method for Quantitative Analysis of an Anticancer Drug in Human Plasma With CE-ESI-TOF-MS. Electrophoresis 32: 1778-1785.

[30] Kapnissi-Christodoulou CP, Zhu X, Warner IM (2003) Analytical Separations in Open-Tubular Capillary Electrochromatography. Electrophoresis 24: 3917-3934.
[31] Mayer BX (2001) How to Increase Precision in Capillary Electrophoresis. J. Chromatogr. A 907: 21-37.
[32] Rosler M, Retz W, Retz-Junginger P, Dennle HJ (1998) Effects of Two-Year Treatment With the Cholinesterase Inhibitor Rivastigmine on Behavioural Symptoms in Alzheimer's Disease. Behav. Neurol. 11: 211-216.
[33] Nicolaou I, Kapnissi-Christodoulou CP (2011) Simultaneous Determination of Nine Acetylcholinesterase Inhibitors Using Micellar Electrokinetic Chromatography. J. Chromatogr. Sci. 49: 265-271.
[34] Pokorna L, Revilla A, Havela J, Patocka J (1999) Capillary Zone Electrophoresis Determination of Galanthamine in Biological Fluids and Pharmaceutical Preparatives: Experimental Design and Artificial Neural Network Optimization. Electrophoresis 20: 1993-1997.
[35] Banks JF (1997) Recent Advances in Capillary Electrophoresis/Electrospray/Mass Sprectrometry. Electrophoresis 18: 2255-2266.
[36] Niessen WMA, Tjaden UR, van der Greef J (1993) Capillary Electrophoresis-Mass Spectrometry. J. Chromatogr. 636: 3-19.
[37] Zheng J, Shamsi SA (2003) Combination of Chiral Capillary Electrochromatography with Electrospray Ionization Mass Spectrometry: Method Development and Assay of Warfarin Enantiomers in Human Plasma. Anal. Chem. 75: 6295-6305.
[38] Frankfort SV, Ouwehand M, van Maanen MJ, Rosing H, Tulner LR, Beijnen JH (2006) A Simple and Sensitive Assay for the Quantitative Analysis of Rivastigmine and its Metabolite NAP 226-90 in Human EDTA Plasma Using Coupled Liquid Chromatography and Tandem Mass Spectrometry. Rapid Commun. Mass Spectrom. 20: 3330-3336.
[39] Lefevre G, Allison M, Ho YY (2008) Pharmacokinetics and Bioavailability of the Novel Rivastigmine Transdermal Patch Versus Rivastigmine Oral Solution in Healthy Elderly Subjects. J. Clin. Pharmacol. 48: 246-252.
[40] Sha Y, Deng C, Liu Z, Huang T, Yang B, Duan G (2004) Headspace Solid-Phase Microextraction and Capillary Gas Chromatographic-Mass Spectrometric Determination of Rivastigmine in Canine Plasma Samples. J. Chromatogr. B 806: 271-276.
[41] Bhatt J, Subbaiah G, Kambli S, Shah B, Nigam S, Patel M, Saxena A, Baliga A, Parekh H, Yadav G (2007) A Rapid and Sensitive Liquid Chromatography–Tandem Mass Spectrometry (LC–MS/MS) Method for the Estimation of Rivastigmine in Human Plasma. J. Chromatogr. B 852: 115-121.
[42] Pommier F, Frigola R (2003) Quantitative Determination of Rivastigmine and its Major Metabolite in Human Plasma by Liquid Chromatography With Atmospheric Pressure Chemical Ionization Tandem Mass Spectrometry. J. Chromatogr. B 784: 301-313.
[43] Enz A, Chappuis A, Dattler A (2004) A Simple, Rapid and Sensitive Method for Simultaneous Determination of Rivastigmine and its Major Metabolite NAP 226-90 in

Rat Brain and Plasma by Reversed-Phase Liquid Chromatography Coupled to Electrospray Ionization Mass Spectrometry. Biomed. Chromatogr. 18: 160-166.

[44] Płotka JM, Biziuk M, Morrison C (2011) Common Methods for the Chiral Determination of Amphetamine and Related Compounds I. Gas, Liquid and Thin-Layer Chromatography. Trends in Analytical Chemistry 30: 1139-1158.

[45] Li L, Zhou S, Jin L, Zhang C, Liu W (2010) Enantiomeric Separation of Organophosphorus Pesticides by High-Performance Liquid Chromatography, Gas Chromatography and Capillary Electrophoresis and their Applications to Environmental Fate and Toxicity Assays. J. Chromatogr. B 878: 1264-1276.

[46] Wang Z, Ouyang J, Baeyens WRG (2008) Recent Developments of Enantioseparation Techniques for Adrenergic Drugs Using Liquid Chromatography and Capillary Electrophoresis: A Review. J. Chromatogr. B 862: 1-14.

[47] Cavazzini A, Pasti L, Massi A, Marchetti N, Dondi F (2011) Recent Applications in Chiral High Performance Liquid Chromatography: A Review. Anal. Chim. Acta 706: 205-222.

[48] Ilisz I, Berkecz R, Péter A (2008) Application of chiral derivatizing agents in the high-performance liquid chromatographic separation of amino acid enantiomers: A review. J. Pharmaceutical and Biomedical Analysis 47: 1-15.

[49] König WA, Ernst K (1983) Application of Enantioselective Capillary Gas Chromatography to the Analysis of Chiral Pharmaceuticals. J. .Chromatogr. A 280: 135-141.

[50] Giuffrida A, Tabera L, González R, Cucinotta V, Cifuentes A (2008) Chiral Analysis of Amino Acids from Conventional and Transgenic Yeasts. J. Chromatogr. B 875: 243-247.

[51] Huang L, Lin JM, Yu L, Xu L, Chen G (2009) Improved Simultaneous Enantioseparation of β-Agonists in CE Using β-CD and Ionic Liquids. Electrophoresis 30: 1030-1036.

[52] Suntornsuk L, Ploygam S (2010) Simultaneous Determination of R-(−)-, S-(+)-Baclofen and Impurity A by Electrokinetic Chromatography. J. Pharmaceutical and Biomedical Analysis 51: 541-548.

[53] Dominguez-Vega E, Sanchez-Hernandez L, Garcia-Ruiz C, Crego AL, Marina ML (2009) Development of a CE-ESI-ITMS Method for the Enantiomeric Determination of the Non-Protein Amino Acid Ornithine. Electrophoresis 30: 1724-1733.

[54] Deñola NL, Quiming NS, Catabay AP, Saito Y, Jinno K (2008) Effects of Alcohols on CE Enantioseparation of Basic Drugs With Native γ-CD as Chiral Selector. J. Sep. Sci. 31: 2701-2706.

[55] Tsioupi DA, Nicolaou I, Moore L, Kapnissi-Christodoulou CP (2011) Chiral Separation of Huperzine A Using CE - Method Validation and Application in Pharmaceutical Formulations. Electrophoresis 33: 516-522.

[56] Mayer S, Schurig V (1992) Enantiomer Separation by Electrochromatography on Capillaries Coated With Chirasil-dex. J. High Res. Chromatogr. 15: 129-131.

[57] Mayer S, Schurig V (1993) Enantiomer Separation by Electrochromatography in Open Tubular Columns Coated with Chirasil-Dex. J. Liq. Chromatogr. 16: 915-931.

[58] Mayer S, Schurig V (1994) Enantiomer Separation Using mobile and Immobile Cyclodextrin Derivatives With Electromigration. Electrophoresis 15: 835-841.

[59] Yang J Hage DS (1994) Chiral Separations in Capillary Electrophoresis Using Human Serum Albumin as a Buffer Additive. Anal. Chem. 66: 2719-2725.

[60] Tanaka Y, Terabe S (2000) Studies on Enantioselectivities of Avidin, Avidin-Biotin Complex and Streptavidin by Affinity Capillary Electrophoresis. Chromatographia 49: 489-495.

[61] Lin JM, Uchiyama K, Hobo T (1998) Enantiomeric Resolution of Dansyl Amino Acids by Capillary Electrochromatography Based on Molecular Imprinting Method. Chromatographia 47: 625-629.

[62] Quaglia M, de Lorenzi E, Sulitzky C, Massolini G, Sellergren B (2001) Surface Initiated Molecularly Imprinted Polymer Films: A New Approach in Chiral Capillary Electrochromatography. Analyst 126: 1495-1498.

[63] Kapnissi CP, Valle BC, Warner IM. (2003) Chiral Separations Using Polymeric Surfactants and Polyelectrolyte Multilayers in Open-Tubular Capillary Electrochromatography. Analytical Chemistry 75: 6097-6104.

[64] Easson, LH, Stedman E (1933) Studies on the Relationship Between Chemical Constitution and Physiological Action. Biochem. J. 27: 1257-1266.

[65] Nevado JJB, Penalvo GC, Sanchez JCJ, Mochon MC, Dorado RMR, Navarro MV (2009) Optimisation and Validation of a New CE Method for the Determination of Lansoprazole Enantiomers in Pharmaceuticals. Electrophoresis 30: 2940-2946.

[66] Li W, Zhao L, Tan G, Sheng C, Zhang X, Zhu Z, Zhang G, Chai Y (2011) Enantioseparation of the New Antifungal Drug Iodiconazole and Structurally Related Triadimenol Analogues by CE with Neutral Cyclodextrin Additives. Chromatographia 73: 1009-1014.

[67] Wedig M, Laug S, Christians T, Thunhorst M, Holzgrabe U (2002) Do we Know the Mechanism of Chiral Recognition Between Cyclodextrins and Analytes? J. Pharmaceutical and Biomedical Analysis 27: 531-540.

[68] Valkó IE, Billiet HAH, Frank J, Luyben KCAM (1994) Effect of the Degree of Substitution of (2-hydroxy)propyl-β-Cyclodextrin on the Enantioseparation of Organic Acids by Capillary Electrophoresis. J. Chromatogr. A 678: 139-144.

[69] Zangara A (2003) The psychopharmacology of huperzine A: an alkaloid with cognitive enhancing and neuroprotective properties of interest in the treatment of Alzheimer's disease. Pharm. Biochem. Behav. 75: 675-686.

[70] Wang J, Warner IM (1994) Chiral Separations Using Micellar Electrokinetic Capillary Chromatography and a Polymerized Chiral Micelle. Anal. Chem. 66: 3773-3776.

[71] Yarabe HH, Billiot E, Warner IM (2000) Enantiomeric Separations by Use of Polymeric Surfactant Electrokinetic Chromatography. J. Chromatogr. A 875: 179-206.

[72] Palmer CP, Tanaka N (1997) Selectivity of Polymeric and Polymer-Supported Pseudo-Stationary Phases in Micellar Electrokinetic Chromatography. J. Chromatogr. A 792: 105-124.

[73] Palmer CP, Terabe S (1997) Micelle Polymers as Pseudostationary Phases in MEKC: Chromatographic Performance and Chemical Selectivity. Anal. Chem. 69: 1852-1860.

[74] Luces CA, Warner IM (2010) Achiral and Chiral Separations Using MEKC, Polyelectrolyte Multilayer Coatings, and Mixed Mode Separation Techniques With Molecular Micelles. Electrophoresis 31: 1036-1043.

[75] Palmer CP, Terabe S (1997) Micelle Polymers as Pseudostationary Phases in MEKC: Chromatographic Performance and Chemical Selectivity. Anal. Chem. 69: 1852-1860.

[76] Shamsi SA, Akbay C, Warner IM (1998) Polymeric Anionic Surfactant for Electrokinetic Chromatography: Separation of 16 Priority Polycyclic Aromatic Hydrocarbon Pollutants. Anal. Chem. 70: 3078-3083.

[77] Dobashi A, Hamada M, Dobashi Y (1995) Enantiomeric Separation with Sodium Dodecanoyl-L-amino Acidate Micelles and Poly(sodium (10-undecenoyl)-L-valinate) by Electrokinetic Chromatography. Anal. Chem. 67: 3011-3017.

[78] Agnew-Heard KA, Sanchez Pena M, Shamsi SA, Warner IM (1997) Studies of Polymerized Sodium *N*-Undecylenyl-L-valinate in Chiral Micellar Electrokinetic Capillary Chromatography of Neutral, Acidic, and Basic Compounds. Anal. Chem. 69: 958-964.

[79] Billiot E, Thibodeaux S, Shamsi S, Warner IM (1999) Evaluating Chiral Separation Interactions by Use of Diastereomeric Polymeric Dipeptide Surfactants. Anal. Chem. 71: 4044-4049.

[80] Shamsi SA, Valle BC, Billiot F, Warner IM (2003) Polysodium *N*-Undecanoyl-L-leucylvalinate: A Versatile Chiral Selector for Micellar Electrokinetic Chromatography. Anal. Chem. 75: 379-387.

[81] Zhu X, Kamande MW, Thiam S, Kapnissi CP, Mwongela SM, Warner IM (2004) Open-Tubular Capillary Electrochromatography/Electrospray Ionization-Mass Spectrometry Using Polymeric Surfactant as a Stationary Phase Coating. Electrophoresis 25: 562-568.

[82] Kapnissi CP, Akbay C, Schlenoff JB, Warner IM (2002) Analytical Separations Using Molecular Micelles in Open-Tubular Capillary Electrochromatography. Anal. Chem. 74: 2328-2335.

[83] Kamande MW, Kapnissi CP, Akbay C, Zhu X, Agbaria RA, Warner IM (2003) Open-Tubular Capillary Electrochromatography Using a Polymeric Surfactant Coating. Electrophoresis 24: 945-951.

[84] He J, Shamsi SA (2009) Multivariate Approach for the Enantioselective Analysis in MEKC-MS: II. Optimization of 1,1′-Binaphthyl-2,2′-Diamine in Positive Ion Mode. J. Sep. Sci. 32: 1916-1926.

[85] He J, Shamsi SA (2009) Multivariate Approach for the Enantioselective Analysis in Micellar Electrokinetic Chromatography-Mass Spectrometry: I. Simultaneous Optimization of Binaphthyl Derivatives in Negative Ion Mode. J. Chromatogr. A 1216: 845-856.

[86] Hou J, Zheng H, Shamsi SA (2007) Separation and Determination of Warfarin Enantiomers in Human Plasma Using a Novel Polymeric Surfactant for Micellar Electrokinetic Chromatography–Mass Spectrometry. J. Chromatogr. A 1159: 208-216.

[87] Simo C, Garcia-Canas V, Cifuentes A (2010) Chiral CE-MS. Electrophoresis 31: 1442-1456.

Chapter 3

Numerical Modelling of Light Propagation for Development of Capillary Electrophoretic and Photochemical Detection Systems

Tomasz Piasecki, Aymen Ben Azouz, Brett Paull,
Mirek Macka and Dermot Brabazon

Additional information is available at the end of the chapter

http://dx.doi.org/10.5772/45764

1. Introduction

Capillary electrophoresis (CE) has been used for over 30 years as an efficient separation technique [1]. These separations are typically preformed using capillaries with internal diameter ranging from 2 μm to 200μm and as the diameter increases above 100μm, a reduction in separation performance is observed [2]. One of the main factors limiting the overall performance of CE separations is Joule heating associated with the passing of the electric current through resistive medium. The high surface-to-volume ratio in a smaller capillary diameter allows for efficient heat dissipation which is beneficial for compound separation, however, simultaneously this has a negative impact on the detection performance due to the corresponding reduction in analyte volume available for detection [3].

Historically when capillaries were first used for CE, they were coated with polyimide and presented excellent robustness but posed problems with optical detection as polyimide is highly absorbing below 550nm preventing optical detection in that spectral range. Such capillaries were striped of the coating to enable the optical detection in UV and low wavelength visible range. However, capillaries striped of their coating are brittle and can easily be damaged making usage often impractical. The introduction of the polytetrafluoroethylene (PTFE) coated capillaries allowed for optical detection within the UV range as well as across the entire visible spectrum. Although the PTFE coated capillaries are not as durable as the polyimide coated capillaries, they are significantly more robust than coatless ones and allow for easier deployment during typical daily laboratory routine.

Absorbance photometry is a detection technique used to determine the concentration of target species in a liquid sample based on interaction between the probe light and species [4]. Typically it is a measurement of the light intensity with and without a sample placed in the light path. A scheme of light intensity measurement with the sample located for detection is presented in Figure 1. The sample transmittance, *T,* is defined as the ratio of the initial light intensity, I_0, to the recorded light intensity, I_1 (Eq. 1). I_0 should be measured with the sample holding cuvette empty. This allows for reflections and potential absorption by the cuvette material to be taken into account during sample measurement. Sample absorbance, *A,* is measured as the negative log of the transmittance (Eq. 2). The cuvette length l is known, as well as species molar absorptivity coefficient α, which is a specific characteristic of every species. Light attenuation along the light path is governed by Beer-Lambert's law where c is molar concentration (Eq. 3) [5]. The method of calculating the actual optical path length is presented elsewhere [6].

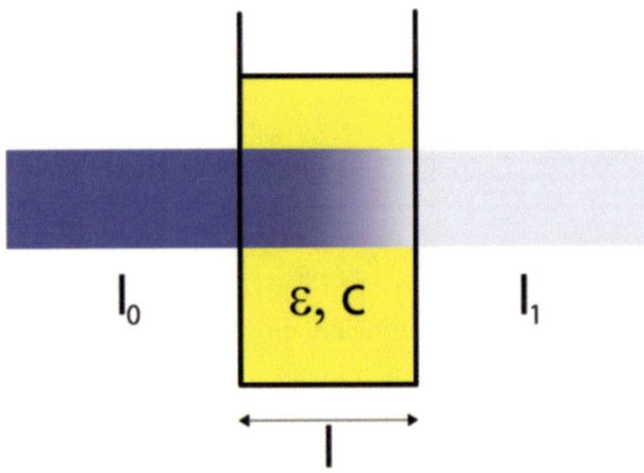

Figure 1. Schematic of light intensity with sample located for photometric detection

$$T = \frac{I_1}{I_0} \tag{1}$$

$$A = -\,log\ T = -log\ \frac{I_1}{I_0} \tag{2}$$

$$I(l) = I_0 \cdot e^{-\alpha c l} = I_0 \cdot e^{-A} \tag{3}$$

Despite excellent performance in analyte separation, CE techniques still constitute a minority of the commercial applications due to detection limitations imposed by popular absorbance photometric detectors. Various developments in CE have occurred such as the development of UV-transparent capillaries which facilitated CE by making it possible to employ the commercially available detectors. In the past different approaches have also been employed to alter the geometry of the detection system in order to improve the detection performance for CE. These include the application of rectangular capillaries to reduce refraction effects on the cylindrical boundaries [2], multi-reflection cells for increased signal intensity [7], and development of Z-shaped flow cells with increased optical pathlength for analyte detection [8]. Some detectors were built employing these methods, but to date the majority of the commercially available absorbance based detectors work on the basis of a single passage of light through the sample contained in a capillary.

To gain a better understanding of light propagation through capillaries and how to improve detection levels, the modelling of the light propagation through capillaries has been previously undertaken. Much of this work was limited to two-dimensional projections and two-layer models (no coating present) taking into account capillary material and inner cavity [9]. Previous numerical simulations of light propagation through capillaries have been reported as a tool for flow-cell design and have been limited to prediction of light path for these specific cases only [10]. The numerical problem of ray-tracing through a capillary is an excellent example where advantage can be taken of the ability to perform numerous calculations quickly on a computer to allow for ray paths, ray path overlaps, and resultant light intensities to be calculated.

This book chapter presents a theoretical study on the light propagation through coated capillaries, focusing on PTFE-coated capillaries. These models can be used to increase the performance of absorbance photometric detection and for associated photochemical applications.

2. General description of the model

The Light propagation and light intensity distribution models were developed using LabVIEW™ 2011 graphical programming environment. Graphic-related work, such as generation of multi-colour maps and reading pixel colours were conducted with Adobe Photoshop CS 3 ver. 10.0.1. Dimension and angle measurements taken from photos were performed using Image J 1.43u software.

Whenever light is incident on the boundary of two transparent dielectrics part of it is reflected and part is transmitted, see Figure 2. The angle of incidence is related with the angle of transmittance by Snell's Law (Eq. 4). In the case of cylindrical symmetry, as in capillaries, it is possible that the exiting ray will not be parallel with the incident, see Figure 3.

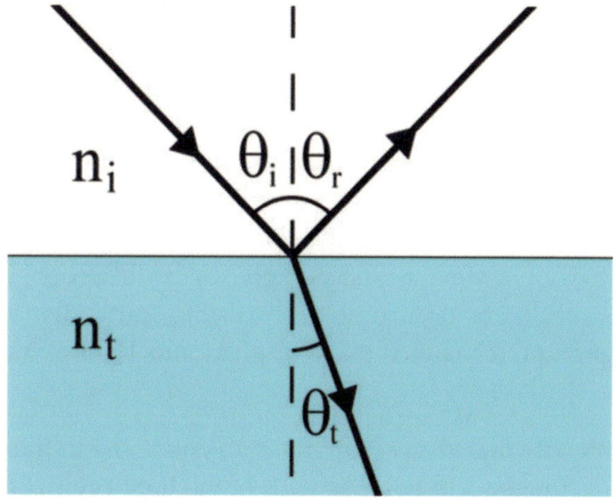

Figure 2. Light incident on boundary of two different dielectrics.

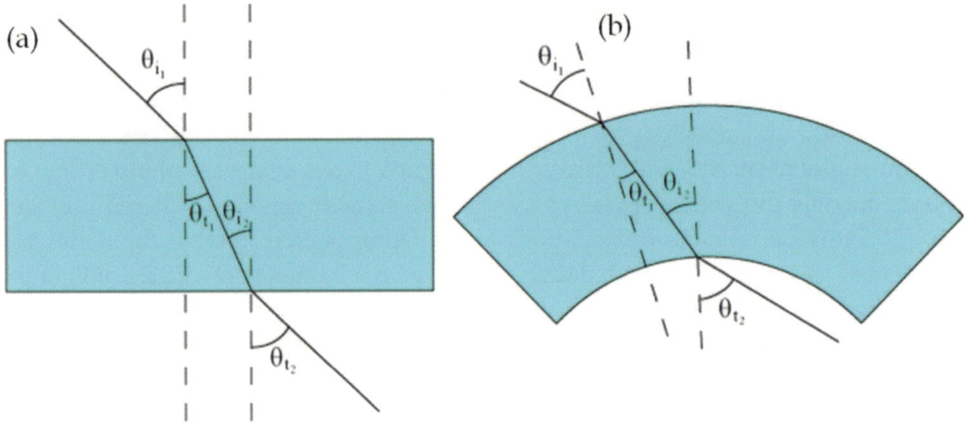

Figure 3. Schematic of light transmission through, (a) a flat transparent material and (b) a cylindrical capillary, showing curvature affects on angle of incidence and transmittance.

$$n_i \cdot sin\,\theta_i = n_t \cdot sin\,\theta_t \tag{4}$$

where n_i is the refractive index of the incident ray transmission medium, θ_i is the angle between incident wavevector and the normal to surface, n_t is the refractive index of the material through with the transmitted ray passes and θ_t is the angle between transmitted wavevector and normal to the surface. Light attenuation inside the absorbing medium is governed by Beer-Lambert law (Eq. 5).

$$I = I_0 \cdot e^{-\alpha c l} \tag{5}$$

where I is the initial light intensity, I_0 is the light intensity behind the sample, α is the molar absorptivity coefficient, c is concentration of the compound and l is length is the absorption path length.

The numerical modelling software was developed to calculate the light ray path through multi-layered cylinders, and the light intensity distribution map through the cylinder cross-sectional area. Light propagation within multimode optical fibres occurs by the phenomena of total internal reflection, where the values of refractive indices and fibre diameter remain within the limits of geometrical optics. The size of the capillary used in this work was comparable with the size of multimode optical fibres. It was assumed that capillary body, coating and bore were perfectly cylindrical and concentric. Only the right half of the capillary cross-section is displayed in the developed model, as the diameter acts as the modelled axis of symmetry and no light ray could propagate through from left to right side. In general such occurrence is possible, but only for higher values of refractive indices approximately twice those of glass and PTFE which were used in this work.

The programmed model calculated the light ray path equations in the Cartesian coordinate system. Separate linear functions to describe each of the light ray path segments were used (for example between air/tube, tube/tube or tube/liquid).

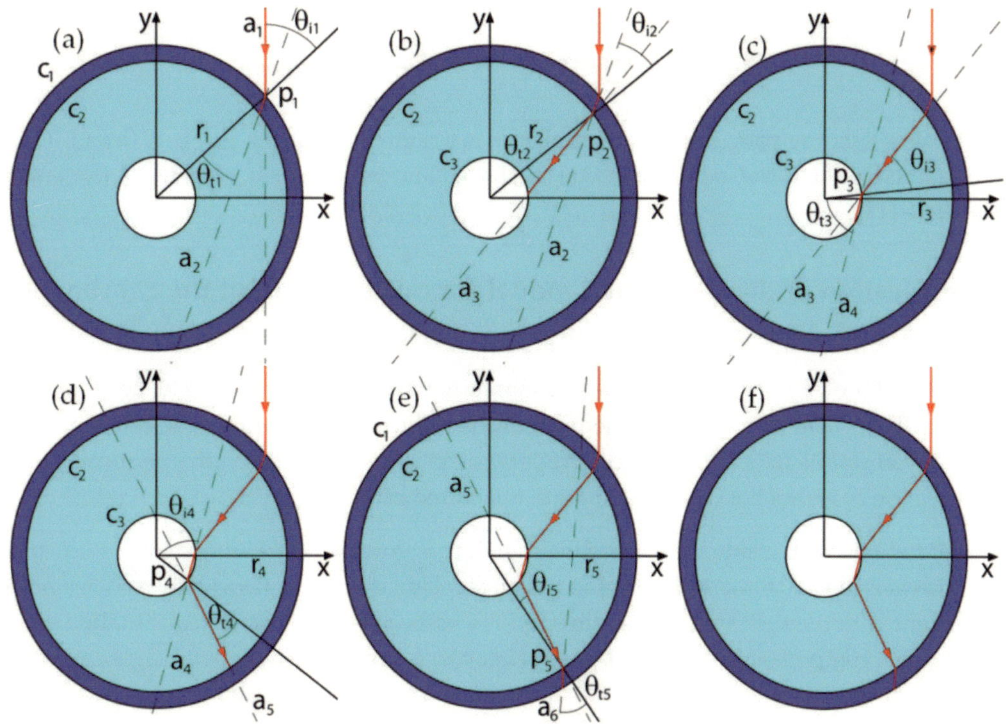

Figure 4. Schematic of reference lines and points, and light ray path calculated by each sub-routine for light passing through a three layer system in which the light ray path is passing through a coated hollow capillary.

Each light path segment was calculated in a separate subroutine calculating the light path in each zone. Incident light was in the form of rays parallel to y-axis, see Figure 4. The first subroutine calculated the coordinates of the light incident from infinity on the air/coating boundary under desired angle. This point of incidence of the ray a_1 on outermost circle c_1 was assigned as p_1. An extended radius r_1 going from (0, 0) through p_1 was drawn for calculation of angle of incidence θ_{i1} and in turn angle of transmission θ_{t1} was calculated from Eq .4, for the use inputted values of refractive indices n_i and n_t. A line a_2 through p_1 was drawn representing the refracted ray in the coating with θ_{t1} as the angle between a_2 and r_1, *ending the first program subroutine,* see Figure 4A. The next subroutine began with calculation of point p_2 (where line a_2 crossed boundary c_2) and the drawing of extended radius r_2 from (0, 0) through p_2. Angle of incidence θ_{i2} was calculated as the angle bounded by r_2 and a_2, see Figure 4B. Angle of transmittance θ_{t2} was calculated from Eq. 4 and the line a_3 was drawn where θ_{t2} was and angle between r_2 and a_3, *ending the second program subroutine,* see Figure 4B. This subroutine was iterated a further three times to calculate light ray paths segments along lines a_4, a_5 and a_6 after refraction on each encountered boundary. These next subroutines of the light path formation are illustrated in Figure 4C to E and the complete generated light path is presented in Figure 4F without reference lines. The entire light path was represented as a sum of individual rays calculated separately according to the symbolic algorithm:

$$r = \sum_{k=1}^{2n} \left(a_k \cap c_k \rightarrow p_k \rightarrow r_k \rightarrow \theta_{i_k} \rightarrow \theta_{t_k} \rightarrow a_{k+1}\right)$$

where r is a light ray path, k is a step number; n is a number of layers; a, c, p, r, θ_i and θ_t are as described earlier. The experiments on model validation and their results are presented elsewhere [11].

3. Application of the developed model for optimisation of the capillary-based optical detection system

The main aim of this part is to present quantitative information about light behaviour in capillary base detection setups. Their highly complicated geometry and multi-layered construction results in unusual optical properties can significantly affect the performance of the detection system if not designed and manufactured carefully.

The light attenuation inside light absorbing medium is governed by Lambert-Beer's law (Eq. 2). However in this form it is assumed that the light that is collected by a detector is concurrent with the light that passed through the absorbing medium. In real situation there is additional component that can seriously affect photometric measurements: the stray light. It can be defined as the light that omits the sample but still is collected by the detector. In such case (Eq. 2) should be written in another form (Eq.4):

$$A = -log\frac{I_A - I_s}{I_0} \qquad (6)$$

where: A is absorbance, I_A is the light intensity of the attenuated light beam, I_S is intensity of the stray light, and I_0 is the initial light intensity. In such case, with increase of the amount of light attenuating medium I_A goes down to zero and Eq. 6 becomes Eq. 7, effectively imposing maximum possible absorbance:

$$A = -log\frac{I_S}{I_0} = const. \qquad (7)$$

Figure 5 presents the maximum possible absorbance recorded depending on the percentage amount of the stray light in the detection system. For the illustration purpose, the relation between concentration and absorbance is set to one. The line labelled "Ideal" represents situation with no stray light at all, giving completely linear response depending on the test compound concentration. The other four curves labelled "10%", "1%", "0.1%" and "0.01%" represent the same response with the respective percentage of the stray light present in the system. This allows visualising the detection limits associated with the imperfectness of the detection system. Also it is noteworthy to compare the deviation from the linearity that is associated with the stray light.

Figure 6 presents the deviation from the ideal response depending on the amount of the stray light. As it can be read from the graph on Figure 6, the deviation is small, and therefore assumption of linearity is valid typically one absorbance unit before the maximum, e.g. for

0.1% of the stray light, the maximum measure absorbance is 3 AU, and the assumption of linearity is valid from 0 to 2 AU.

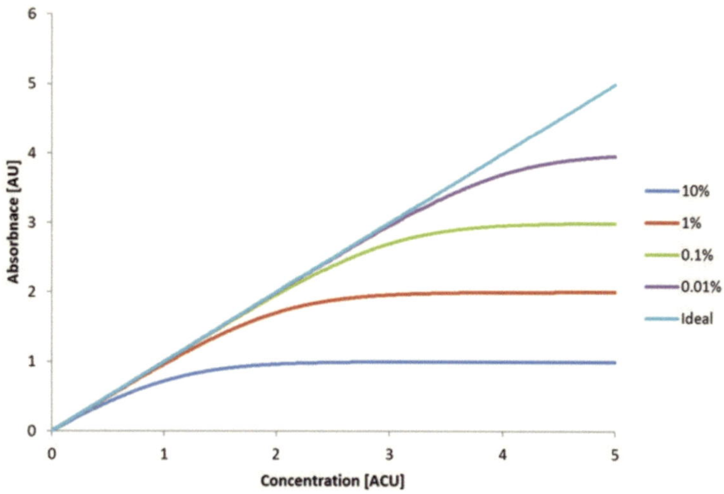

Figure 5. A comparison of different amounts of stray light present in the detection system and how it imposes detecion limits. Concentrations are scaled in arbitrary concentration units (ACU). Absorbance scaled in absorbance units (AU).

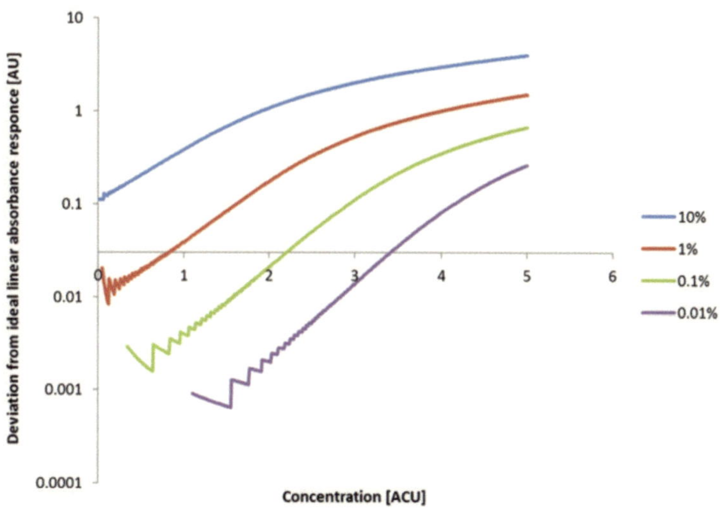

Figure 6. A comparison of the deviation from the linear response depending on the amount of stray light in the system. Concentration scaled in arbitrary concentration units (ACU). Absorbance scaled in absorbance units (AU).

For numerous reasons it is convenient to employ the same capillary for detection after any reactions or separations. The cylindrical geometry of the capillary combined with its multi-layer structure creates a quite complex medium for the light to propagate through and eventually can significantly alter the light path. Figure 7 presents a simplified image of possible light refractions on different boundaries (air/coating, coating/fused silica body, body/content). In this figure only light perpendicular to the capillary symmetry axis is portrayed. In real situation light coming out from an uncollimated light source can be incident under any angle. This situation can be compared with a capillary illuminated by a laser.

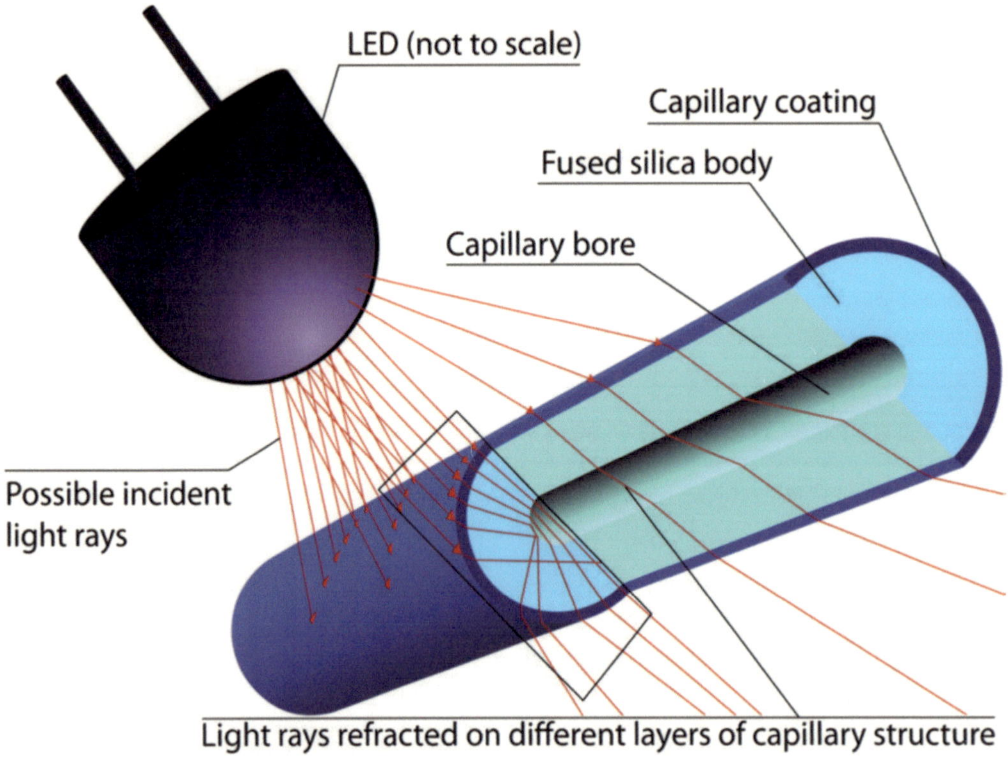

Figure 7. Schematic of a capillary cross-section illuminated by a beam of parellel rays.

In order to prevent undesired light to reach the detector the light source and the detector should be comparable with the capillary bore internal diameter. This can be achieved by installing an aperture or by using an optical fibre to illuminate the capillary and collect the light for detection. Such application of optical fibres has been reported already several years ago [12], [13]. Application of any of these two approaches has an impact on the overall performance of the detection system. Light going through an aperture undergoes diffraction and light coming out of an optical fibre is divergent with the spread angle confined by numerical aperture (NA) of the fibre itself. The NA of 0.22 gives cone of acceptance with maximum possible angle of 25.4°, being as well the maximum angle at which the light will exit the fibre. Figure 8 present a comparison of the light intensity that can interact with capillary bore content depending on the distance at which a 100μm collecting optical fibre of with NA of 0.22 for four different maximum incidence cones. Zero degrees is equivalent of an ideally parallel beam, while 25 degrees is an equivalent of the light coming out of an optical fibre. Light intensity loss is the effect of the capillary geometry only and the capillary coating, capillary body and the theoretical solution filling the capillary bore are assumed to be completely transparent. As it can be seen the light coming out of the optical fibre has the angle of divergence of 25 degrees. This results that only over 22% of the delivered light intensity can reach the capillary bore, interact with its content and be collected by similar 100μm diameter optical fibre. That number decreases with distance as light after propagating through the entire capillary becomes even more divergent, similarly to a ball lens.

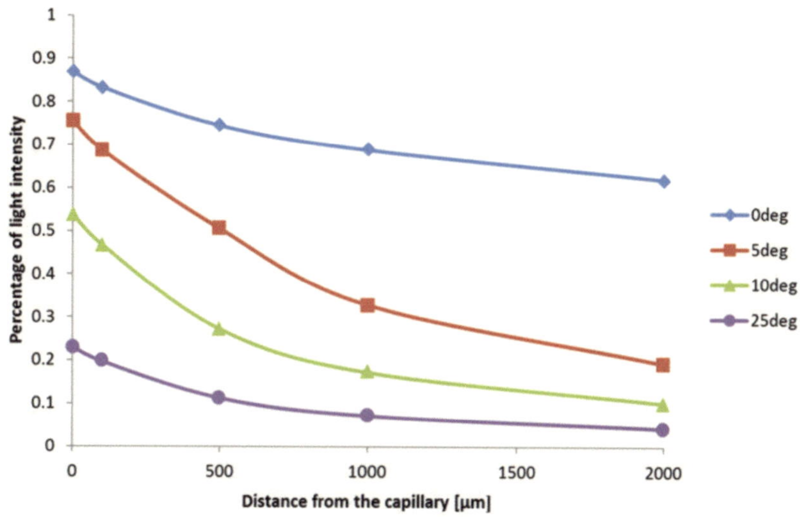

Figure 8. Relation between the percentage of light passing though the capillary bore, comming from a 100 μm fibre, and the distance between the collecting 100 μm optical fibre and the capillary. These results were calculated for incident beams with 0, 5, 10 and 25 degrees divergent cone of illumination.

The light intensity loss associated with the divergent illuminating beam is not the only concern. At the same time due to the capillary geometry it is possible that the light exiting the illuminating fibre will enter the capillary, propagate through omitting the capillary bore completely and exit it in a manner that can still enter the collecting optical fibre. Such light would contribute toward the stray light and reduce overall efficiency of the detection system. Figure 9 illustrates the percentage amount of the stray light that can enter collecting optical fibre under an angle below the numerical aperture depending on a distance between the capillary and the end of the fibre.

For the collecting fibre perfectly touching the capillary the amount of stray light is zero. However already at distance of 0.5mm this goes up to 40% if the capillary is illuminated with an optical fibre as well. The observed maximum for 10 degrees divergent incident beam (Figure 9 - green line) followed by drop of stray light to zero a 2mm from capillary is a result of the capillary geometry – light omitting the capillary bore after passing the coating and exiting the capillary becomes divergent and eventually is not present within the central region where the collecting optical fibre is located. However the value of 10 degrees divergent incident beam is possible theoretically it is unlikely to occur in the real experimental setup.

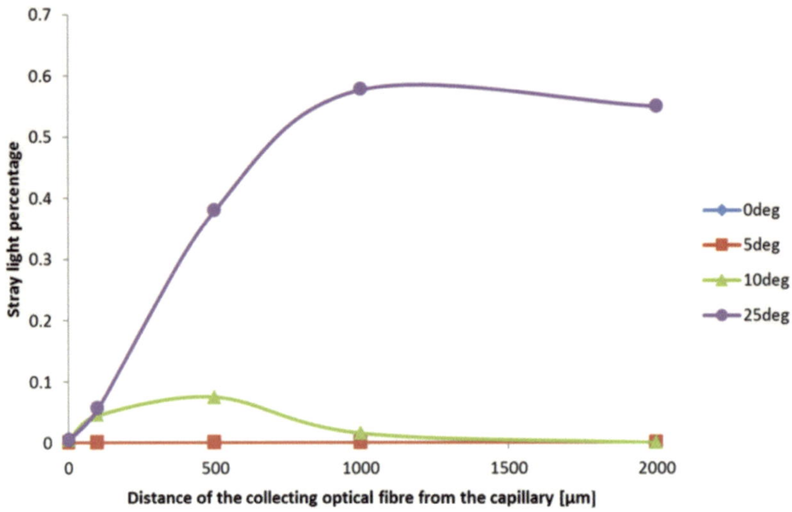

Figure 9. Percentage of stray light collected by a 100μm optical fibre related to its distance from capillary and for four different values of the incident beam divergence angle: 0, 5, 10 and 25 degrees.

The divergence value of 25 degrees as the divergence of the light exiting fused silica optical fibre was selected for other comparison – what diameter of the illuminating and collecting optical fibre should be used to maximise the amount of light interacting with the capillary content while minimising the amount of stray light. The values were modelled for 100μm internal diameter capillary. Two different approaches were tested:

- wide-area illumination and small area collection, attempting to collect only the light passing through the capillary bore.
- narrow-area illumination attempting only to illuminate the capillary bore and collecting as much light as possible

The results are summarised in Figure 10 and Figure 11. Figure 10 presents the collected light intensity that passed through the capillary bore and could interact with its content depending on the distance of the collecting fibre from the capillary.

Figure 11 presents percentage amount of stray light associated with the respective setting. The first number represents the illuminating optical fibre diameter in micrometres, and the second represents the collecting fibre diameter in micrometres as well.

The highest overall light intensity of the light able to interact with the capillary content was achieved with 50μm illuminating and 200μm collecting fibres. However at the position when maximum light intensity is collected the amount of stray light was unsatisfactory (close to 20%). The large diameter of the collecting fibre, twice of the internal capillary diameter, was large enough for the light to propagate around the capillary bore and still enter the fibre under a sufficient angle to stay within the optical fibre.

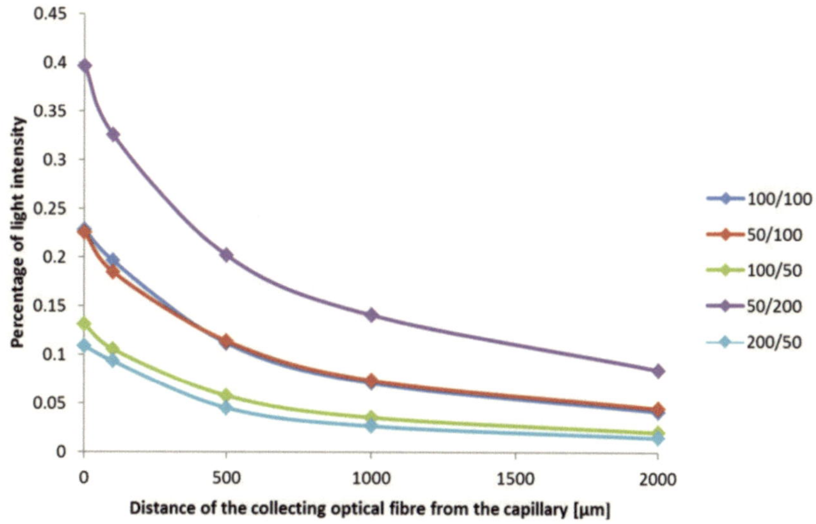

Figure 10. The amount of the initial light intensity that propagates through the capillary bore and subsequently is collected compared for different combinations of diamteres of the illuminating fibre (the first number) and the collecting fibre (the second number)

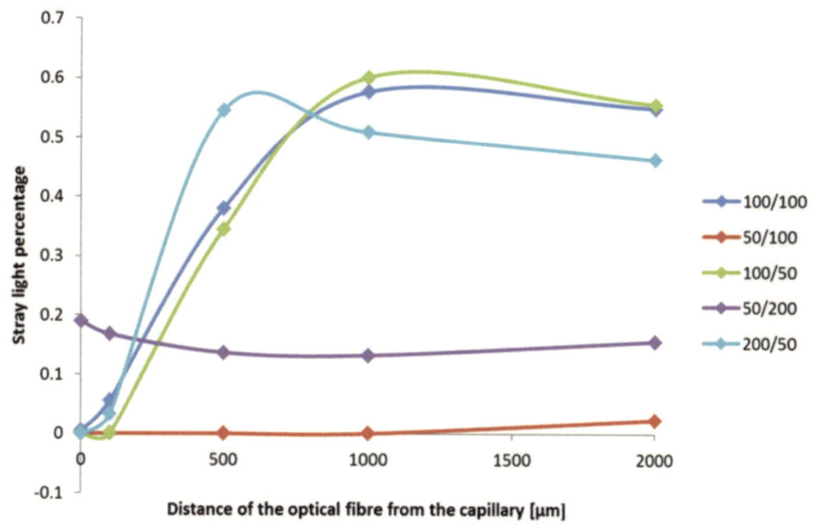

Figure 11. The amount of the stray light collected, compared for different combinations of diamteres of the illuminating fibre (the first number) and the collecting fibre (the second number)

Reducing the diameter of the collecting fibre to 100µm helped avoid stray light even when the distance between the collecting fibre and the capillary was around 1mm. This result can be very helpful as it allows for the high efficiency of the light collection combined with the minimisation of the stray light due to imprecise assembly of the detection system.

Combination of 100µm illuminating and 100µm collecting fibres gave good results in terms of the available light intensity, but even a slightest misalignment of the collecting optical fibre (0.2mm distance between can result in large amount of stray light delivered to the detector. Other combinations (100μm and 200μm illuminating fibres and 50μm collecting fibre) produced significantly lower available light intensity combined with high vulnerability to stray light introduction due to non-perfect assembly of the optical system.

In order to design a capillary based photo detection system the following occurrences should be considered: effect of stray light, amount of light intensity within the capillary bore, location of illuminating and collecting optical fibres and diameters thereof, as well as their angle of incidence. When considering all of these factors in conjunction, a complex hyper-plane of possible solutions presents itself which cannot be simply understood or calculated. It is therefore valuable to have a numerical model which can simultaneously take into account these factors. Such a model can also be used for other applications. An example of this is presented in the following sections for the application of photopolymerisation.

4. Optimisation of the photochemical reactions in capillaries

The main aim of this part is to present qualitative and semi-quantitative information about light behaviour in photopolymerisation setups. The capillary geometry and its multi-layered construction resulting in unusual and unexpected optical properties should be better understood as it allows to improve the work efficiency with capillaries. As explained in later part of this article many of described problems occur only in polytetrafluoroethylene (PTFE) coated capillaries.

For photoinitiated polymerisation in UV-range polyimide coated capillaries cannot be used because of polyimide high absorptivity below 550 nm. PTFE-coated capillaries were selected as PTFE is transparent in UV region. During several experiments it was noted that position of LED versus capillary play significant role in shape and quality of formed polymer. When LED was placed too close to the capillary monolith was growing distance of several millimetres under photomask. It was also noted that edges of formed polymer were uneven and blurry. If LED tilted there was noticeable difference in growth of one end of monolith. Observed problems in polymerisation setup are presented schematically on Figure 12. A series of experiments was conducted to repeat mentioned observation and approach them qualitatively and semi-quantitatively. The experimental conditions for photopolymerisation reactions are presented elsewhere [14].

A series of experiments were conducted in order to evaluate the degree of monolith growth under the masked region of a capillary where direct light should not enter. Observed results in experiments with photopolymerisation using same mixture and light sources showed that monolith can grow few millimetres under the mask. Such growth is considered undesired. Table 1 and Figure 14 shows average of three results of photoinitiated polymerisation of monolith with measured distance of the part formed under the photomask – in a region where direct light should not reach and therefore polymerisation would not be anticipated or well controlled. There was a clear trend of increased length of monolith grown under the mask as the LED was placed closer to the capillary. Also when the LED was tilted by an angle of 45° growth was significantly larger that side compared to when the LED was held perpendicularly. The light source was located directly over the edge of the photomask as shown in Figure 13.

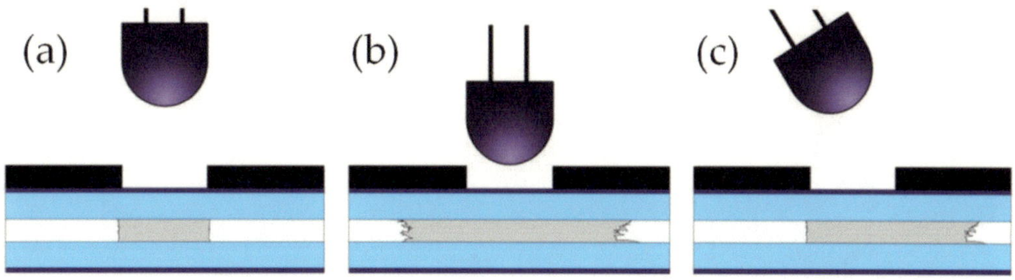

Figure 12. Schematic illustration of observed problems with monoliths obtained by photoinitiated polymerisation: (a) ideal monolith, (b) monolith growing up to several milimeters under the photomask in both direction when LED was too close to the capillary, (c) monolith growing on one end when the LED was tilted.

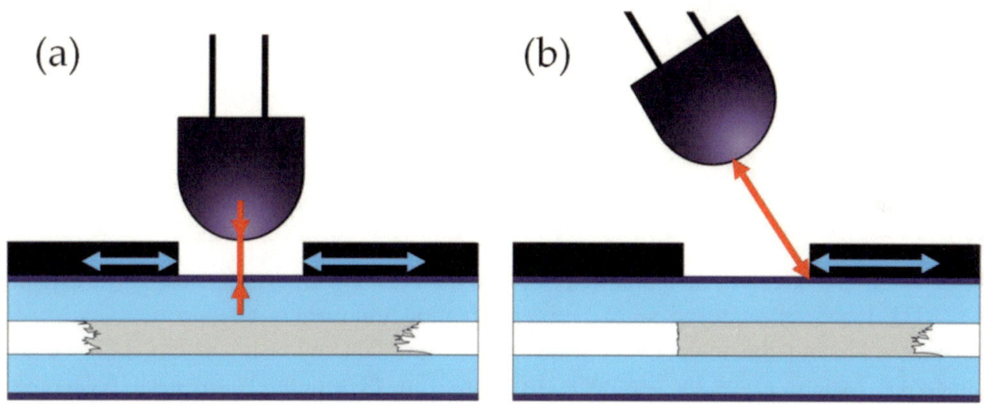

Figure 13. (a) schematic of the LED placement versus the capillary for perpendicular illumination and (b) for illumination under 45° angle. Red arrow marks distance from capillary to the LED, blue arrows marks distance of the monolith growth.

LED perpendicular		**LED tilted by 45°**	
LED distance from capillary	**Growth length**	**LED distance from capillary**	**Growth length**
0 cm	1.1 mm	0 cm	2.3 mm
1 cm	0.7 mm	1 cm	1.9 mm
2.5 cm	0.5 mm	2.5 cm	1.7 mm
4 cm	0 mm	4 cm	1.4 mm

Table 1. Monolith growth depending on light source position.

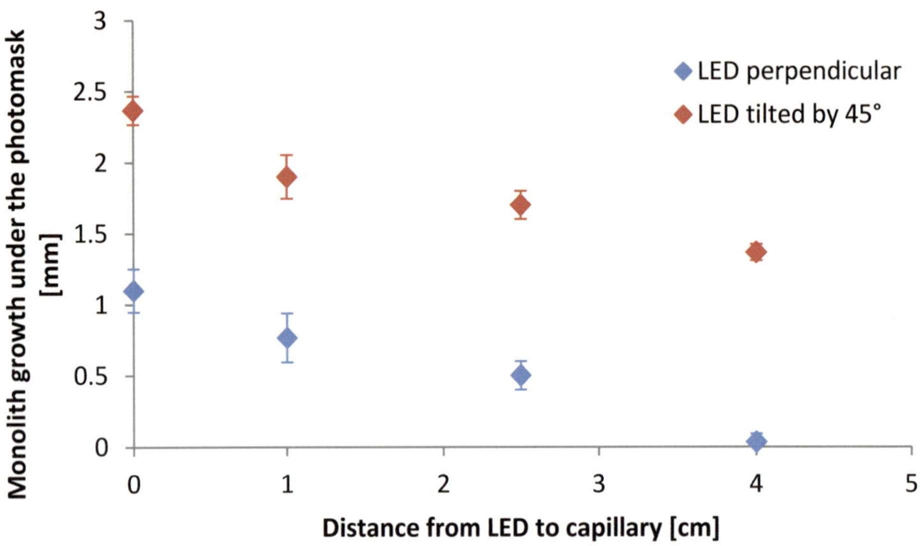

Figure 14. Relation between distance from the light source and monolith growth distance beyond the photomask covering for the case of the LED shining perpendicularly to the capillary and at an angle of 45° to the capillary.

Previously described theoretical model for light propagation was used in order to understand what is happening during photopolymerisation with light, solution and monolith. Tilting of the LED and thus delivering significantly more light to one side of the monolith suggest that capillary geometry plays significant role.

When the light is incident on a boundary of two dielectrics with different dielectric constant (e.g. PTFE/fused silica, fused silica / polymerization mixture) a portion of it undergoes reflection reducing the intensity of transmitted light. The intensity of the reflected and transferred light is given by Fresnel's equations (Eq. 8) separately for each polarisation of light:

$$R_s = \left[\frac{sin(\theta_i - \theta_t)}{sin(\theta_i + \theta_t)}\right]^2 R_p = \left[\frac{tan(\theta_i - \theta_t)}{tan(\theta_i + \theta_t)}\right]^2 \tag{8}$$

The uncollimated light is mixture of all possible polarisations, that is combination of linear polarisations and can be separated for *p* polarisation (vector of the electric field is parallel to plane of incidence) and *s* polarisation (vector of the electric field is perpendicular to plane of incidence). Graph on Figure 15 shows percentage of light being reflected on boundary PTFE/fused silica (light coming from medium with lower refractive index to medium with higher one) versus angle of incidence for both polarisations calculated from Eq. 8.

The light that reflects multiple times on dielectric boundary quickly loses intensity. The developed numerical model showed that incidence angles that are present in a capillary at the fused silica/PTFE boundary illuminated from outside do not provide significant reflectance and light is quickly transmitted outside of the capillary. Figure 16 provide information with upper limits of the light intensity that can be reflected by the capillary assuming total lack of absorption at this stage.

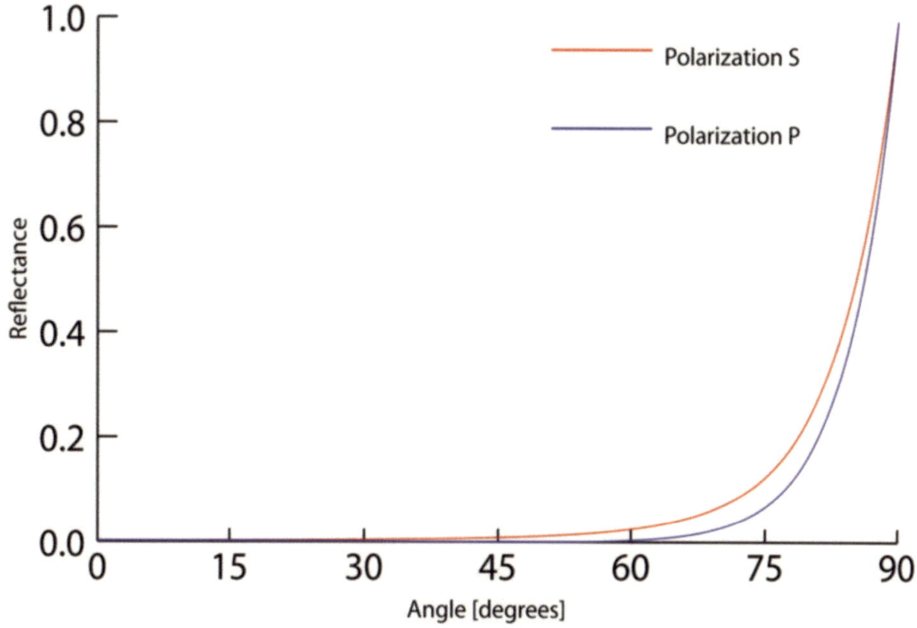

Figure 15. Theoretical reflectance of s and p polarised light incident on boundary PTFE/fused silica versus the angle of incidence; calculated from Eq. 8.

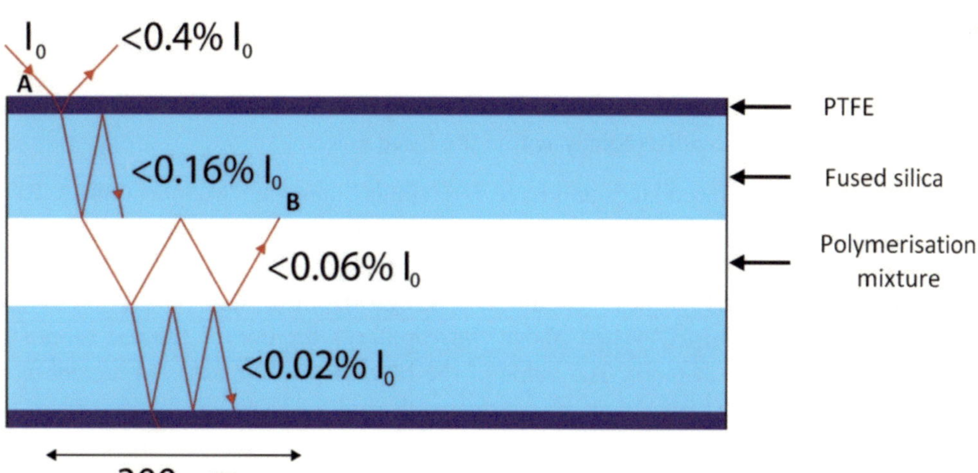

Figure 16. Schematic of the light propagation by multiple reflections inside PTFE-coated fused silica capillary with upper limit of the initial intensity I_0 (no more than). Dark blue – PTFE, light blue – fused silica, white – polymerisation mixture, red – light path

Figure 16 shows schematic cross-section of capillary with sample light ray. After two reflections (first on boundary fused silica/polymerisation mixture, second on boundary fused silica/PTFE) not more than 0.16% of intensity from point A is delivered to point B. The real value of the light intensity in point A is already lower than I_0 due to reflection on boundary air/PTFE and boundary PTFE/fused silica, but these effects are neglected and rounded to I_0 is this discussion. After three reflections intensity drops to not more than 0.06% of initial, and after four to not more than 0.02%. Dimensions of the capillary and refractive indices of PTFE, fused silica and polymerisation mixture are enforcing angles in further reflections and transmittances – after around 300 μm from initial point of illumination total delivered light intensity is below 0.2% of the initial light intensity I_0 delivered to the capillary is available. This calculations are based on assumption that all materials (air, PTFE, fused silica and polymerisation mixture) are completely transparent and do not absorb any light. Because their transmission coefficients are below 1 total amount of light available will be significantly lower, making impossible to penetrate distances observed in the experiments.

The size of a standard capillary is comparable with size of multimode optical fibres, where geometrical optics is sufficient to explain observations and calculate results. Capillaries and coating have form of coaxial cylinders. The basic principle of optical (multimode) waveguide is total internal reflection (TIR). This phenomenon is occurring when dielectric with refractive index n_1 (e.g. fused silica (FS)) is covered with dielectric with refractive index $n_2 < n_1$ (e.g. PTFE, $n_{PTFE} < n_{FS}$). The light wave incoming from medium with higher refractive index to boundary can undergo TIR provided angle of incidence is high enough. The graph on Figure 17 shows dependence of intensity of reflected light versus angle of incidence (for fused silica/PTFE boundary).

In order to observe total internal reflection for light incident on fused silica/PTFE boundary, angle of incidence must be higher than 64.8°, otherwise some light will be transmitted through the boundary resulting with loss of the light intensity. The developed numerical model showed that no angle higher than 44.7° is available in any part of the capillary meaning that no light introduced to capillary from source placed above the capillary can achieve angle sufficient to reflect totally within the fused silica.

Capillaries are made of fused silica and have very similar diameters to optical fibres. PTFE has lower refractive index than fused silica. Initially a hypothesis about the capillary acting as an optical waveguide was discussed to explain occurrences of monolith growth under the photomask. There were two major contradictions: light coming from outside of cone of acceptance (i.e. from source located above the capillary) for optical fibre it cannot be transmitted over longer distance. The value of the highest possible angle of incidence on boundary fused silica/PTFE for light interacting with capillary content obtained from the developed model is significantly lower than required for the TIR. Graph on Figure 17 shows that for angles of incidence below 45°, reflectance is very low, and light is mostly transmitted through the boundary.

The last possibility was reflection on boundary PTFE/air – the highest possible ratio of refractive indices. In order to observe TIR on this boundary the incidence angle of light must not be lower than 49.25°[1]. The highest incidence angle on fused silica/PTFE boundary calculated using light propagation model gives angle of 44.57°. Although amount of light reflected on that boundary is around 10% it does not satisfy the condition to observe total internal reflection and could not explain growth of few millimetres.

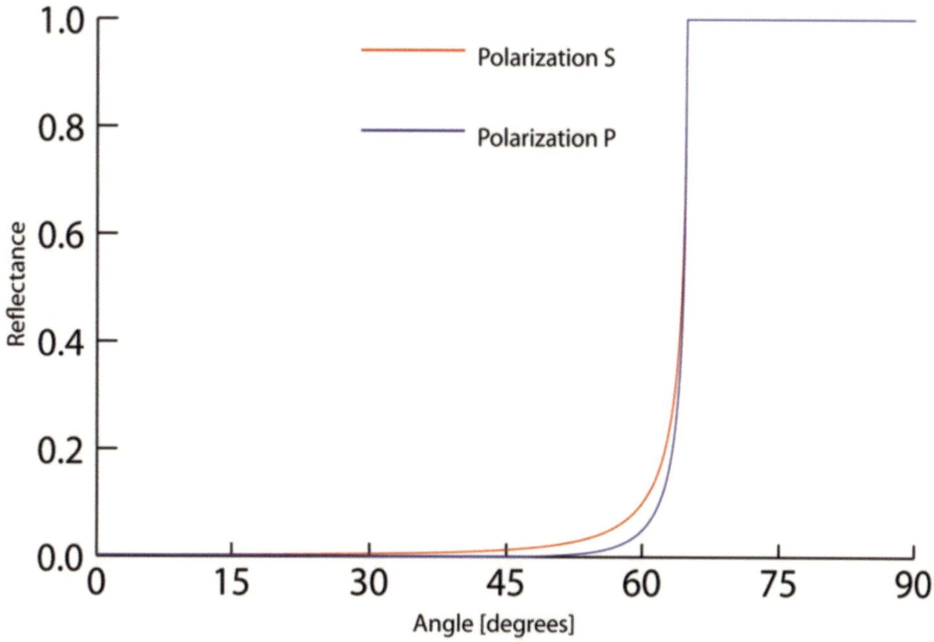

Figure 17. Theoretical reflectance on boundary fused silica/PTFE versus the angle of incidence.A reflectance value of one (starting at 64.8°) shows where total internal reflection occurs and no light is transmitted through the boundary (from Eq. 8).

These attempts mentioned above to explain observed result were based on a static system with constant time-independent properties. A capillary with on-going polymerisation reaction is a dynamic system, which changes its physical and optical properties in time. A hypothesis that a monolith forming inside capillary is changing optical properties of the setup during the polymerisation was posed. To prove this hypothesis, new photographs of capillary filled with polymerisation mixture and monolith were taken to show the transmission of light when the LED was shining on the capillary. The capillary was installed vertically above the digital microscope. In order to prevent any undesired light, a black cardboard separated the microscope objective from the rest of the setup and photos were taken in total darkness. Any possible openings near capillary wall were covered with a sealant. Schematic of that setup is shown on Figure 18.

It can be clearly seen in the Figure 19a that light was transmitted through the fused silica. The distance from the light source to the microscope was 20 mm to prevent any other discussed method than waveguiding to propagate light toward the end of capillary. To confirm that the

[1] Calculated from Eq. 22, refractive index of PTFE n_{PTFE}=1.32 and refractive index of air n_{air}=1

observed effect has nothing to do with collimation of the light the experiment was repeated using 532 nm green laser as light source. Result is shown on Figure 20. Black spots are effect of destructive interference of laser beam with itself after multiple reflections inside fused silica. LED light is non-collimated thus the interference effect was not observed.

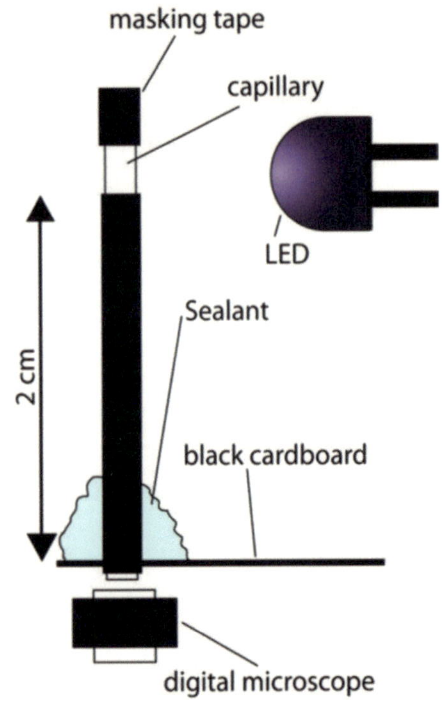

Figure 18. Schematic of experimental setup for observing light waveguiding inside the capillary with monolith.

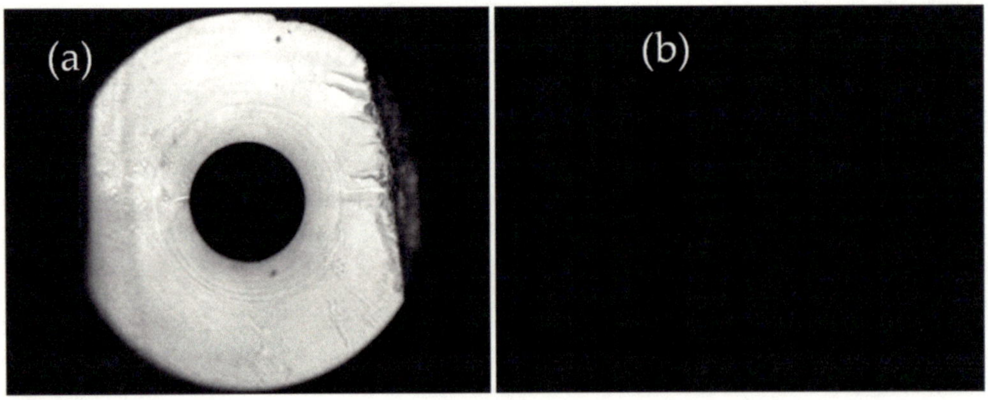

Figure 19. (a) Capillary with monolith and (b) empty capillary.
Image collected by digital microscope using setup from Figure 18.

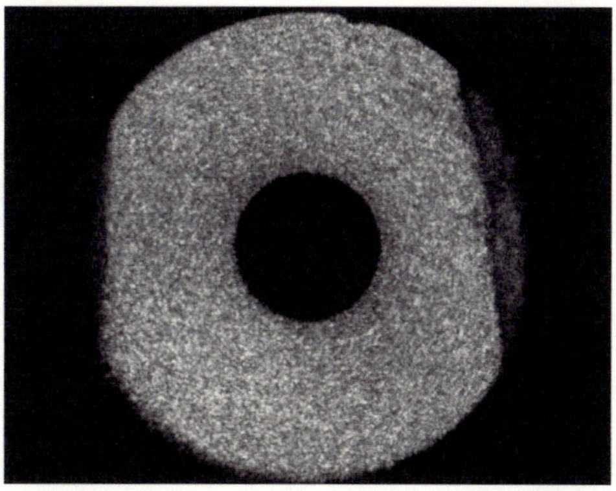

Figure 20. Image observed when 532 nm green laser is used as light source (setup same as in Figure 18).

The monolith inside the capillary has a very irregular surface. Moreover the refractive index of a polymer is higher than of the fused silica. The incident light was scattered on the monolith surface (polymerisation mixture/monolith boundary) and due to the morphology of the monolith surface it was scattered in all directions (Figure 21). This type of reflection is called diffuse reflection or diffuse scattering. In this situation light can be reflected under an angle sufficient to undergo total internal reflection on boundary fused silica/PTFE. These angles are not available for light that is not a subject to diffuse reflection. The refractive index of polymethacrylic polymer is higher than fused silica and ranges from 1.472 to 1.506 [15]. Light reflected diffusively in one part of monolith can propagate through the monolith, cross the boundary monolith/fused silica, and then remain in the TIR regime in regions where no monolith is present, effectively turning capillary into an optical waveguide.

Also diffuse scattering allows photons entering fused silica under angles higher than for those coming directly from light source. Wherever total internal reflection is occurring, on opposite side of boundary an evanescent field appears strong enough to initiate photopolymerisation. Figure 22 shows a schematic of that principle.

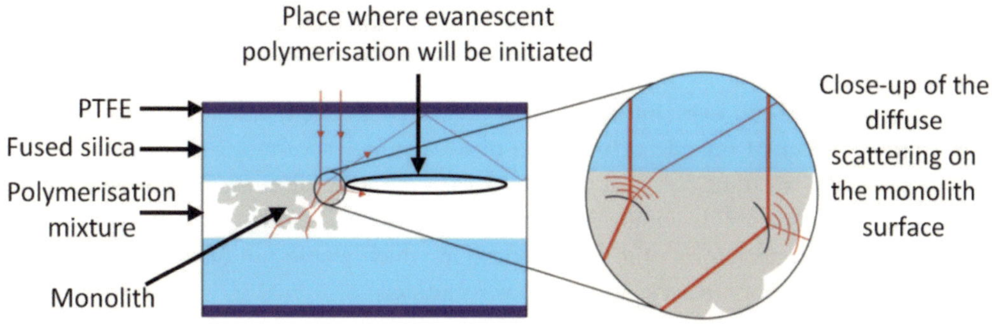

Figure 21. Schmatic of the diffuse scattering of incident light on the porous surface of the monolith that has formed inside the capillary.

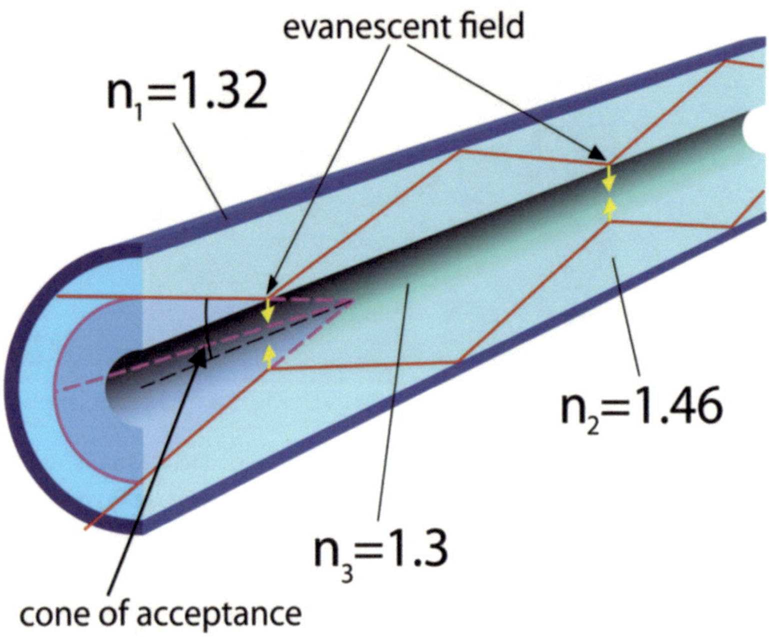

Figure 22. Schematic showing formation of the evanescent field (yellow arrow) outside dielectric inside which light undergoes total internal reflection. Dark blue – PTFE, light blue – fused silica, transparent – polymerisation mixture.

5. Conclusions

The developed numerical model was used for investigation of the capillary optical properties during the photopolymerisation reaction. One of the important goals of this work was determining the relation between alignment of the light source and the shape and size of produced polymerised monoliths. Previous observations suggested that light transmission through the capillary fused silica body occurs in a similar manner as in an optical fibre. The numerical model was required to determine the conditions necessary for total internal reflection of the light used to initiate the polymerisation reaction, which in consequence would lead to waveguiding of the light through the capillary and enabled evanescent polymerisation in an obscured region under the photomask.

From the modelling and experimental results presented above, it can be seen that in cases where light is incident outside of the cone of acceptance, it cannot be confined by total internal reflection and thus does not propagate for significant distance. The presence of the photopolymerised monolith within the capillary, provides a surface for diffuse reflection, which in turn presents angles of internal reflection which would not otherwise be present. This enables the capillary to act as an optical waveguide.

The minimum required angle to observe total internal reflection without monolith was not present in the system. This result is concurrent with basic knowledge of the optical fibres: the light must be introduces into the waveguide under a sufficient angle to remain within total internal reflection regime. If the initial angle of incidence is higher than required the beam will eventually escape the waveguide. This hypothesis was tested using an empty capillary and a capillary filled with polymerisation mixture, which was illuminated perpendicularly. A photodetector was used to record the light emitted though the capillary cross-section. The results in both cases showed that there was no propagation of light along the capillary.

During this investigation the dynamics of the polymerisation process which are affected by the system optical properties was studied. The polymerisation process and the corresponding formation of a monolith inside the capillary would be expected to affect the optical properties of the whole system. An organic monolith is a highly porous, amorphic structure, that when illuminated reflects a significant amount of light. The presence of the monolith formed during the reaction changes the optical properties of the capillary that may no longer be regarded as a uniform cylindrical structure with constant refractive indices through the cross-section.

The monolith inside the capillary bore provided a surface for diffuse light reflection. The monolith's surface morphology allowed for presence of angles that are normally unavailable inside the capillary system, but with monolith waveguiding of the light becomes possible. The light can be transmitted along the capillary for significant distances (measured distance was 20 mm) utilising the total internal reflection mechanism. When light is entering the capillary walls as in Figure 22 and undergoes total internal reflection within fused silica, an evanescent field will appear within the coating and the internal boundary of the capillary bore which is strong enough to initiate the reaction of photopolymerisation. Increased distance between the capillary and the light source significantly reduces the amount of light reflected from the monolith surface and thus the amount of light that can propagate through the capillary.

Author details

Tomasz Piasecki, Aymen Ben Azouz and Dermot Brabazon
School of Mechanical Engineering and Manufacturing, Dublin City University, Dublin, Ireland
Irish Separation Science Cluster, Dublin, Ireland

Brett Paull and Mirek Macka
School of Chemistry, University of Tasmania, Hobart, Australia

6. References

[1] *Capillary and microchip electrophoresis: Challenging the common conceptions.* Breadmore, M. C. 2012, Journal of Chromatography A, Vol. 1221, pp. 42– 55.

[2] *Rectangular Capillaries for Capillary Zone Electrophoresis.* Tsuda, T., Sweedler, J. V., Zare, R. N. 1990, Analytical Chemistry, Vol. 62, pp. 2149-2152.

[3] *Capillary Electrophoresis.* Kuhr, W. G. 1990, Analytical Chemistry, Vol. 62, pp. 403R-414R.

[4] *Enhancement of detection sensitivity for indirect photometric detection of anions and cations in capillary electrophoresis.* Johns, C., Macka, M., Haddad, P. R. 2003, Electrophoresis, Vol. 24, pp. 2150-2167.

[5] Skoog, D. A., Holler, F. J., Crouch, S. R. *Principles of instrumental analysis.* Belmont : Thomson Books/Cole, 2007. ISBN 13: 978-0-495-12570-9.

[6] *Practical method for evaluation of linearity and effective pathlength of on-capillary photometric detectors in capillary electrophoresis.* Johns, C., Macka, M., Haddad, P. R., King, M., Paull, B. 2001, Journal of Chromatography A, Vol. 927, pp. 237-241.

[7] *Nanoliter-Scale Multireflection Cell for Absorption Detection in Capillary Electrophoresis.* Wang T., Aiken, J. H., Huie, C. W., Hartwick, R. A. 1991, Analytical Chemistry, Vol. 63, pp. 1372-1376.

[8] *Optical Improvements of a Z-Shaped Cell for High-Sensitivity UV Absorbance Detection in Capillary Electrophoresis.* Moring, S., E., Reel, R. T. 1993, Analytical Chemistry, Vol. 65, pp. 3454-3459.

[9] *On-column capillary flow cell utilizing optical wave-guide for chromatographic applications.* Bruno, A. E., Gassmann, E., Pericles, N., Anton, K. 1989, Analytical Chemistry, Vol. 61, pp. 876-883.

[10] *The pigtailing approach to optical detection in capillary electrophoresis.* Bruno, A. E., Maystre, F., Krattiger, B., Nussbaum, P., Gassmann, E. 1994, Trends in Analytical Chemistry, Vol. 13, pp. 190-198.

[11] *Numerical model for light propagation and light intensity distribution inside coated fused silica capillaries.* Piasecki, T., Macka, M., Paull, B., Brabazon, D. 2011, Optics and Lasers in Enginnering, Vol. 49, pp. 924-931.

[12] *Fiber-optic-based UV–visible absorbance detector for capillary electrophoresis, utilizing focusing optical elements.* Lindberg, P., Hanning, A., Lindberg, T., Roeraade, J. 1998, Journal of Chromatography A, Vol. 809, pp. 181-189.

[13] *Capillary Electrophoresis Detector Using a Light Emitting Diode and Optical Fibres.* Butler, P. A. G., Mills, B., Hauser, P. 1997, Analyst, Vol. 122, pp. 949-953.

[14] *UV-LED photopolymerised monoliths.* Abele, S., Nie, F. Q., Foret, F., Paull, B., Macka, M. 2008, Analyst, Vol. 133, pp. 864-866.

[15] Brandrup, J., Immergut, E. H. *Polymer Handbook.* Ney York : John Wiley & Sons, 1975.

Chapter 4

Protein Electrophoresis in Saliva Study

Elsa Lamy, Ana R. Costa, Célia M. Antunes,
Rui Vitorino and Francisco Amado

Additional information is available at the end of the chapter

http://dx.doi.org/10.5772/48586

1. Introduction

Saliva started for been less studied than other body fluids, but in the last years it has being receiving an increased attention. Until now, more than 2000 different proteins and peptides have been identified in whole saliva and salivary glandular secretions [1]. From these, more than 90% derive from the secretion of the three pairs of "major" salivary glands (parotid, submandibular and sublingual glands). The remaining 10% derives from "minor" salivary glands and from extra-glandular sources, namely gingival crevicular fluid, mucosal transudations, bacteria and bacterial products, viruses and fungi, desquamated epithelial cells, and food debris [2].

Saliva secretion is mainly under autonomic nervous system regulation. Sympathetic and parasympathetic stimulation have different effects on the flow rate and composition of saliva secreted. Whereas parasympathetic stimulation results in the production of a high volume of saliva with low protein concentration, stimulation of the sympathetic branch of the autonomic nervous system is responsible for the secretion of a small amount of saliva with increased protein concentration. Besides this distinctive characteristic, and inversely to what is observed for the majority of body systems, the effects of parasympathetic and sympathetic innervations are not antagonic but rather exert relatively independent effects in which the activity of one branch may synergistically augment the effect of the other [3,4]. Despite the thought of an exclusive nervous regulation, recent in vivo animal experiments indicate a short-term endocrine regulation of salivary glandular activities as well [5-9].

The primordial function of saliva is to aid in the moistening and preprocessing of food, aiding in deglutition. Besides this, other important functions exist for saliva, which can generally be grouped in digestive (and ingestive) and protection [10]. For digestive (and ingestive) purposes, saliva contains enzymes, including proteases, lipases and glycohydrolases, which initiate partial break-down of food components. Among these enzymes, alpha-amylase is by far the enzyme present in higher amounts. There are also salivary proteins involved in food perception, such as: salivary PRPs (proline-rich proteins), which bind dietary polyphenols (mainly tannins) and are involved in astringency perception [11]; carbonic anhydrase VI, suggested to influence bitter taste sensitivity [12]; and alpha-amylase, involved in sweet taste sensitivity [13]. Additionally, recent studies suggest changes in saliva composition induced by taste [14-16], reinforcing the potential of saliva in food perception and ingestive choices. Concerning protection role, several different salivary proteins have been identified, namely mucins, acidic PRPs, statherins, among many others. For example, salivary proteins adsorbed to the enamel surface form the enamel pellicle, which helps to protect teeth [17]. It is also relevant to point the presence of proteins with more than one function, and the sharing of the same function by different families of proteins. This functional redundancy may help to ensure that a given function is always present under a broader range of physiological conditions [18, 19]. An in-depth analysis of saliva proteome, including the posttranslational modifications can therefore provide a valuable resource for saliva function research.

The high potential of saliva as a source of biomarkers was one of the main responsible for the great interest in this fluid. Several analytes are present in saliva in amounts that relates to blood, with the great advantage of being collected using simple and non-invasive methods. Proteomic techniques such as two-dimensional electrophoresis (2-DE), 2D-liquid chromatography/mass spectrometry (2D-LC/MS), matrix-assisted laser desorption ionization-time of flight mass spectrometry (MALDI-TOF/MS), and surface enhanced laser desorption/ionization time-of-flight mass spectrometry (SELDI-TOF/MS), have been used in saliva studies. Based on those techniques, potential salivary biomarkers for diseases such as Sjögren syndrome [20], diabetes mellitus [21] and some different cancers [22] have been suggested. 2-DE has been one technique of choice for the global analysis and initial profiling of salivary proteins, being used as a first step for protein separation, followed by MS or tandem MS (MS/MS) [23].

Whereas the salivary proteins from humans have deserved substantial attention, both in terms of identification and characterization, as well as functional properties, animal saliva has been much less studied. However, interest for the latter is being increasing, due to the convenience on the use of animal models for diverse pathological and physiological conditions and due to the potential of this fluid for disease diagnostic and for understanding behavioral and physiological processes, important in animal production. Moreover, a multidisciplinary approach that integrates knowledge about salivary proteins in the animal kingdom (most important in mammals) and draws comparisons to possible functions in humans would be valuable.

The following sections will give an overview about the use of electrophoresis in saliva studies. Methodological issues and the major advantages and limitations for the use of this technique in human and animal saliva studies will be presented. We will finish the chapter by presenting alternatives to electrophoresis for the study of salivary proteome.

2. Applications of electrophoresis for saliva proteome characterization

Even before the advent of proteomics, electrophoresis was frequently used for salivary protein separation. Sodium dodecyl sulfate polyacrylamide gel electrophoresis (SDS-PAGE) [24], PAGE in non-denaturing conditions [25], isoelectric focusing [26], two-dimensional electrophoresis (2-DE) [27], capillary electrophoresis (CE) [28,29] and free flow electrophoresis [30] have all been used in saliva studies, with different purposes.

One-dimensional gel electrophoresis under denaturing conditions has several advantages: virtually all proteins are soluble in SDS, allowing their separation; it covers a relatively high range of molecular masses (from 10000 to 300000 Da) and allows the possibility of extremely acidic and basic proteins to be visualized [31]. Moreover, SDS-PAGE has the advantage of having a low sensitivity to salt concentration. However, by separating proteins only based in their molecular masses, only limited information is obtained, since each of the bands present in the gels is frequently constituted by different proteins. Limitation in the number of proteins separated also occur using IEF, which separates proteins only based in their charges.

2-DE takes advantages of the two different properties of proteins (molecular masses and isoelectric points), allowing the separation and visualization, in a gel matrix, of a considerable number of different proteins. This technique, which originates from the work of O'Farrell and Klose in the 1970's [32,33] became very useful for the study of complex protein mixtures, such as saliva. Besides the very high separation capability of 2-DE, ensuring well-resolved protein maps with more than 2000 protein spots, this technique has also the advantage of mapping posttranslational modifications. Coupled to mass spectrometry, for protein identification, 2-DE has been considerably used in several different samples, including saliva [27,34]. Using this approach, the protein spots observed in 2-DE gels are subsequently digested using a protease (usually trypsin), with the resultant digest products analyzed by mass spectrometry. 2-DE can be used to compare expression levels of proteins in related samples, such as those from altered experimental conditions, allowing the response of classes of proteins to be determined. This approach has been suceffuly used in a number of saliva studies [e.g. 35].

Salivary peptides and proteins have been analyzed by a variety of CE approaches (reviewed in [36]. The term CE, although often used as shorthand for capillary zone electrophoresis, refers to a family of related techniques, all based on the performance of the separations in narrow-bore capillaries across which an electric field is applied. The types of CE that have the greatest potential in proteomics (commonly used together with mass spectrometry) are capillary zone electrophoresis and capillary isoelectric focusing [37]. The use of CE to analyze saliva proteome, have been described and allowed the profiling and identification of several salivary proteins [38]. Recently this technique was successfully used for the identification of salivary profiles in cancer diagnosis [29].

The existence of different electrophoretic methodologies, allowing separations based on different protein characteristics is of great utility in proteomics, contributing for the resolution of complex samples such as saliva. The electrophoretic methods described are complementary of each other and since all have advantages and limitations, the choice for each one will depend on the objectives of the study and on the salivary proteins of interest.

3. Methodological issues related to saliva proteome analysis

The concept of salivary proteome is related to the creation of a salivary protein catalogue, where information that can be further used for several different purposes (e.g diagnostics, physiological status) can be placed [39]. In this context, it is important that the results obtained from different laboratories can be compared. Accurate examination of salivary components requires optimal collection, processing and storage conditions. Moreover, most of the studies aim comparisons among pathological/physiological conditions, and as such it is essential that the differences obtained are not due to external factors. To avoid changes in the protein and peptide composition from salivary secretions, standardized salivary sampling protocols, processing and storage conditions need to be applied [23,39,40].

3.1. Sample collection, processing and storage

One of the particularities of saliva is its capacity of rapidly adapt to different conditions. This is related to the fact of salivary secretion being mainly regulated by the two branches of the autonomic nervous system (both sympathetic and parasympathetic), with only a minor regulation from hormonal origin. Such type of regulation results in variations in composition according to the stimulus. Circadian rhythm [41], gender [42], drugs [43], exercise [44], among others, are factors that change salivary flow rate and saliva composition. Additionally to variations in the composition from each salivary gland cell type, the contribution of each individual salivary gland to the total fluid is not the same. For example, minor salivary glands and submandibular glands have an important contribution at rest, whereas in response to strong stimuli during feeding is the parotid contribution that becomes dominant [45]. Concerning daytime, the flow rate of resting and stimulated saliva is higher in the afternoon than in the morning, with the peak occurring in the middle of the afternoon. The salivary protein concentration also follows this diurnal pattern [2]. Eating is another strong stimulus for the secretion of saliva, and as such the interval between feeding and saliva collection influences salivary flow rate, viscosity and protein composition.

The points referred above are important for the selection of the collection method: with or without stimulation. For unstimulated saliva collection the draining method is usually choose. Accordingly, saliva is allowed to drip off the lower lip to a tube maintained on ice. On the other hand, stimulated saliva is frequently obtained after parafilm mastication, or after sour taste stimulation [46].

Advantages and disadvantages exist for both approaches. Unstimulated saliva collection is several times preferred, as stimulated saliva contains a diluted concentration of several proteins, which may be of interest. However, it is difficult to have saliva completely free of stimulation: due to the high range of stimulus influencing salivary secretion, even small variations in collection conditions, among different saliva donors (e,g light intensity, temperature, emotional status, or others) may be sufficient to induce differences in results [39].

Stimulated saliva collection is generally used to have higher volumes of sample. In some pathological conditions (e.g Sjögren's syndrome, xerostomia) or post-radiation, stimulation of salivary secretion may be the only way of obtaining adequate amounts of saliva samples for analysis. However, difficulties may exist to make uniform the intensity and duration of the stimulus and the secretion of certain proteins may be affected by the duration of stimulation. For example, with prolonged stimulation of salivary flow, certain glycoproteins may be incompletely glycosylated [39].

Another issue in regard with saliva sampling is the origin of the collected secretions, i.e. whether it is glandular or whole saliva. Glandular fluid can be obtained through the use of adapted collection devices, for both parotid and submandibular/sublingual secretions [47]. Through this, changes in salivary proteins by exogenous enzymes are avoided, since the fluid is collected before it reaches the mouth.

Most of the studies on saliva in general have been obtained from human saliva. Concerning animal saliva less is known and, although methodologies of collection, processing and storage are based on the ones reported for humans, in most cases it is difficult to obtain unstimulated saliva from animals. Only in domestic ruminants, which produce considerably high volumes of saliva daily, in a continuous flow, it is possible to obtain whole saliva samples by simply collecting the mouth flow into a large beaker [48]. Moreover, this method only works with domesticated animals that are used to be handled by man.

For all the animals that do not produce large amounts of saliva, or for specific investigation purposes, stimulation may be necessary. Under controlled experimental conditions, the collection of stimulated saliva can be initiated either mechanically or chemically. Cotton rolls (e.g. Sallivetes) are currently used for animal saliva collection for being both practical and efficient in getting this fluid. Animals chew for some time, being saliva production stimulated by mastication. The cotton roll moistened by the saliva is further centrifuged to release the fluid. There is the possibility of some salivary proteins such as mucins and/or other potential biomarkers irreversibly adsorb to the cotton roll, resulting in their loss. However, to our knowledge, studies to elucidate this aspect are lacking. Chemical (pharmacological) stimulation is sometimes used for saliva collection in animals with low amounts of saliva. For example, in small rodents, it is very difficult to collect sufficient amounts for analysis without stimulation and the use of parasympathetic agonists (e.g. pilocarpine) appears to be effective [49]. This type of stimulant is referred to increase the volume production without changing the proportion of the proteins secreted. Others types of chemical stimulants may be used. For example, sympathetic agonists, such as isoproterenol, induce the synthesis and secretion of granular proteins from salivary glands, thereby increasing salivary protein concentration and changing the salivary protein profile [50].

While whole saliva can be collected using the already mentioned cotton-rolls, or through direct aspiration from the mouth [49], glandular saliva collection can be achieved using catheters inserted into the ducts of the gland of interest. Although this method is invasive, the level of invasion is minimal, and once the collection period is finished, the catheter can easily be removed without causing any damage to the animal. This method was used to study sheep and goat parotid salivary proteomes [34,51]. Collection of glandular saliva has the advantage of accessing the secretion product of certain glands, as well as obtaining a clear saliva sample with virtually no interfering compounds, such as food debris or microorganisms. Glandular collection is most important for studies that aim on unraveling the effects of particular factors on the secretion from individual glands [e.g. effects of dietary constituents on parotid saliva [35]. However, this approach, if not carried properly, might have the disadvantage of faulty catheter insertion which causes leakage of plasma proteins into the salivary secretions due to disruption of epithelial integrity.

Wherever in humans or animals, the collection method of choice, as well as the origin of fluid selected, will affect the outcomes and thus should depend on the objectives of the study and on the specific group of proteins of interest.

In complex protein mixtures, such as saliva, sample preparation and fractionation constitute one of the most crucial processes for proteome study. None of the currently available proteomic techniques allow the analysis of the entire proteome in a single step. In body fluids such as saliva, proteins have different physicochemical characteristics and a wide concentration dynamic range and, as such, fractionation is essential. In this context, electrophoresis may also be useful for that purpose. Preparative isoelectric focusing using free flow electrophoresis is an example of the methods available for sample fractionation and which has been used in saliva [30]. Before 2-DE separation, free flow electrophoresis has the advantage of generating different fractions, which can be independently run. For example, a first separation using isoelectric free flow electrophoresis results in fractions with different pI ranges. Each of the resulting fractions can be further separated according to charge, by using narrow pH ranges, in the first dimension allowing a more detailed picture of the protein profile [30].

Saliva contains proteins present in high levels, and numerous low abundant proteins, for which analysis may be of interest. The observation of the latter in electrophoretic gels is obscured by the presence of the high-abundant proteins. When salivary proteome is used for disease diagnostic purposes, the analysis of low abundant proteins is almost always necessary, since most of potential salivary biomarkers are present in relative low amounts. The major components of saliva are mucins, proline-rich glycoproteins, amylase and some antimicrobial proteins that include agglutinin, lisozyme, lactoferrin, immunoglobulins, histatins and defensins [27]. The protein alpha-amylase contributes to almost 60% of total salivary proteome [52] and its depletion allows the analysis of less abundant proteins. Salivary alpha amylase depletion can be achieved through elution of samples from starch columns to deplete this protein specifically [53].

One of the main concerns when working with protein sample is to avoid undesirable alterations during the several steps that go from collection to final analysis. One characteristic of saliva is that many salivary proteins enter post-translational modifications

(PTMs), namely glycosylation, phosphorylation, sulfation and proteolysis. From the considerable amount of glycoproteins present in saliva, mucins represent an important group, which is responsible for bacterial agglutination and lubrication of the oral cavity tissues. Phosphoproteins also exist in a considerable amount of salivary proteins, with several diverse roles [54]. Such modifications start in the acinar cells of salivary glands, and continue when saliva enters the mouth, mainly due to the presence of host- and bacteria-derived enzymes, what results in additional protein modifications [55].

Since these forms of PTMs are responsible for many functions of this fluid, a particular attention should be directed to saliva collection processing and storage, in order to minimize proteolysis, de-glycosylation and de-phosphorylation. Different research groups have employed different methods for avoiding proteolysis, de-glycosylation and de-phosphorylation. The addition of 0,2% trifluoroacetic acid to saliva after collection has been one protocol used [56]. In a recent study, it was observed that the addition of protease inhibitors to saliva may allow its storage at 4°C for approximately two weeks, without significant degradation [53]. Nonetheless, some authors report that not even an inhibitor cocktail can prevent all protein degradation [57]. Additionally, it was suggested to be possible to keep the samples at room temperature (for a period of about two weeks), without considerable changes in salivary proteome occurred, only by adding ethanol, [53]. Nevertheless, working on ice for no longer than one hour and subsequent storage of samples at -80°C has been considered a safe and practical handling protocol [57-59]. Nonetheless, long time storage, as well as freeze-thaw cycles can induce protein precipitation, in particular from low molecular mass components [57]. In any case, little research has been directed on ways of minimizing degradative processes, and this is clearly needed [39].

3.2. Staining procedures and PTMs in-gel analysis

Protein visualization is necessary for quantitative and qualitative analysis. In electrophoresis this is achieved through reversible or irreversible binding of a colored organic or inorganic chemical to the protein. An ideal staining procedure would be the one with a very low detection limit, an optimal signal to noise ratio, a wide dynamic range and a wide linear relationship between the quantity of protein and the staining intensity, and compatible to mass spectrometry [60]. However, such an ideal stain does not exist.

Silver, Coomassie Brilliant Blue (CBB) and fluorescent staining are the most frequently used methodologies. Silver staining presents a high sensitivity, making possible the visualization of proteins present in amounts as low as 1 ng [61]. Silver staining techniques are based upon saturating gels with silver ions, washing the less tightly bound metal ions out of the gel matrix and reducing the protein-bound metal ions to form metallic silver. In 2-DE, Silver staining is regularly used due to its potential for the visualization of low intensity protein spots. However, it presents a narrow dynamic range and the tendency of the dye to stain differently based on amino acid composition and PTMs. Moreover, by detecting low levels of protein, each of the stained spots may have not sufficient amounts o protein for subsequent analysis by mass spectrometry.

CBB is a disulfonated triphenylmethane textile dye. CBB staining presents a linear dynamic range and a moderately sensitivity. As such, CBB dyes are suitable for protein quantitative analysis, which is necessary in proteomics analysis. Moreover, this staining technique is compatible with mass spectrometry. Two modifications of CBB exist: R-250 and G-250. In acidic solutions the dye sticks to the amino groups of the proteins by electrostatic and hydrophobic interactions. Inversely to silver stain, CBB is not so extremely sensitive, thus being necessary to load higher amounts of proteins in the gel. Consequently, CBB stained spots contains considerable amounts of proteins, suitable for mass spectrometry analysis.

Saliva samples present particularities that should be considered for silver and CBB staining. Parotid saliva contains a considerable amount of proline-rich proteins (PRPs), which are difficult to stain with silver since they present low amounts of amino acids containing sulfur, necessary for the binding of silver ions [61]. On the other hand, when stained with CBB R-250 they usually present a violet-pink stain, which allows differentiating them from all the other salivary proteins that stain blue. A destain protocol of 10% acetic acid, instead of the common 10% acetic acid/ 10% methanol is generally used for this purpose [62].

The fluorescent dyes were more recently developed, presenting a high sensitivity and a linear dynamic range, and in this way, being advantageous relatively to common silver staining. For example, SYPRO staining technique has been used in several salivary proteomic studies [e.g. 63]. This is a novel, ruthenium-based fluorescent dye. SYPRO Red and Orange, bind to the detergent coat surrounding proteins in SDS denaturing gels, thus, staining in such gels is not strongly selective for particular polypeptides [64]. SYPRO stains are compatible with mass spectrometry and can be used in combination with other staining techniques, for detection of PTMs, such as glycosylations or phosphorylations.

DIGE (Difference Gel Electrophoresis) technology represented an improvement in 2-DE based proteomics, allowing more accurate comparisons among samples. Its convenience is also true for saliva samples [65]. DIGE is based on the modification of proteins, before electrophoresis, by attaching a fluorescent labeling. Cyanine-bases dyes (Cy2, Cy3 and Cy5) are used with this purpose. These dyes label the amine group of protein lysines specifically and covalently to form an amide. A different cyanine-based dye is added to each individual sample. The dyes are designed to have the same molecular mass and charge to ensure that proteins common to both samples have the same relative 2-DE mobility. The samples are mixed and resolved in a single 2-DE gel. The proteins from the different dyes are visualized by alternatively illuminating the gel at different wavelengths. With this technique, it is possible to avoid some inter-gel variation.

PTMs, such as glycosylation and phosphorylation, can be accessed in electrophoretic gels through staining procedures. As it was referred before, several different salivary proteins suffer PTMs, which confer their characteristic functions. Consequently, their visualization in electrophoretic gels is valuable. Detection of phosphoproteins can be achieved using different procedures. Recently, Pro-Q Diamond was developed for phosphoproteins staining. This affords wide specificity and high sensitivity. Phosphoserine, phosphothreonine and phosphotyrosine containing proteins are detected [66].

O- and N- glycosylated proteins are abundant in saliva. Some of the most well studied salivary glycoproteins are mucins (MUC5B and Muc7) and proline-rich glycoproteins. The most classical procedure for glycoproteins staining is periodic-acid Schiff (PAS) with the protocol that allows detection of glycoproteins in gels being adapted from protocols of histochemistry. However, it does not present a high sensitivity, resulting in the need of high levels of protein load [67]. Another possible method is the more recently developed stain Pro-Q Emerald that allows glycoprotein detection, presenting approximately 50-fold more sensitivity than PAS [68] very useful when amount of sample is limited.

3.3. Protein identification

Proteome studies may be performed with two main different but somewhat complementary purposes: to characterize a particular sample (e.g organism, cell, fluid), or to compare samples from different experimental conditions. Independently of which of these two is the focus, protein identification is a central aspect. In proteomic studies, electrophoresis is frequently coupled to mass spectrometry technologies. Proteins selected in the gels are excised and subjected to enzymatic in-gel digestion. In this process it is important that each of the excised spots (or bands) present the amount of protein sufficient for MS analysis. So, the choice of the staining methodology should take this into consideration, as well as the need for the use of dyes compatible with MS (referred in previous section).

After proteins being identified, MS identifications need further to be validated, by Western blotting, through the use of antibodies to the proteins of interest. This is only possible for proteins for which commercial antibodies are available. There are some particularities of saliva samples that may be considered for Western blotting. Using this technique, only the protein spots (or bands) that react with the antibody are visualized. When comparison among different samples is to be made, it is important to have the same load of protein in all lanes or 2-DE gels. Different loads may result in erroneous results. The existence of internal controls, i.e., proteins for which the levels are proportional to the amount of total protein loaded in the gels, is important and is commonly used to circumvent putative differences in protein loads [e.g 69]. The simultaneous use of a primary antibody for such internal control and the use of a primary antibody for the protein of interest may allow adjustments: by comparing the intensity of these internal controls, the relative amount of protein run for each sample can thus be estimated. For saliva studies such strategy is not possible, since it is not yet known one salivary protein which relative amount to total protein content remains constant. One way to circumvent this limitation is through the staining of the membrane, allowing the relative evaluation of protein content, before incubation with the primary antibody [70]. With this procedure, it is possible to visualize the several lanes and consequently the expression of the protein of interest may be compared as a percentage of the total band intensity. The reversible Ponceau red staining [71] is the standard procedure, despite its low sensitivity (detection limit in the range of 1 μg/spot).

Although in human saliva, protein identification can be generally performed with success, for animal saliva this is not always the case. One of the major limitations in animal proteomic studies is the lack of complete and annotated genome and protein sequences for a great number of species, making salivary protein identification challenging. As consequence there is the need for search in other related species databases, at the cost of eventually producing high number of false positive results [72]. Moreover, the inexistence of commercial antibodies for most of animal proteins makes validation of identifications more difficult. Nevertheless, recent advances in sequencing the genomes of various domestic animals (cattle, pigs and sheep) are increasing the ability to identify salivary proteins in these animal species.

4. Advantages and limitations of electrophoresis for saliva proteome analysis

Electrophoresis has been widely used in saliva characterization. Despite the advances in the knowledge of salivary protein composition that SDS-PAGE and IEF allowed, the development of 2-DE and its application to the study of saliva contributed for a great advance in the comprehension of this body fluid. 2-DE has played a major role in the birth and developments of proteomics, although it is no longer the exclusive separation tool used in this field. Nevertheless, 2-DE continues to be essential in proteomics studies. Apart from the great advantage arising from its efficiency in resolving a high number of proteins in complex samples such as saliva and allowing the visualization of PTMs, 2-DE method has also the advantage of being a method relatively inexpensive (at least comparatively to most of the other techniques used in proteomics).

Although the great advantages of electrophoresis for the study of salivary protein composition, saliva presents several particularities that need to be consider, and that will limit the use of this technique. One of these particularities is the considerable ionic content that saliva presents, which is important to account for when IEF or 2-DE is to be used in protein separation; since separation according to electrical charge will occur, the existence of charged compounds, or salts, in the sample may cause interference. Lowering of ions and salt content may be achieved using ultracentrifugation membranes [e.g. 51], or through protein acid precipitation (e.g. 10% TCA/90% acetone), in which the precipitate is subsequently re-suspended in a buffer. Both of these procedures will also increase the protein concentration, which may constitute an advantage since saliva samples may sometimes be too diluted (for example, after parasympathetic stimulation). In any case some protein loss will inevitably occur.

Another characteristic of saliva that limits its study by electrophoresis is the considerable amount of mucins present. These are glycoproteins with high molecular masses, which present particular physychochemical characteristics that difficult analysis using electrophoretic separation. Moreover, the presence of mucins also limits the study of other salivary proteins, since mucins form complexes with several proteins, interfering in the analysis of the latter. Human salivary glands secrete two types mucins: oligomeric mucin (MG1) with molecular mass above 1 MDa and monomeric mucin (MG2) with molecular mass of 200–250 kDa, which together represent about 26% of total proteins from saliva [73].

The high molecular mass of these molecules impede them to migrate through the polyacrylamide matrix, resulting in their deposition in the top of the gel [74]. The exact mucin content is dependent on the proportion of contribution from the different major salivary glands for the total saliva. As such, the type of stimulation will influence the amount of salivary mucins present. Stimuli that increase sublingual and submandibular saliva result in higher amount of these proteins than when parotid glands are the major contributors.

The presence of high amounts of mucins confers a high viscosity to saliva, what has practical drawbacks in sample preparation for analysis, such as difficulties in sample pipetting. Usage of denaturing conditions, such as buffers containing 4-6M guanidine hydrochloride (GdmCl), or the reducing agent dithiothreitol (DTT) can diminish the viscosity of mucous salivary secretions [75]. However, this effect is achieved at expense of effects in the structure of proteins, and should be avoided in studies where the maintenance of such structure is necessary (for example when studying salivary complexes). The centrifugation of saliva samples, which is one of the first approach usually performed in saliva preparation, also aids in the removing of mucins. However, mucins and other glycoproteins are frequently involved in protein complexes with other salivary proteins, namely amylase, statherin and PRPs, resulting in particular protein losses, when the pellet is discharged.

Another limitation for the study of saliva using electrophoresis, which is common to the majority of body fluids, is the high diversity in the levels of the different protein species. As it was mentioned in section 3.1., a few salivary proteins are present in high amounts, whereas many others appear in low levels. Depletion methods are needed, in order to visualize the low abundant proteins, which may be of particular importance. If considering the important application of saliva as source of disease biomarkers, and knowing that many of those potential biomarkers are not secreted by the salivary glands, it is essential to have access to the low abundant proteins, such as metabolic enzymes with antimicrobial activity (e.g. lysozyme and lactoperoxidase).

Finally, saliva contains several low-molecular mass components that have important functions, namely important bactericidal activity (e-g- histatins, defensins) [76]. The fact that electrophoresis does not allow the separation of these compounds constitutes another limitation in the application of electrophoresis to saliva characterization.

5. Alternatives to electrophoresis for saliva proteome analysis

Through this chapter the use of electrophoresis in saliva proteome analysis has been emphasized. 2-DE and protein MS represent an integrated technology by which several thousand proteins can be separated, quantified and identified. And, as has been being referred, this approach has been considerably used in saliva proteome studies. However, although the advantages of 2-DE, it does not allow the study of the complete proteome. Moreover, as it was stated before, 2-DE have drawbacks which include poor gel-to-gel reproducibility, and the requirement of relatively large amounts of sample, as well as extensive labor and a considerable time required. As such, other proteomic techniques are valuable for the study of saliva proteome.

With the introduction of high-throughput LC coupled to tandem MS (MS/MS), the study of complex systems moved towards a bottom-up proteomics analysis, where complex protein samples are digested and the generated peptides, which are separated by high pressure liquid chromatography (HPLC), are introduced into a mass spectrometer for fragmentation and sequencing to identify and quantify the parent proteins. However, LC-MS analysis of highly complex proteomic samples remains a challenging endeavor [77]. With this approach, tryptic cleavage generates multiple peptides per protein so that proteomic samples typically consist of hundreds of thousands of peptides. To date, no separation method is capable of resolving so many components in a single analytical dimension prior to the MS analysis. Thus, many research efforts have focused on the development of a more sensitive multidimensional liquid chromatography (MudPIT) with higher peptide separation power [21,22]. Initially, large scale shotgun proteomics was defined as an ion exchange chromatography (specifically strong cation exchange-SCX) coupled to reverse phase (RP) and mass spectrometry [78]. Nowadays, alternative configurations to SCX with RP have also been investigated and include the use of anion exchange chromatography and RP, affinity chromatography (AC) and RP, isoelectric focusing (IEF) and RP, capillary electrophoresis (CE) [79,80]. Recently, a RPRP system was proposed for proteomics where the first RP column uses a pH of 10 and the second RP column uses a pH of 2.6 [81,82]. This later approach yielded higher proteome coverage when compared with the SCX-RP approach [81,83].

These methodologies have been applied to saliva proteome characterization aiming the extension of salivary proteins catalogue and further comparison within other biological fluids. For instance, the use of a classical MudPit approach through combination of SCX-RP, allowed the identification of more than 100 proteins [84]. In other experiments a shotgun approach using only LC-MS/MS resulted in the identification of more than 300 proteins [85]. Different chromatographic combinations have been succeeded [30,80,86] which in conjugation to instrumentation advances conducted to the identification of more than 3000 different components in saliva. Other approaches were also performed aiming the reduction of saliva sample complexity through the utilization of combinatorial chemistry derived hexapeptide libraries (Proteominer from BioRad) which lead to the identification of more 2300 different proteins [1]. This has been later used for PTM characterization mainly N-linked salivary glycoproteins and their glycosylation sites [87]. Other methodologies aimed at characterizing salivary glycoproteins consist in enrichment procedures based on affinity chromatography with lectins [88] acting as a first dimension. In fact, saliva is a rich source of both N- and O-linked glycoproteins, which play an important role in the maintenance of oral health and protection of teeth [68]. In line with the characterization of the most abundant PTMs in salivary proteins, namely phosphorylation, other systems were also developed. For instance, Salih et al [89] developed an enrichment procedure based on chemical derivatization using dithiothreitol (DTT) leading to the identification of 65 phosphoproteins. In a different approach, combining hexapeptide libraries, immobilized metal ion affinity chromatography, SCX and RP, 217 unique phosphopeptides sites were positively identified representing 85 distinct phosphoproteins [54].

As stated above, and similarly to other bodily fluids, saliva contains several protein species of low molecular weight which comprise around 40-50% of the total secreted protein content [90]. Albeit particular functions can be attributed to the major peptide classes, several questions about their precise role in oral cavity remain unclear. A strategy based on acidic precipitation or passing the saliva supernatant through a defined cut-off filters and LC-MS/MS have been widely adopted to perform the characterization of this low molecular weight fraction [91-107]. Behind the identification of several fragments deriving from those major peptide classes, several PTMs were also assigned. For instance, novel N- and O-glycosylation sites were identified in PRPs [108] as well as S-Glutathionyl, S-cysteinyl and S-S 2-mer recently identified in cystatin B [109].

More than identification of proteins, many studies aim the evaluation of protein expression under different purposes and pathophysiological conditions. In fact, quantitative proteome profiling is key for comparative analysis of proteins from normal and diseased patients, as similar proteins may be present in both states but at significantly different concentrations. Without quantitative information, the value of these differentially abundant proteins as biomarkers may be overlooked. In the pursuit of these goals, gel-free approaches using stable-isotope tagging or label free (based on spectral counting) have been used for comparative analysis involving salivary samples. In these approaches, the typical flowchart starts by protein digestion being, in case of isotope labeling, derivatized with respective isotope, mixed and analysed simultaneously. Depending on the approach, up to 8 different samples (iTRAQ-8plex, Absciex) can be compared at same time. As an example, Streckfus et al. [110] evaluated the salivary protein expression in patients with breast cancer using a iTRAQ approach identifying 55 proteins that were common to both cancer stages in comparison to each other and healthy controls while there were 20 proteins unique to Stage IIa and 28 proteins that were unique to Stage IIb. In case of label free, eluted peptides are aligned in terms of retention time and comparative analysis will be based on spectral counting [111]. For instance, Ambatipudi et al. [112] by MudPit and label free evaluated the aging effect in the abundance of human female parotid salivary proteins where extensive age associated changes in the abundance of many of salivary proteins were noted, especially for proteins associated with host defense mechanisms.

6. Concluding remarks

The proteome of human saliva received considerable attention in the last years. However, the improvement of the methods to study and characterize salivary proteins and/or the changes in its profile is still an issue since the identification of potential biomarkers for several pathological and physiological conditions is yet to be established. Moreover, a complete knowledge on the importance of each salivary protein in oral environment is not completely understood. Additionally, the growing interest in animal saliva, both due to their value as models for humans, as well as for veterinarian and production purposes, justifies the efforts to develop new protocols adapted to the particularities of these samples.

Despite the existence of limitations, electrophoresis continues to be an essential tool in the study of salivary proteome. It constitutes the bases for the separation of several different components, allowing a summary characterization and also providing a purification step prior the application of more selective and commonly more expensive methods. Nevertheless, enhanced methodologies for sample fractionation and processing might be useful to circumvent some of the limitations for the study of this fluid by electrophoresis. Profiling such a fluid that rapidly changes according stimulus and where some of its constituents interact with each other is challenging. Improved approaches will be necessary to cope with the challenges in understanding of these interactions, their functions and health consequences – the so called interactome – which will be the future in saliva characterization and biomarker identification.

Author details

Elsa Lamy*
ICAAM – Institute of Mediterranean Agricultural and Environmental Sciences, University of Évora, Évora, Portugal
QOPNA, Mass Spectrometry Center, Department of Chemistry, University of Aveiro, Aveiro, Portugal

Ana R. Costa
ICAAM – Institute of Mediterranean Agricultural and Environmental Sciences, University of Évora, Évora, Portugal
Department of Chemistry, University of Évora, Évora, Portugal

Célia M. Antunes
ICAAM – Institute of Mediterranean Agricultural and Environmental Sciences, University of Évora, Évora, Portugal
Department of Chemistry, University of Évora, Évora, Portugal
Center for Neuroscience and Cell Biology, University of Coimbra, Portugal

Rui Vitorino and Francisco Amado
QOPNA, Mass Spectrometry Center, Department of Chemistry, University of Aveiro, Aveiro, Portugal

Acknowledgement

Authors acknowledge the financial support from FCT - Fundação para a Ciência e a Tecnologia – Science and Technology Foundation (Lisbon, Portugal) of the Ministry of Science, Technology and Higher Education (Post-doctoral grant SFRH/BPD/63240/2009 Elsa Lamy) and by FEDER Funds through the Operational Programme for Competitiveness Factors - COMPETE and National Funds through FCT - Foundation for Science and Technology under the Strategic Projects PEst16C/AGR/UI0115/2011 and PEst-C/QUI/UI0062/2011.

* Corresponding Author

7. References

[1] Bandhakavi S, Stone MD, Onsongo G, Van Riper SK, Griffin TJ (2009) A dynamic range compression and three-dimensional peptide fractionation analysis platform expands proteome coverage and the diagnostic potential of whole saliva. J. Proteome Res. 8: 5590–5600.

[2] Castagnola M, Cabras T, Iavarone F, Fanali C, Nemolato S, Peluso G, Bosello SL, Faa G, Ferraccioli G, Messana I (2012) The human salivary proteome: a critical overview of the results obtained by different proteomic platforms. Expert Rev Proteomics.9: 33-46.

[3] Emmelin N (1987) Nerve interactions in salivary glands. J Dent Res. 66: 509-517.

[4] Huang AY, Castle AM, Hinon BT, Castle D 2001 Resting (basal) secretion of proteins is provided by the minor regulated and constitutive-like pathways and not granule exocytosis in parotid acinar cells. J Biol Chem. 276: 22296-22306.

[5] Cevik-Aras H, Ekstrom J (2006) Pentagastrin-induced proteinsynthesis in the parotid gland of the anaesthetized rat, and its dependence on CCK-A and -B receptors and nitric oxide generation. Exp Physiol. 91: 673-679.

[6] Cevik-Aras H, Ekstrom J (2006) Cholecystokinin- and gastrin induced protein and amylase secretion from the parotid gland of the anaesthetized rat. Regul Pept. 134: 89-69.

[7] Cevik-Aras H, Ekstrom J (2008) Melatonin-evoked in vivo secretion of protein and amylase from the parotid gland of the anaesthetised rat. J Pineal Res. 45: 413-421.

[8] Cevik-Aras H, Godoy T, Ekstrom J (2011) Melatonin-induced protein synthesis in the rat parotid gland. J Physiol Pharmacol. 62: 95-99.

[9] Loy F, Diana M, Isola R, Solinas P, Isola M, Conti G, Lantini MS, Cossu M, Riva A, Ekström J (2012) Morphological evidence that pentagastrin regulates secretion in the human parotid gland. J Anat. doi: 10.1111/j.1469-7580.2012.01489.x

[10] Ruhl S (2012) The scientific exploration of saliva in the post-proteomic era: from database back to basic function. Expert Rev Proteomics. 9: 85-96.

[11] Soares S, Vitorino R, Osório H, Fernandes A, Venâncio A, Mateus N, Amado F, de Freitas V (2011) Reactivity of human salivary proteins families toward food polyphenols. J Agric Food Chem. 59: 5535-5547.

[12] Padiglia A, Zonza A, Atzori E, Chillotti C, Calò C, Tepper BJ, Barbarossa IT (2010) Sensitivity to 6-n-propylthiouracil is associated with gustin (carbonic anhydrase VI) gene polymorphism, salivary zinc, and body mass index in humans. Am J Clin Nutr. 92: 539-545.

[13] Mandel AL, Peyrot des Gachons C, Plank KL, Alarcon S, Breslin PA (2010) Individual differences in AMY1 gene copy number, salivary α-amylase levels, and the perception of oral starch. PLoS One. 5: e13352.

[14] Dsamou M, Palicki O, Septier C, Chabanet C, Lucchi G, Ducoroy P, Chagnon MC, Morzel M (2012) Salivary protein profiles and sensitivity to the bitter taste of caffeine. Chem Senses. 37: 87-95.

[15] Neyraud E, Sayd T, Morzel M, Dransfield E (2006) Proteomic analysis of human whole and parotid salivas following stimulation by different tastes. J Proteome Res. 5: 2474-2480.

[16] Harthoorn LF, Brattinga C, Van Kekem K, Neyraud E, Dransfield E (2009) Effects of sucrose on salivary flow and composition: differences between real and sham intake. Int J Food Sci Nutr. 60: 637-46.

[17] Vitorino R, Calheiros-Lobo MJ, Duarte JA, Domingues P, Amado F (2006) Salivary clinical data and dental caries susceptibility: is there a relationship? Bull Group Int Rech Sci Stomatol Odontol. 47: 27-33.

[18] Nieuw Amerongen AV, Veerman EC (2002) Saliva - the defender of the oral cavity. Oral Dis. 8: 12-22.

[19] Huq, NL, Cross KJ, Ung M, Myroforidis H, Veith PD, Chen D, Staton D, He H, Ward BR, Reynols EC (2007) A review of the salivary proteome and peptidome and saliva-derived peptide therapeutics. Int J Res Ther. 13: 547-564.

[20] Baldini C, Giusti L, Ciregia F, Da Valle Y, Giacomelli C, Donadio E, Ferro F, Galimberti S, Donati V, Bazzichi L, Bombardieri S, Lucacchini A (2011) Correspondence between salivary proteomic pattern and clinical course in primary Sjögren syndrome and non-Hodgkin's lymphoma: a case report. J Transl Med. 9: 188.

[21] Caseiro A, Vitorino R, Barros AS, Ferreira R, Calheiros-Lobo MJ, Carvalho D, Duarte JA, Amado F (2012) Salivary peptidome in type 1 diabetes mellitus. Biomed Chromatogr. 26: 571-82. doi: 10.1002/bmc.1677

[22] Xiao H, Zhang L, Zhou H, Lee JM, Garon EB, Wong DT (2012) Proteomic analysis of human saliva from lung cancer patients using two-dimensional difference gel electrophoresis and mass spectrometry. Mol Cell Proteomics. 11: M111.012112

[23] Al-Tarawneh SK, Border MB, Dibble CF, Bencharit S (2011) Defining salivary biomarkers using mass spectrometry-based proteomics: a systematic review. OMICS. 15: 353-361

[24] da Costa G, Lamy E, Capela e Silva F, Andersen J, Sales Baptista E, Coelho AV (2008) Salivary amylase induction by tannin-enriched diets as a possible countermeasure against tannins. J Chem Ecol. 34: 376-387

[25] Fontanini D, Capocchi A, Saviozzi F, Galleschi L (2007) Simplified electrophoretic assay for human salivary alpha-amylase inhibitor detection in cereal seed flours. J Agric Food Chem. 55: 4334-4339.

[26] Chisholm DM, Beeley JA, Mason DK (1973) Salivary proteins in Sjögren's syndrome: separation by isoelectric focusing in acrylamide gels. Oral Surg Oral Med Oral Pathol. 35: 620-630,

[27] Vitorino R, Lobo MJ, Ferrer-Correira AJ, Dubin JR, Tomer KB, Domingues PM, Amado FM (2004) Identification of human whole saliva protein components using proteomics. Proteomics. 4: 1109-1115.

[28] Pobozy E, Czarkowska W, Trojanowicz M (2007) Determination of amino acids in saliva using capillary electrophoresis with fluorimetric detection. J Biochem Biophys Methods. 69: XIII-XXIII.

[29] Sugimoto M, Wong DT, Hirayama A, Soga T, Tomita M (2010) Capillary electrophoresis mass spectrometry-based saliva metabolomics identified oral, breast and pancreatic cancer-specific profiles. Metabolomics. 6:78-95.
[30] Xie H, Rhodus NL, Griffin RJ, Carlis JV, Griffin TJ (2005) A catalogue of human saliva proteins identified by free flow electrophoresis-based peptide separation and tandem mass spectrometry. Mol Cell Proteomics. 4: 1826-1830.
[31] Pandey A, Mann M (2000) Proteomics to study genes and genomes. Nature 405: 837-846.
[32] Klose J 1975 Protein mapping by combined isoelectric focusing and electrophoresis of mouse tissues. A novel approach to testing for induced point mutations in mammals. Humangenetik 26: 231–243.
[33] O'Farrell, PH 1975 High resolution two-dimensional electrophoresis of proteins. J Biol Chem. 250: 4007–4021.
[34] Lamy E, da Costa G, Santos R, Capela E Silva F, Potes J, Pereira A, Coelho AV, Sales Baptista E (2009) Sheep and goat saliva proteome analysis: a useful tool for ingestive behavior research? Physiol Behav. 98: 393-401.
[35] Lamy E, da Costa G, Santos R, Capela e Silva F, Potes J, Pereira A, Coelho AV, Baptista ES (2011) Effect of condensed tannin ingestion in sheep and goat parotid saliva proteome. J Anim Physiol Anim Nutr (Berl). 95: 304-312.
[36] Lloyd DK (2008) Capillary electrophoresis analysis of biofluids with a focus on less commonly analyzed matrices. J Chromatogr B Analyt Technol Biomed Life Sci. 866: 154-166.
[37] Simpson DC, Smith RD (2005) Combining capillary electrophoresis with mass spectrometry for applications in proteomics. Electrophoresis. 26: 1291-1305.
[38] Guo T, Lee CS, Wang W, DeVoe DL, Balgley BM (2006) Capillary separations enabling tissue proteomics-based biomarker discovery. Electrophoresis. 27: 3523-3532.
[39] Siqueira WL, Dawes C (2011) The salivary proteome: challenges and perspectives. Proteomics Clin Appl. 5: 575-579.
[40] Castagnola M, Picciotti PM, Messana I, Fanali C, Fiorita A, Cabras T, Calò L, Pisano E, Passali GC, Iavarone F, Paludetti G, Scarano E (2011) Potential applications of human saliva as diagnostic fluid. Acta Otorhinolaryngol Ital. 31: 347-357.
[41] Dawes C (1972) Circadian rhythms in human salivary flow rate and composition. J Physiol 220:529-545.
[42] Inoue H, Ono K, Masuda W, Morimoto Y, Tanaka T, Yokota M, Inenaga K (2006) Gender difference in unstimulated whole saliva flow rate and salivary gland sizes. Arch Oral Biol. 51: 1055-1060.
[43] Scully C (2003) Drug effects on salivary glands: dry mouth. Oral Dis. 9: 165-176.
[44] Usui T, Yoshikawa T, Orita K, Ueda SY, Katsura Y, Fujimoto S, Yoshimura M. Changes in salivary antimicrobial peptides, immunoglobulin A and cortisol after prolonged strenuous exercise. Eur J Appl Physiol. 111: 2005-2014.
[45] Humphrey SP, Williamson RT (2001) A review of saliva: Normal composition, flow and function. J Prosthet Dent. 85:162-169.
[46] Navazesh M (1993) Methods for collecting saliva. Ann N Y Acad Sci. 694:72-77.

[47] Veerman EC, van den Keybus PA, Vissink A, Nieuw Amerongen AV (1996) Human glandular salivas: their separate collection and analysis. Eur J Oral Sci. 104: 346-352.

[48] Mau M, Kaiser TM, Südekum KH (2010) Carbonic anhydrase II is secreted from bovine parotid glands. Histol Histopathol. 25:321-329.

[49] Lamy E, Graça G, da Costa G, Franco C, E Silva FC, Baptista ES, Coelho AV (2010) Changes in mouse whole saliva soluble proteome induced by tannin-enriched diet. Proteome Sci. 8:65.

[50] Gorr SU, Venkatesh SG, Darling DS (2005) Parotid secretory granules: crossroads of secretory pathways and protein storage. J Dent Res. 84: 500-509.

[51] Lamy E, da Costa G, Capela e Silva F, Potes J, Coelho AV, Baptista ES (2008) Comparison of electrophoretic protein profiles from sheep and goat parotid saliva. J Chem Ecol. 34:388-397.

[52] Deutsch O, Fleissig Y, Zaks B, Krief G, Aframian DJ, Palmon A (2008) An approach to remove alpha amylase for proteomic analysis of low abundance biomarkers in human saliva. Electrophoresis 29: 4150-4157.

[53] Xiao H, Wong DT (2012) Method development for proteome stabilization in human saliva. Anal Chim Acta. 722: 63-69.

[54] Stone MD, Chen X, McGowan T, Bandhakavi S, Cheng B, Rhodus NL, Griffin TJ (2011) Large-scale phosphoproteomics analysis of whole saliva reveals a distinct phosphorylation pattern. J Proteome Res. 10: 1728-1736.

[55] Helmerhorst EJ, Oppenheim FG (2007) Saliva: a dynamic proteome. J Dent Res. 86: 680-693.

[56] Castagnola M, Inzitari R, Fanali C, Iavarone F, Vitali A, Desiderio C, Vento G, Tirone C, Romagnoli C, Cabras T, Manconi B, Sanna MT, Boi R, Pisano E, Olianas A, Pellegrini M, Nemolato S, Heizmann CW, Faa G, Messana I (2011) The surprising composition of the salivary proteome of preterm human newborn. Mol Cell Proteomics. 10: 1-14.

[57] Thomadaki K, Helmerhorst EJ, Tian N, Sun X, Siqueira WL, Walt DR, Oppenheim FG (2011) Whole-saliva proteolysis and its impact on salivary diagnostics. J Dent Res. 90: 1325-1330.

[58] Schipper R, Loof A, de Groot J, Harthoorn L, Dransfield E, van Heerde W (2007) SELDI-TOF-MS of saliva: methodology and pre-treatment effects. J Chromatogr B Analyt Technol Biomed Life Sci 847:45-53.

[59] Schipper RG, Silletti E, Vingerhoeds MH (2007) Saliva as research material: Biochemical, physicochemical and practical aspects. Arch Oral Biol. 52: 1114-1135.

[60] Westermeier R, Marouga R (20059 Protein Detection Methods in Proteomics Research. Bioscience Reports 25; DOI: 10.1007/s10540-005-2845-1

[61] Patton WF (2002) Detection technologies in proteome analysis. J Chromatogr B Analyt Technol Biomed Life Sci. 771(1-2):3-31.

[62] Beeley JA, Sweeney D, Lindsay JC, Buchanan ML, Sarna L, Khoo KS (1991) Sodium dodecyl sulphate-polyacrylamide gel electrophoresis of human parotid salivary proteins. Electrophoresis12: 1032-1041.

[63] Yao Y, Berg EA, Costello CE, Troxler RF, Oppenheim FG (2003) Identification of protein components in human acquired enamel pellicle and whole saliva using novel proteomics approaches. J Biol Chem. 278: 5300-5308.
[64] Steinberg TH, Jones LJ, Haugland RP, Singer VL (1996) SYPRO orange and SYPRO red protein gel stains: one-step fluorescent staining of denaturing gels for detection of nanogram levels of protein. Anal Biochem. 239: 223-237.
[65] Zhang L, Xiao H, Karlan S, Zhou H, Gross J, Elashoff D, Akin D, Yan X, Chia D, Karlan B, Wong DT (2010) Discovery and preclinical validation of salivary transcriptomic and proteomic biomarkers for the non-invasive detection of breast cancer. PLoS One. 5: e15573
[66] Steinberg TH (2009) Protein gel staining methods: an introduction and overview. Methods Enzymol. 463: 541-563.
[67] Miller I, Crawford J, Gianazza E (2006) Protein stains for proteomic applications: which, when, why? Proteomics. 6: 5385-5408.
[68] Sondej M, Denny PA, Xie Y, Ramachandran P, Si Y, Takashima J, Shi W, Wong DT, Loo JA, Denny PC (2009) Glycoprofiling of the Human Salivary Proteome. Clin Proteomics. 5: 52-68.
[69] Costa AR, Capela e Silva F, Antunes CM, Cruz-Morais J (2011) Key role of AMPK in glucose-evoked Na,K-ATPase modulation. Diabetologia. 54 [Supplement1]; S196.
[70] Nakamura K, Tanaka T, Kuwahara A, Takeo K (1985) Microassay for proteins on nitrocellulose filter using protein dye-staining procedure. Anal Biochem. 148: 311-319.
[71] Salinovich O, Montelaro RC (1986) Comparison of glycoproteins by two-dimensional mapping of glycosylated peptides. Anal Biochem. 157: 19-27.
[72] Johnson RS, Davis MT, Taylor JA, Patterson SD (2005) Informatics for protein identification by mass spectrometry. Methods 35: 223-236.
[73] Zalewska A, Zwierz K, Zółkowski K, Gindzieński A (2000) Structure and biosynthesis of human salivary mucins. Acta Biochim Pol. 47: 1067-1079.
[74] Tabak LA 1990 Structure and function of human salivary mucins. Crit Rev Oral Biol Med. 1:229-234.
[75] Schipper RG, Silletti E, Vingerhoeds MH (2007) Saliva as research material: biochemical, physicochemical and practical aspects. Arch Oral Biol. 52:1114-1135.
[76] Amado F, Lobo MJ, Domingues P, Duarte JA, Vitorino R (2010) Salivary peptidomics. Expert Rev Proteomics. 7: 709-721.
[77] Boersema PJ, Mohammed S, Heck AJ (2008) Hydrophilic interaction liquid chromatography (HILIC) in proteomics. Anal Bioanal Chem. 391: 151-159.
[78] Nagele E, Vollmer M, Horth P, Vad C (2004) 2D-LC/MS techniques for the identification of proteins in highly complex mixtures. Expert Rev Proteomics. 1: 37-46.
[79] Zhou F, Sikorski TW, Ficarro SB, Webber JT, Marto JA (2011) Online nanoflow reversed phase-strong anion exchange-reversed phase liquid chromatography-tandem mass spectrometry platform for efficient and in-depth proteome sequence analysis of complex organisms. Anal Chem. 83: 6996-7005.

[80] Guo T, Rudnick PA, Wang W, Lee CS, Devoe DL, Balgley BM (2006) Characterization of the human salivary proteome by capillary isoelectric focusing/nanoreversed-phase liquid chromatography coupled with ESI-tandem MS. J Proteome Res. 5: 1469-1478.

[81] Manadas B, Mendes VM, English J, Dunn MJ (2010) Peptide fractionation in proteomics approaches. Expert Rev Proteomics. 7: 655-663.

[82] Stephanowitz H, Lange S, Lang D, Freund C, Krause E (2012) Improved Two-Dimensional Reversed Phase-Reversed Phase LC-MS/MS Approach for Identification of Peptide-Protein Interactions. J Proteome Res. 11: 1175-1183.

[83] Manadas B, English JA, Wynne KJ, Cotter DR, Dunn MJ (2009) Comparative analysis of OFFGel, strong cation exchange with pH gradient, and RP at high pH for first-dimensional separation of peptides from a membrane-enriched protein fraction. Proteomics. 9: 5194-5198.

[84] Wilmarth PA, Riviere MA, Rustvold DL, Lauten JD, Madden TE, David LL (2004) Two-dimensional liquid chromatography study of the human whole saliva proteome. J Proteome Res. 3: 1017-1023.

[85] Hu S, Xie Y, Ramachandran P, Ogorzalek Loo RR, Li Y, Loo JA, Wong DT (2005) Large-scale identification of proteins in human salivary proteome by liquid chromatography/mass spectrometry and two-dimensional gel electrophoresis-mass spectrometry. Proteomics. 5: 1714-1728.

[86] Denny P, Hagen FK, Hardt M, Liao L, Yan W, Arellanno M, Bassilian S, Bedi GS, Boontheung P, Cociorva D, Delahunty CM, Denny T, Dunsmore J, Faull KF, Gilligan J, Gonzalez-Begne M, Halgand F, Hall SC, Han X, Henson B, Hewel J, Hu S, Jeffrey S, Jiang J, Loo JA, Ogorzalek Loo RR, Malamud D, Melvin JE, Miroshnychenko O, Navazesh M, Niles R, Park SK, Prakobphol A, Ramachandran P, Richert M, Robinson S, Sondej M, Souda P, Sullivan MA, Takashima J, Than S, Wang J, Whitelegge JP, Witkowska HE, Wolinsky L, Xie Y, Xu T, Yu W, Ytterberg J, Wong DT, Yates JR 3rd, Fisher SJ (2008) The Proteomes of Human Parotid and Submandibular/Sublingual Gland Salivas Collected as the Ductal Secretions. J Proteome Res. 7: 1994-2006.

[87] Bandhakavi S, Van Riper SK, Tawfik PN, Stone MD, Haddad T, Rhodus NL, Carlis JV, Griffin TJ (2011) Hexapeptide libraries for enhanced protein PTM identification and relative abundance profiling in whole human saliva. J Proteome Res. 10(3): 1052-1061.

[88] Ferreira JA, Daniel-da-Silva AL, Alves RM, Duarte D, Vieira I, Santos LL, Vitorino R, Amado F (2011) Synthesis and Optimization of Lectin Functionalized Nanoprobes for the Selective Recovery of Glycoproteins from Human Body Fluids. Anal Chem. 2011;83: 7035-7043.

[89] Salih E, Siqueira WL, Helmerhorst EJ, Oppenheim FG (2010) Large-scale phosphoproteome of human whole saliva using disulfide-thiol interchange covalent chromatography and mass spectrometry. Anal Biochem. 407: 19-33.

[90] Amado FM, Vitorino RM, Domingues PM, Lobo MJ, Duarte JA (2005) Analysis of the human saliva proteome. Expert Rev Proteomics. 2: 521-539.

[91] Huq NL, Cross KJ, Ung M, Myroforidis H, Veith PD, Chen D, et al. A review of the salivary proteome and peptidome and saliva-derived peptide therapeutics. Int J Pept Res Ther. 2007;13:547-64.

[92] Hu S, Loo JA, Wong DT (2006) Human body fluid proteome analysis. Proteomics. 6: 6326-6353.

[93] Schlesinger DH, Hay DI, Levine MJ (1989). Complete primary structure of statherin, a potent inhibitor of calcium phosphate precipitation, from the saliva of the monkey, Macaca arctoides. Int J Pept Protein Res. 34: 374-380.

[94] Oppenheim FG, Xu T, McMillian FM, Levitz SM, Diamond RD, Offner GD, Troxler RF (1988) Histatins, a novel family of histidine-rich proteins in human parotid secretion. Isolation, characterization, primary structure, and fungistatic effects on Candida albicans. J Biol Chem. 263: 7472-7477.

[95] Hay DI, Bennick A, Schlesinger DH, Minaguchi K, Madapallimattam G, Schluckebier SK (1988) The primary structures of six human salivary acidic proline-rich proteins (PRP-1, PRP-2, PRP-3, PRP-4, PIF-s and PIF-f). Biochem J. 255: 15-21.

[96] Shomers JP, Tabak LA, Levine MJ, Mandel ID, Ellison SA (1982) Characterization of cysteine-containing phosphoproteins from human submandibular-sublingual saliva. J Dent Res. 61:764-767.

[97] Kauffman D, Wong R, Bennick A, Keller P (1982) Basic proline-rich proteins from human parotid saliva: complete covalent structure of protein IB-9 and partial structure of protein IB-6, members of a polymorphic pair. Biochemistry.;21: 6558-62.

[98] Isemura S, Saitoh E, Sanada K (1982) Fractionation and characterization of basic proline-rich peptides of human parotid saliva and the amino acid sequence of proline-rich peptide P-E. J Biochem. 91: 2067-2075.

[99] Isemura S, Saitoh E, Sanada K (1980) The amino acid sequence of a salivary proline-rich peptide, P-C, and its relation to a salivary proline-rich phosphoprotein, protein C. J Biochem. 87: 1071-1077.

[100] Bennick A (1977) Chemical and physical characterization of a phosphoprotein, Protein C, from human saliva and comparison with a related protein A. Biochem J. 163: 229-239.

[101] Bennick A (1975) Chemical and physical characteristics of a phosphoprotein from human parotid saliva. Biochem J. 145: 557-567.

[102] Hay DI, Oppenheim FG (1974) The isolation from human parotid saliva of a further group of proline-rich proteins. Arch Oral Biol. 19: 627-632.

[103] Azen EA, Oppenheim FG (1973) Genetic polymorphism of proline-rich human salivary proteins. Science. 180: 1067-1069.

[104] Oppenheim FG, Hay DI, Franzblau C (1971) Proline-rich proteins from human parotid saliva. I. Isolation and partial characterization. Biochemistry. 10: 4233-4238.

[105] Vitorino R, Barros A, Caseiro A, Domingues P, Duarte J, Amado F (2009) Towards defining the whole salivary peptidome. Proteom Clin Appl. 3: 528-540.

[106] Helmerhorst EJ, Sun X, Salih E, Oppenheim FG (2008) Identification of Lys-Pro-Gln as a novel cleavage site specificity of saliva-associated proteases. J Biol Chem. 283: 19957-19966.

[107] Vitorino R, Barros AS, Caseiro A, Ferreira R, Amado F (*in press*) Evaluation of different extraction procedures for salivary peptide analysis. Talanta.

[108] Vitorino R, Alves R, Barros A, Caseiro A, Ferreira R, Lobo MC, Bastos A, Duarte J, Carvalho D, Santos LL, Amado FL (2010) Finding new posttranslational modifications in salivary proline-rich proteins. Proteomics. 10: 3732-3742.

[109] Cabras T, Manconi B, Iavarone F, Fanali C, Nemolato S, Fiorita A, Scarano E, Passali GC, Manni A, Cordaro M, Paludetti G, Faa G, Messana I, Castagnola M (2012) RP-HPLC-ESI-MS evidenced that salivary cystatin B is detectable in adult human whole saliva mostly as S-modified derivatives: S-Glutathionyl, S-cysteinyl and S-S 2-mer. J Proteomics. 75: 908-913.

[110] Streckfus CF, Storthz KA, Bigler L, Dubinsky WP (2009) A Comparison of the Proteomic Expression in Pooled Saliva Specimens from Individuals Diagnosed with Ductal Carcinoma of the Breast with and without Lymph Node Involvement. J Oncol. 2009:737619.

[111] Matros A, Kaspar S, Witzel K, Mock HP (2009) Recent progress in liquid chromatography-based separation and label-free quantitative plant proteomics. Phytochemistry. 72: 963-974.

[112] Ambatipudi KS, Lu B, Hagen FK, Melvin JE, Yates JR (2009) Quantitative analysis of age specific variation in the abundance of human female parotid salivary proteins. J Proteome Res. 8: 5093-5102.

Chapter 5

Electrophoresis of Myocardial Cells

Ying Zhou

Additional information is available at the end of the chapter

http://dx.doi.org/10.5772/53508

1. Introduction

Mammals are composed of a large number of surface charged cells. The cell structure and functions in different tissues and organs are different. Through the method of cell electrophoresis the information of the cell surface structure can be obtained and is valuable for the function study of cells, tissues and organs. Many studies have been reported on the fluidic electric phenomena of cells (Ertan & Rampling, 2003; Aki et al., 2010; Brown et al., 1985; Pimenta & de Souza, 1982), but quite limited on the separation of myocardial cell electric phenomena possibly due to the short of myocardial cell electrophoresis technology to conduct the experiments. The methods only observing through separation and investigation of myocardial cell surface complex sugars and plasma membrane phospholipid composition on the cell contraction within the ion flow (inward ionic current) may not actually take the cell membrane and membrane structure as a whole but insularly highlight the single component in achieving cardiac function. Integration is not equal to the simple sum of the single components. The heart is the vital organs of humans and animals, its interfacial electric phenomena (such as electrocardiogram) and its response to the pacemaker have revealed the hints on the close relationship between the myocardial cell membrane structure and function of the heart (Podrid et al., 1995). For this reason, electric phenomenon of the myocardial cells was systematically studied. Although electrophoresis does not directly measure the cell surface charge density and zeta potential, it can help to find a trace to elucidate the mentioned relationship by exploring the classic knowledge of colloidal particles and to set up a electric double layer model for insight into the structural characteristics of the myocardial cell surface and their variations. It opens a way to realize the mechanism in respect of the cardiac function. We found that the myocardial cell is rich and complex internal and external membrane structure, and the distribution and variation of the adsorbed ions on cell surfaces are completely different from the colloidal particles, these should be the basis for completion of the cardiac function.

2. Experimental

2.1. Cell preparation

Ventricular myocardial cells were isolated from adult Sprague-Dawley rats (2–3 months old, weight 225–300 g) using standardized enzymatic techniques (Zhou et al., 2000). Freshly isolated single cells were stored in Tyrode's solution containing (in mM) 137 NaCl, 5.4 KCl, 1.2 $MgCl_2$, 1 $NaH2PO_4$, 1 $CaCl_2$, 20 glucose, and 20 HEPES (pH 7.4). (Fig. 1)

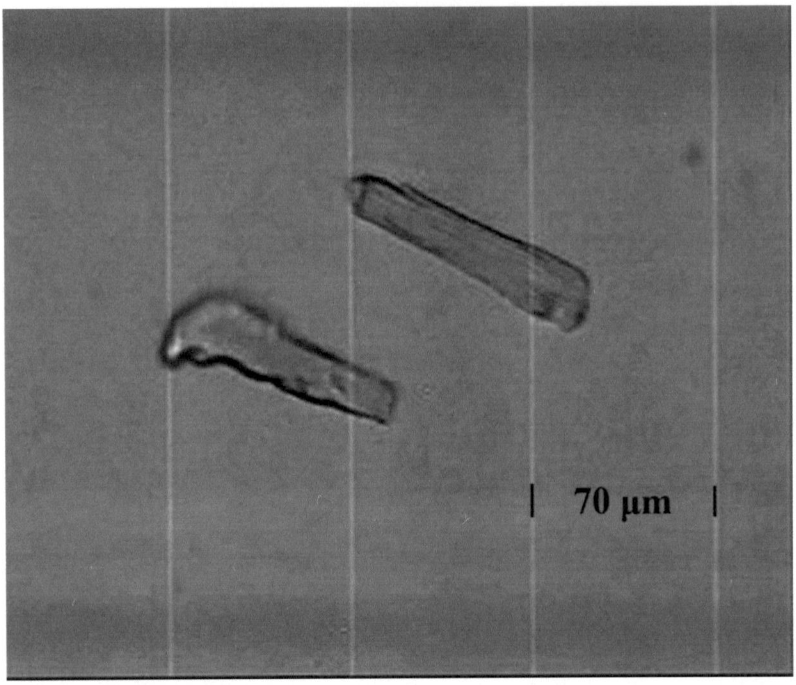

Figure 1. Photo of isolated myocardial cells under low power lens.

2.2. Preparation of cell suspensions

Cell suspensions were prepared in isotonic solutions composed of an aqueous glucose solution and 60%, 30%, 15%, 10%, 7.5%, 5% and 2.5% (v/v) of a stock solution of 145 NaCl and 2.97 $CaCl_2$ (referred to in Table 1). To adjust the transmembrane potential, which is determined primarily by the K^+ transmembrane equilibrium potential (Fozzard et al., 1991), K^+ was added at a required concentration depending on the potential needed which can be calculated using the Nernst equation: $E_K = -\frac{RT}{F}\ln\left(\frac{c_{Kin}}{c_{Kout}}\right)$, where E_K is the K^+ transmembrane equilibrium potential, R is the universal gas constant, T the absolute temperature, F the Faraday, and c_{Kin} and c_{Kout} are the concentration of K^+ inside and outside the cell, respectively (Liu, 2005). Aqueous Dextran 40 (free of ion, transmembrane potential not known) was from Shijiazhuang No.4 Pharmaceutical (China). A proper amount of hydroxypropyl methyl cellulose (HPMC) was added into all suspensions to adjust the viscosity of the final suspensions at 3.8 mPa·s (24°C, Table.1).

2.3. Removal of sialic acid from myocardial cells

Myocardial cells were suspended, at 0.5%(v/v), in the Tyrode's solution containing 0.25U/mL neuraminidase (Sigma Chemical, St. Louis, Mo) at pH 6.0, continually shaken at 37°C for 90 minutes, and then washed with the Tyrode's solution (Post, 1992) .

2.4. Procedure of electrophoresis

The prepared myocardial cells were washed three times with the suspending medium and then suspended in the same medium at 0.1% (v/v) measured at 24°C. The used media were collected in Table 1 where the No. 10-14 solutions were used for the suspension of enzyme-treated myocardial cells while the No. 1-9 for non-treated cells. The suspended cells were loaded on a cell electrophoresis system (Beijing Warder Biomedicine Instrument Company, China) and electrophoresed, within 2-3 minutes, at 2-7V/cm depending on the ionic strength of the suspension. Only the perfect cells were recorded. Each sample solution was determined in parallel for 12 times. All the measurements were finished within eight hours after the myocardial cells were isolated.

№	Ionic strength mM	Glucose mM	Dextran mM	E_K [d] mV	HPMC mM	Viscosities mPa.s	HEPES mM	pH
1	154~163 [a]	5		-96.8~-61.9	0.57	3.8	20	2.5~7.35
2	93.6~102.6 [a]	108.1		-96.8~-61.9	0.54	3.8	20	2.5~7.35
3	48.3~57.3 [a]	180.1		-96.8~-61.9	0.52	3.8	20	2.5~7.35
4	25.7~34.7 [a]	214.5		-96.8~-61.9	0.5	3.8	20	5.0~7.35
5	18.1~27.1 [a]	227.1		-96.8~-61.9	0.5	3.8	20	5.0~7.35
6	14.3~23.3 [a]	233.4		-96.8~-61.9	0.5	3.8	20	5.0~7.35
7	10.6~19.6 [a]	239.7		-96.8~-61.9	0.5	3.8	20	5.0~7.35
8	6.8~15.8 [a]	246		-96.8~-61.9	0.5	3.8	20	5.0~7.35
9	151 [b]	8		0	0.57	3.8	20	5.0~7.35
10		252	1.5		0.2	3.8	20	5.0
11	12.6 [c]	239.7		-83.9	0.5	3.8	20	5.0
12	20.1 [c]	227.1		-83.9	0.5	3.8	20	5.0
13	156 [c]	5		-83.9	0.57	3.8	20	5.0
14	151 [b]	8		0	0.57	3.8	20	5.0

a) Total-3~12mM KC=96%NaCl+4%$CaCl_2$.

b) 140mM KCl+5mM NaCl+3mM $CaCl_2$.

c) Total-5mM KCl=96%NaCl+4%$CaCl_2$.

d) Theoretical value.

Table 1. The suspensions used in this study.

2.5. Note

2.5.1. Cell separation should be to reduce the loss of the charged matter on cell surface

a. Single collagenase (collagenase II, 1mg/ml) can be used as the first choice for myocardial cell separation enzyme; b. enzymatic digestion time of myocardial tissue control in about 15 minutes (not more than 20 minutes), c. the same experimental animals age and enzyme digestion time should be consistent.

2.5.2.

Select the electrophoresis tank stationary layer as the level of observation and measurement of the electrophoretic velocity, to ensure the accuracy of the measured values.

2.5.3.

When the high ionic strength suspensions were studied, the electrophoresis voltage is better selected an proper high value in compromising with the electrophoresis time (making the cells migrate at <10μm • s-1).

2.5.4.

The experiment should be completed within 6-8 hours after the cell separation, to ensure the normal activity of the cells.

3. Data processing and interpretation

Myocardial cells showed different features under an electric field depending on the nature of the suspensions used. Although the cells may contract at high ionic strength or may not at low ionic strength, the impact of the contraction on the electrophoretic speed did not affect the analysis and judgment of experimental results measured from different conditions. Commonly, myocardial cells show negative charges in the electric field. But their electrophoretic speed may have obvious difference even in same a suspension, suggesting that the composition of the charged layer on myocardial cell membrane is variable. In fact, cells were possibly affected by enzyme action (time-dependent), mechanical damage and so on in the preparation. However, this also did not impact on the judgment of variation tendency of the cell mobility.

In the case that the type and proportion of adions on cell surface were maintained, the cell electrophoretic mobility (EPM) showed a zigzag increase (Fig. 2A). At an ionic concentration from 9.7 to 33.2mM, no electromigration of some cells were observed at some conditions, mostly at pH 5.0 and 6.0, rare at pH 7.35 (Table 2). However, reversed electromigration has not yet been observed. The variation of the mobility was largely dependent of the suspension pH. The mobility was found to be in common greater at low pH (2.5-4.5) than at high pH (4.5-7.35) with the minimum at pH 5.0 (Fig. 2C). The mobility commonly decreased as pH increased but changed to fast increase when pH value was above 7.0. The curves of the EPM against pH values under the conditions of the same transmembrane potential and different ionic strength was basically the same as (Fig. 3).

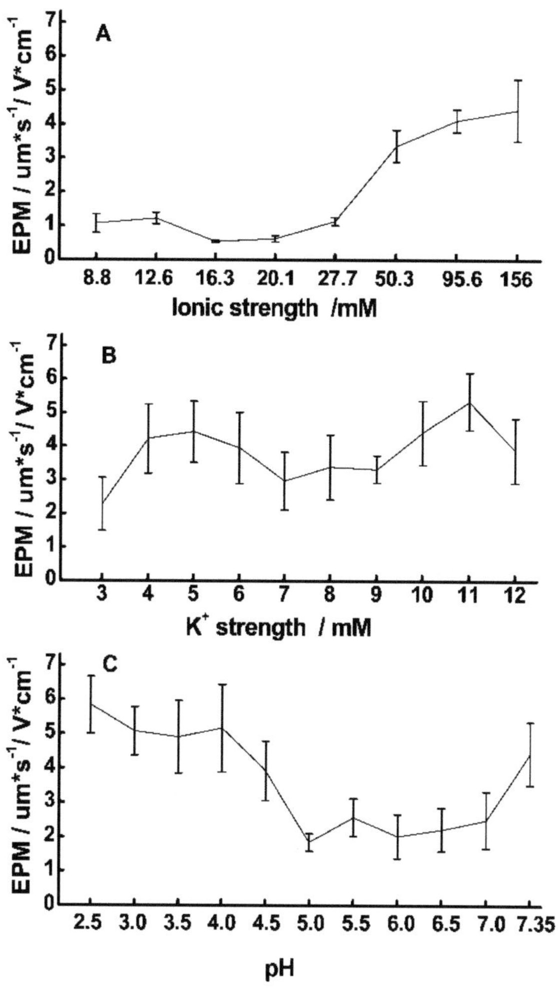

Figure 2. A: Zigzag increase of EPM measured in the suspensions of 5 mM K^+, pH 7.35 and ionic strength 8.8-156mM ($P<0.01$). B: an obvious zigzag increase of EPM in the suspension of 5mM K^+ at 156mM ionic strength and pH 2.5-7.35 ($P<0.01$). C: increasing tendency of EPM in the suspensions of ionic strength 151-160mM, pH 7.35 and K^+ strength 3-12mM ($P<0.01$).

The zigzag increase of cell mobility was also measured when the transmembrane potential or K^+ increased (from 3 to 12mM K^+, Fig. 2B). This suggests that the electromigration of myocardial cells may be imposed largely by the transmembrane potential. Such a zigzag variation of mobility was found in different conditions with somewhat similar features except for some point as shown in Figure 4. Myocardial cells tended to die at the ionic strength < 6.7mM and pH < 5.0 or at the ionic strength <151mM and pH < 2.5. The death could largely be avoided when the cells were suspended in Dextran solution at pH≥5.0.

K$^+$ strength	Ionic strength	pH		
mM	mM	5.0	6.0	7.35
3	10.4	4	0	5
	25.2	0	6	0
4	11.4	0	6	0
5	12.4	0	4	0
	16.1	0	5	0
6	9.7	5	0	0
8	11.7	4	0	0
	15.4	7	0	0
9	16.4	3	0	0
	20.1	0	4	0
	23.8	0	6	0
10	13.7	6	0	0
	21.1	0	4	0
	24.8	4	0	0
11	25.8	0	4	0
	33.2	0	7	0
12	19.4	6	0	0

Table 2. Cells with zero mobility found in various experimental groups

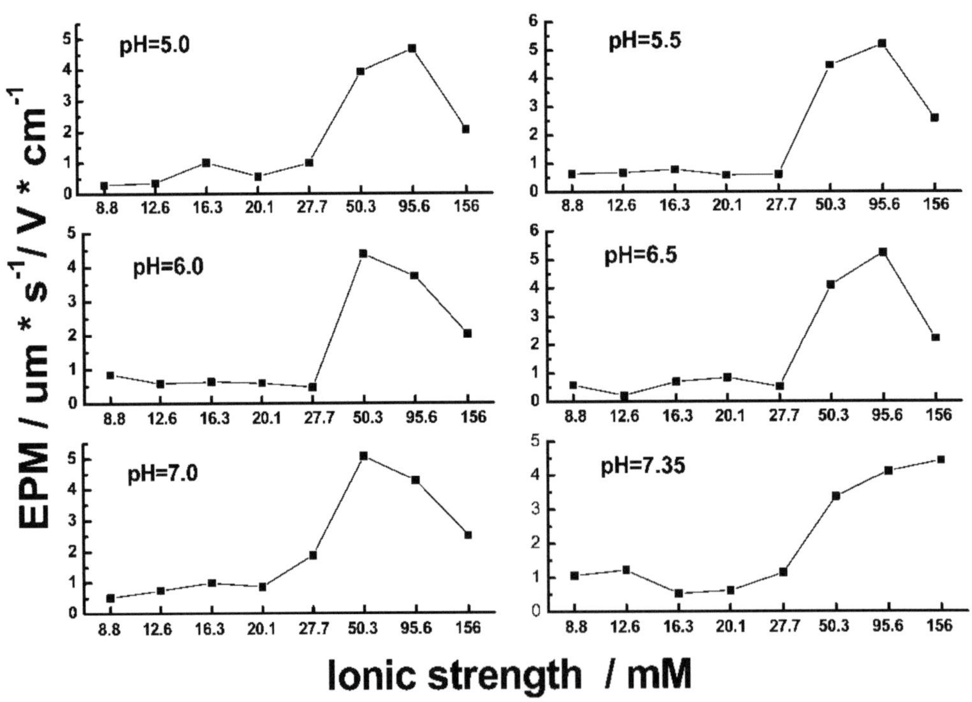

Figure 3. Plot of EPM against ionic strength measured in the suspensions of 5 mM K$^+$ and 8.8-156mM ionic strength at pH values as shown in the figure.

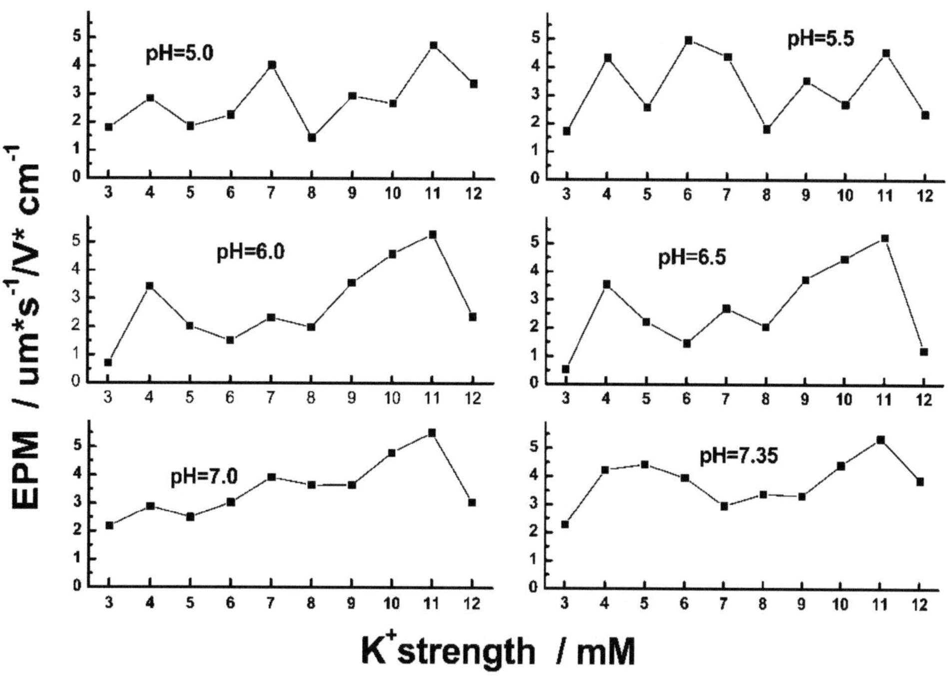

Figure 4. Plot of EPM against the concentration of K^+ from suspensions with ionic strength of 151-160mM at pH values as shown in the figure.

At zero transmembrane potential (140 mM K^+), the variation tendency of cell mobility along with the pH change was still the same but the range changed to 3.09 - 6.00, the highest and the lowest value appeared at the pH value of 6.5 and 6.0. At a transmembrane potential of approximately - 83.9mV (5 mM K^+) and in 153mM ionic strength suspension, the cell mobility generally increased with an exception at pH7.35 (Fig. 5).

After the surface sialic acid was cut off, the mobility of myocardial cells decreased at pH 5.0 as expected (Fig. 6, the first three bars). The decreasing extent was found to depend on the ionic strength. Unexpectedly, the mobility increased when they were suspended in dextran medium (Fig. 6, the 5th bars) or at transmembrane potential of 0 (Fig. 6, the 4th bars).

In conclusion, the zigzag mobility variation (over a range from 0 to 8.67) of myocardial cells was observed for the first time, and the variation was found to depend not only on the suspension's pH and ionic strength but also on K^+ or transmembrane potential.

4. Electric double layer on myocardial cell surface

4.1. Characteristics of myocardial cell mobility

This aberrant change could have four characteristics: First, the cell mobility changes in parallel to the ionic strength and becomes large as the charged layer shrinks. Second, the undulant mobility repeats in a certain range as the surface negative electric field is enhanced or weakened. Third, the cell mobility rises abnormally as a component of the comprehensive surface negative electric field decreases to its minimum or is equal zero (at pH 2.5 or zero volt of transmembrane potential), Fourth, zero mobility appeared several times, especially under the conditions of relatively low ionic strength where the mobility should originally be slow (diffusion layer shrinks).

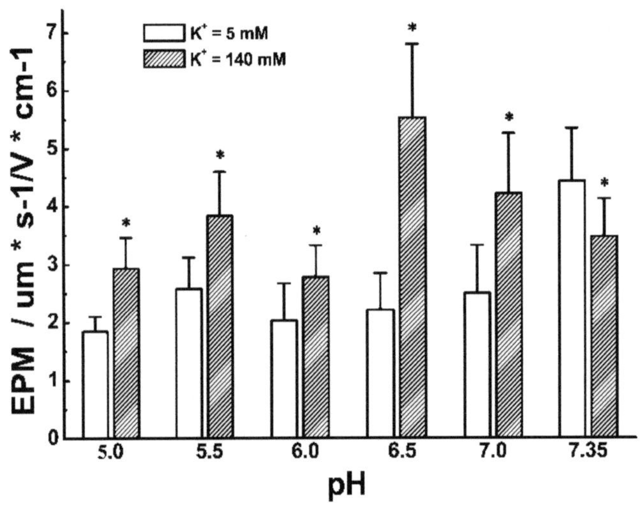

Figure 5. Plot of EMP against pH from the suspension of 140 mM K^+ (transmembrane potential was zero). EMP at pH 7.35 decreased compared with that at 5mM K^+ (transmembrane potential was -83.9mV) and 153mM ionic strength, and the others were increased (*: $P<0.01$).

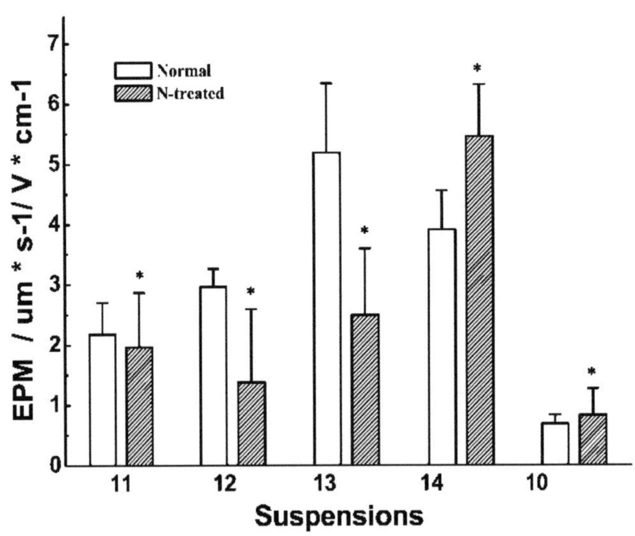

Figure 6. Effects of neuraminidase treatment on EMP of myocardial cells. EPM were measured in suspensions at pH 5.0 and ionic strengths of 12.4mM (Table 1, № 11), 19.8mM (Table 1, № 12) and 153mM(Table 1, № 13), 140 mM K^+ (Table 1, № 14) and Dextran 40 (Table 1, № 10), respectively (*: $P<0.01$).

4.2. Composition of the cell surface electric field

The fact that K^+ strength influenced on the cell EPM demonstrated that the transmembrane potential participation should be considered to impose on the surface negative electric field of myocardial cells, this conforms to the theoretic expectation. However, when the transmembrane potential changes to zero, the characteristics of the EPM of myocardial cells are still different from the common particles. After removing of the sialic acid on the surface of cells, and at pH 5.0, which is lower than the isoelectric point of phospholipids (pI=6.5±) (Yamada et al., 1981) and changed the isoelectric point of membrane protein (Braun et al., 2007; Popot & Engelman, 2000; Cho & Stahelin, 2005), or when the transmembrane potential is zero (at 148mM ionic strengths), the EPM showed a strange rise, they may have very complex contribution to the surface charges.

First, the cell mobility changes in parallel to the ionic strength and becomes large as the charged layer shrinks. The role of transmembrane potential is complex, there should be a complex negative electric field from intracell which concerns with the charged layer but has an effect different from the charged layer. It enhances the half layer of the adsorbed ions and reduces the diffusion layer at the same time or causes redistribution of the absorbed surface ions, making the cell mobility changed zigzag. Second, the undulant mobility repeats in a certain range as the surface negative electric field is enhanced or weakened. This is just an effect of the complex negative electric field inside the cells: the adsorption layer of cell surface varies as the charged layer does, causing part of the ions entering from the diffusion layer. Third, the cell mobility rises abnormally as a component of the comprehensive surface negative electric field decreases to its minimum or is equal zero (at pH 2.5 or zero volt of transmembrane potential), implying that the effect of the complex intracellular negative electric field is reduced as the charged layer shrinks significantly. Fourth, zero mobility appeared several times, especially under the conditions of relatively low ionic strength where the mobility should originally be slow (diffusion layer shrinks). The intracellular complex negative electric field act thus in a limited space within the diffusion layer. The zero mobility concerns only with a zero diffusion layer, not necessarily a zero value for cell surface negative electric field and the adsorption layer. Therefore, the negative electric field of myocardial cell surface is composed by the surface sialic acid, plasmalemma of phospholipids, proteins and transmembrane potential (complex intracellular negative electric).

Each component of negative electric field on the myocardial cell surface has different influence on adions because its amount, isoelectric point and affect characteristics are not the same. Sialic acid on the extracellular surface is about 50 nm away the plasma membrane proteins, phospholipids and transmembrane potential, and in between the negative charges sparsely distributes, including some penetrated adsorbed ions (Langer, 1978). More specifically, in myocardial cells there are many of the same or similar membrane structures besides the negative electric field (transmembrane potential). The surface area of these structures is about 1/3 or more of the cell surface area. The special structure and composition make the electric double layer of the myocardial cell surface different from the general particles: There exist the intersection of the surfaces and inside/outside infiltration of matter, and unique or even strange changes of the distribution characteristics of the cell surface adsorbed ions and mobility. The so-called electric double layer structure is thus not sterling, may be better called an aberrant electric double layer.

4.3. Analysis of mobility change

The one way declining of the cellular mobility after removal of only the extracellular surface sialic acid, which is different from the zigzag mobility vibration, suggests the removal of sialic acid has almost no impact on the complex intracellular negative electric field. At zero transmembrane potential of zero, the mobility was not found to have an obvious relationship with the phospholipid isoelectric point as the pH varied, implying that the pH value may influence on all the membrane composition of sialic acid, protein, phospholipids, and the intracellular complex negative electric field. The mobility is a result of multiple integral ation. Thus the plasma membrane phospholipid may have only weak contribution to the distribution change of adsorbed ions on a myocardial cell surface. After complete removal of the effect of the transmembrane potential, the mobility almost allways increased except for at pH7.35. This indicates the significant effect of the transmembrane potential changes on the intracellular complex negative electric field: The more is the transmembrane potential reduced, the weaker becomes effect of intracellular complex negative electric field and the less are the ions entering from the diffusion layer into the adsorption layer. (In case of pH7.35 there may be other charged components enhancing the effect). The fact that removal of sialic acid and plasma membrane phospholipids and suppression of the transmembrane potential could cause high mobility at high ionic strength, which should also be an integral result, reveals that these factors have some effect on the intracellular complex negative electric field.

By these imitated in vivo, broadly varied suspension conditions, the myocardial cells were shown to have a complex aberrant electric double layer structure on the membrane surfaces. The related surface electric field is a cellular character integrated from the complex combinations of membrane sugars, phospholipids, proteins and transmembrane potential (intracellular complex negative electric field). The ion layer is composed of a large adsorption layer and a large cyclicly varied diffusion layer. This is a special structure, making mobility and its related features completely different the general colloidal particles. Interestingly, the contraction and Ca^{2+} influx (slow inward calcium current) of a living myocardial cell surface, which is chared only negatively, is not significantly affected after removal of the negative sialic acid composition (Yee et al., 1991; Langer & Nudd, 1983). This is due to the contribution of other powerful components in addition to the sialic acid. The aberrant electric double layer structure, which is also distinguished from non-excitable cells, is dependent on the variation of each complex membrane composition such as sugars, membrane proteins, phospholipids and transmembrane potential (intracellular complex negative electric field). The change may be the normal physiological performance but can also be a result of pathology. The charged layer in combination with the intracellular complex negative electric field connects the intra and extra cell environments as a whole body. This point is very important for not only realizing the myocardial cell function (contraction) but also for studying the cardiovascular disease mechanisms.

5. Challenges and prospects of myocardial cells electrophoresis

Animal and human cells are a complex system. According to the basic essence of systems biology research--structure determines function we believe that the study of cell electric phenomena is a key point to crack the hard shells of some diseases and related mechanism. This may unroll all the true information on finding the formation and changing mechanism and inside/outside cellular connecting routes of the cell surface electric double layer, which should be different from the ordinary colloidal particles and on the integration of the double layer's characters into the overall cellular structure and natural biological behaviors. In fact, the occurrence of some diseases is not only caused by one abnormal gene or protein by the unusual results of the electric double layer structure of the cell membrane surface. Therefore, the electric double layer structure of cell membranes or surfaces should be a coherent part in biological researches, and the related methods and materials selected should be able to imitate body conditions as much as possible during performing the investigation of cell electric phenomena. It is also critical to establish cell model capable of re-diplaying the cell functions.

Author details

Ying Zhou
PLA 309 Hospital, Beijing, P.R. China

Acknowledgement

I thank Professor Chen Yi for guidance and help.

6. References

Ertan, N. Z. & Rampling, M.W. (2003) Effect of ionic strength of buffer on the measurement of erythrocyte electrophoretic mobility. Med Sci Monit; 2003 Oct, 9 (10): BR378-81. ISSN: 1234-1010.

Aki, A.; Nair, B. G.; Morimoto, H.; Kumar, D. S. & Maekawa, T. (2010) Label-free determination of the number of biomolecules attached to cells by measurement of the cell's electrophoretic mobility in a microchannel. PLoS One, 2010,5 (12):e15641. ISSN: 1932-6203

Brown, K. A.; Wolstencroft, R. A.; Booth, C. G. & Dumonde, D. C. (1985) A reappraisal of the macrophage electrophoretic mobility (MEM) test for the measurement of lymphokine activity. *J Immunol Methods*. 82, (2) :189-98. ISSN: 0022-1759

Pimenta, P. F. & de Souza, W. (1982) Surface charge of eosinophils. Binding of cationic particles and measurement of cellular electrophoretic mobility. *Histochemistry,* 74(4):569-76, ISSN: 0301-5564

Podrid, P. J.; Kowey, P. R. & Zoll, P. M. (1995) *Cardiac arrhythmia mechanism, diagnosis and management,* Williams & Wilkins, ISBN: 0-7817-2486-4, Baltimore.

Zhou, Y. Y.; Wang, S. Q.; Zhu, W. Z.; Chruscinski, A.; Kobilka, B. K.; Ziman, B.; Wang, S.; Lakatta, E. G.; Cheng,H. & Xiao, R. P. (2000) Culture and adenoviral infection of adult mouse cardiac myocytes: methods for cellular genetic physiology. *Am J Physiol Heart Circ Physiol,* 279, H429-36. ISSN: 0363-6135

Fozzard, H. A.; Haber, E.; Jennings, R. B.; Katz, A. M & Morgan, H. E. (1991) *The Heart and Cardiovascular System, Scientific Foundations.* Raven Press, ISBN 0-88167-747-7, New York. p1-30,1091-1119.

Liu, T. F., Cardiomyocyt Electrophysiology. People's Health Press, Beijing 2005,11-16, ISSN: 7-117-06548-6, Beijing.

Post, J. A. (1992) Removal of sarcolemmal sialic acid residues results in a loss of sarcolemmal functioning and integrity. *Am J Physiol,* 263, H147-52. ISSN: 0363-6135

Yamada, K.; Sasaki, T. & Sakagami, T. (1981) Measurement of isoelectric points of phospholipid exchange proteins by gel isoelectric focusing. *The Tohoku Journal of Experimental Medicine.* 135, 37-42. ISSN: 1349-3329

Braun, R. J.; Kinkl, N.; Beer, M. & Ueffing, M. (2007) Two-dimensional electrophoresis of membrane proteins. *Anal Bioanal Chem,* 9, 1033-45. ISSN: 1618-2642

Popot, J. L. & Engelman, D. M. (2000) Helical membrane protein folding, stability, and evolution. *Annu Rev Biochem,* 69, 881-922. ISSN: 0066-4154

Cho, W. & Stahelin, R. V. (2005) Membrane-protein interactions in cell signaling and membrane trafficking. *Annu Rev Biophys Biomol Struct.* 34, 119–151. ISSN: 1056-8700

Langer, G. A. (1978) The structure and function of the myocardial cell surface. *Am J Physiol.* 1978, 4,H461-468.

Yee, H. F. Jr.; Kuwata, J. H. & Langer, G. A. (1991) Effects of neuraminidase on cellular calcium and contraction in cultured cardiac myocytes. J *Mol Cell Cardiol,* 23,175-185. ISSN: 0022-2828

Langer, G. A. & Nudd, L. M. (1983) Effects of cations, phospholipases, and neuraminidase on calcium binding to "gas-dissected" membranes from cultured cardiac cells. *Circ Res,* 53, 482-90. ISSN: 0009-7330.

Chapter 6

Isozymes: Application for Population Genetics

Vibeke Simonsen

Additional information is available at the end of the chapter

http://dx.doi.org/10.5772/45875

1. Introduction

Electrophoresis as such has been applied for many purposes, among these, analysis of isozymes or isoenzymes. As the designation indicates, this is a group of enzymes having the same catalytic ability, i.e. they catalyse the same chemical reaction. Due to the fact that enzymes are proteins and built from amino acids, they possess an electric change - so isozymes are enzymes with the same catalytic ability, but different electric charge. The charge will cause the enzyme to migrate in an electric field, the migration rate may be faster or slower, depending on the carrying media (see section 3 on electrophoresis) and the power applied. In order to detect the enzyme, the substrate for the particular enzyme and needed co-factors are added, so that the enzymatic reaction can take place. An array of enzymatic detection methods based on for instance histochemical or fluorogenic procedures are known (Manchenko 1994). The revealed pattern is called a zymogram (Fig.1).

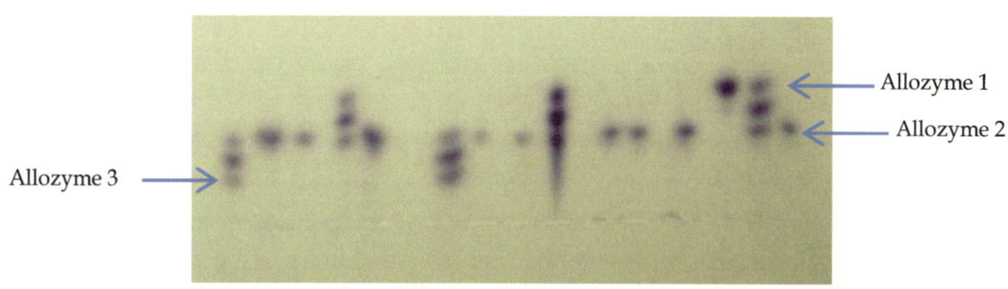

Figure 1. Zymogram of the enzyme glucosephosphateisomerase from 15 individual sprat eggs, dimeric enzyme with three allozymes. For further interpretation, see section 5

Proteins are gene products, which mean that they are an easy way to get information on genetic variation. If more zones of a particular enzyme are revealed in the same zymogram, then the enzyme consists of isozymes. If one zone is interpreted as being the product of one gene, it may reveal different alleles of the gene and, hence, this particular isozyme consists of various allozymes as shown in Fig. 1. Interpretation of the zymograms leads to estimation of allelic frequencies, which is the fundament of population genetics under the assumption that the genes are inherited in accordance with the Mendelian laws. Another important feature of enzymes is that they may consist of one, two, three or more polypeptides. As the enzymes perform the same reaction needed in the metabolism of the organism and that metabolism more or less is universal, the molecular structure is universal. That means that if an enzyme is a dimeric enzyme, built by two polypeptides, this structure will be found regardless of plant species, insect species or any other organism studied, see section 5.1 on interpretation.

1.1. Procedures needed for studying isozymes

When studying isozymes, five major procedures must be considered: Sampling and storage, preparation of the samples for electrophoresis, electrophoretic procedure, detection procedure for the isozymes, and the interpretation of the zymograms.

2. Samples

As mentioned above, one of the key observations in population genetics is the detection of alleles used for estimation of allelic frequencies. In order to achieve a reliable estimate of allelic frequencies, many individuals must be analysed, at least 30 in order to minimise the variance of the estimate of the allelic frequencies (see e.g. Hedrick 1983). When studying vulnerable species, for instance species with low population size, the method for sample collection has to be considered. Two methods for sample collection are considered:

None destructive method where the individual can continue life

Destructive method which will cause the death of the individual

2.1. None destructive methods

None destructive methods require that the samples used for detecting isozymes are removed from the individuals at site of collection. In the case of plants, the samples may consist of leaves or seeds collected from individual plants. For animals the samples may consist of a small amount of blood that can be removed without killing the animal, and with special equipment it may also be possible to remove small biopsies or hair samples. If hair is pulled out, the small tissue sample on the tip of the hair can be used for analysis of isozymes.

2.2. Destructive methods

As mentioned, destructive methods lead to the organism being killed. In this case, it has to be considered how to utilise the samples in an optimal way in order to reduce further collection. When dealing with plants, in the most cases the destructive method can be avoided. However, when using seeds, this will result in a reduction in the next generation, so collecting seeds is a partially destructive method. However, seedlings as samples for analysis of isozymes are often a good choice due to the soft material compared to the fully grown plant. Many small invertebrates have a size that only allows one analysis, e.g. collembolans, but other invertebrates, e.g. earthworms, can support several analyses - however, the collecting method will be destructive. Many fish species when caught will die immediately, so removing biopsies will not save the fish. Analysing fish eggs is a destructive method which will affect the next generation. However, many fish species produce a huge number of eggs, so the actual damage may be small, depending on the vulnerability of the species.

2.3. Transportation

Depending on the species, the samples may either be frozen or kept in a cold box with ice when transporting the samples back to the laboratory. It is crucial that the samples are kept cold in order to avoid destruction of the proteins and, hence, the enzymes. Furthermore, re-freezing the samples will also destroy the proteins, so the number of times for freezing/thawing the samples has to be minimised. Leaves or seeds can easily be transported in a cool bag with cooling blocks if the transportation can be done in one day in the temperate climate zone. If the material is frozen, transportation in a cool bag or a thick foam box with cooling blocks may be sufficient to keep the samples cold, so the activity of the enzymes is not reduced significantly. Frozen fingerlings of Indian major carps were transported this way from Bangladesh to Denmark and later analysed for isozymes with success, see Fig. 2. Transportation time was about 24 hours. For longer storage, the tissue samples have to be kept either at -80°C or in liquid nitrogen. For many organisms, storage of none treated samples are better for storage than solutions of the samples, e.g. a small tissue sample in a plastic bag will do better than a solution of the tissue during storage, see preparation of samples, section 2.4.

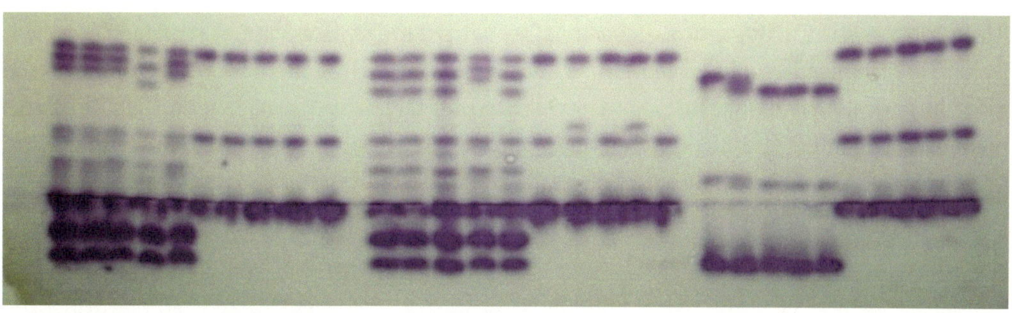

Figure 2. Zymogram of the enzyme glucosephosphateisomerase from Indian major carps, from left to right: 5 hybrids between mrigal and rohu, 5 hybrids between catla and rohu, 5 hybrids between catla and mrigal, 5 rohu, 5 mrigal and 5 catla

2.4. Preparation of samples

Preparation of samples for the electrophoretic procedure may be done in various ways, depending on the organism and the electrophoretic procedure chosen. The simplest way is to use a drilling machine with a glass or plastic rod. A small about of a buffer solution, often the gel buffer, is added and the sample with the buffer is treated with the drilling machine. When dealing with very tough samples, a small amount of sand can be added. After this procedure, the samples are centrifuged and the supernatant is used for the application procedure. It is important that the samples are kept cold during the process in order to minimise the loss of enzyme activity. More advanced milling machines are available and they give a better grinding of the material. Depending on the organism, it may be an advantage to add various chemicals to the grinding buffer, and especially for plants an array of chemicals are suggested (Soltis & Soltis 1990). However, seedlings seem to do just as well with a buffer as grinding solution. Sonication may also be a method for squeezing the material as is treatment with liquid nitrogen.

Depending on the application method, it may be necessary to add sucrose or other chemicals that increase the density of the sample solution. This is often used when the samples are loaded on the surface of the gel, either on a small piece of filter paper (e.g. Whatman No. 3MM) or as a droplet (1-2 μL). Various application templates may be used. It must be mentioned that fish eggs, which are stored in 40% sucrose (Paaver, pers. comm.), can be picked up with a forceps after the sucrose with the eggs is spread on a piece of filter paper, thereby removing the liquid. A single egg can be picked up and placed on a piece of filter paper (size 4mm x 6mm) soaked with 5 μL 40% sucrose, another piece of filter paper also soaked with 40% sucrose is placed on the top, and the whole sandwich is squeezed with a spatula. One piece of paper may be used immediately and the other piece may be stored in an Eppendorf tube (0.5 mL) at -80°C and used later for electrophoresis.

3. Electrophoresis

Electrophoresis is the migration of electrically charged molecules in an electric field. In order to establish an electric field which can be handled, a supporting media is needed. The media may be paper, agar or agarose, starch, cellulose acetate or polyacrylamide. Each of these media has advantages and disadvantages, so one has to select a media which suits to the organism that has to be analysed. Among the media mentioned, I have no experience with paper or cellulose acetate.

The electrophoresis may be performed either vertically or horizontally. Regardless of orientation which you are using, you must ensure that the gel is kept cold during the procedure. The electrophoretic procedure develops heat, and the heat may destroy the activity of the enzyme. In general, the set-up consists of buffer trays which carry the electrodes providing the electric field and provides contact to the carrying media and the gel. When using a horizontal set-up, the contacts may be filter paper, cloths or sponges, which establish the electric field across the gel. It must be mentioned that some enzymes are sensitive to the material from which the contact is made, but ususally filter paper, Whatman quality, will not harm the enzymes. A horizontal set-up is shown in Fig. 3.

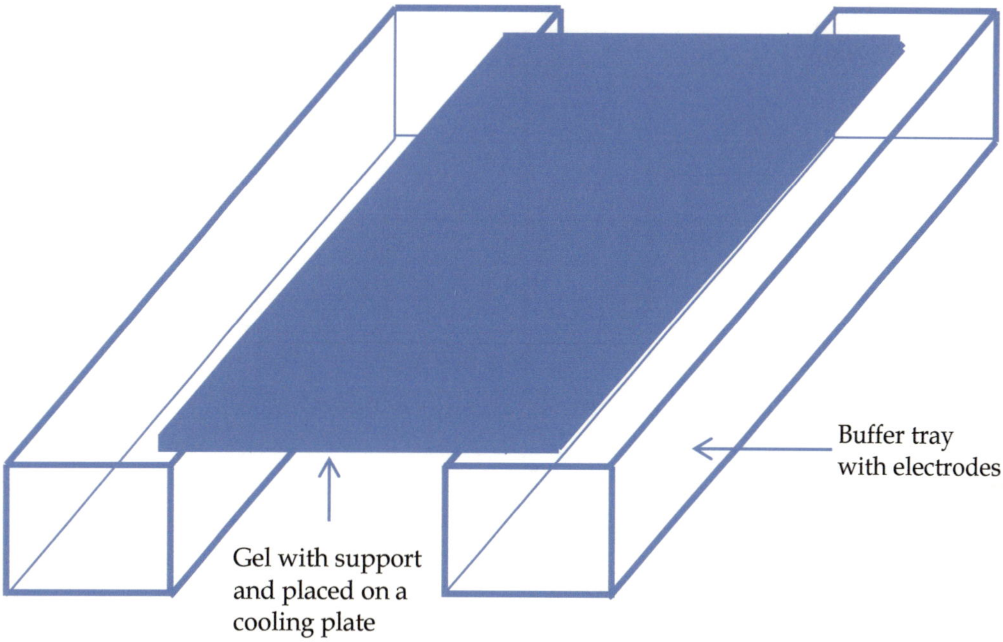

Figure 3. Drawing of horizontal set-up

3.1. Electrophoresis performed with agar or agarose as supporting media

The two methods described in this paragraph are horizontal set ups. Years back, certain purified agars were adequate for doing electrophoresis of e.g. haemoglobin, but nowadays various agaroses are better suited for the electrophoretic procedure. The pore size formed by the polymers is dependent on the concentration of agar/agarose. Due to the fact that there has been a high demand for high quality agarose for separating DNA fragments, many varieties of agarose have been developed, among these also products suitable for proteins.

3.1.1. Electrophoresis performed with agar or agarose on microscope slides

A simple set-up is to use microscope slides for support of the media. The agar/agarose solution (1%) is cooked in water bath or in microwave oven with the buffer used for electrophoresis and the cleaned slides are coated with 2 mL agar/agarose solution, using a pipette. The samples are applied in a slit (one slit, one sample) made by filter paper placed vertically on the slide, see Fig. 4, or holes punched out by suction with a capillary pipette. Depending on the method used, one to four samples may be applied to one microscope slide and the electrophoresis is performed immediately after the application. An array of buffers may be used, for suggestions, see Richardson et al. (1986).

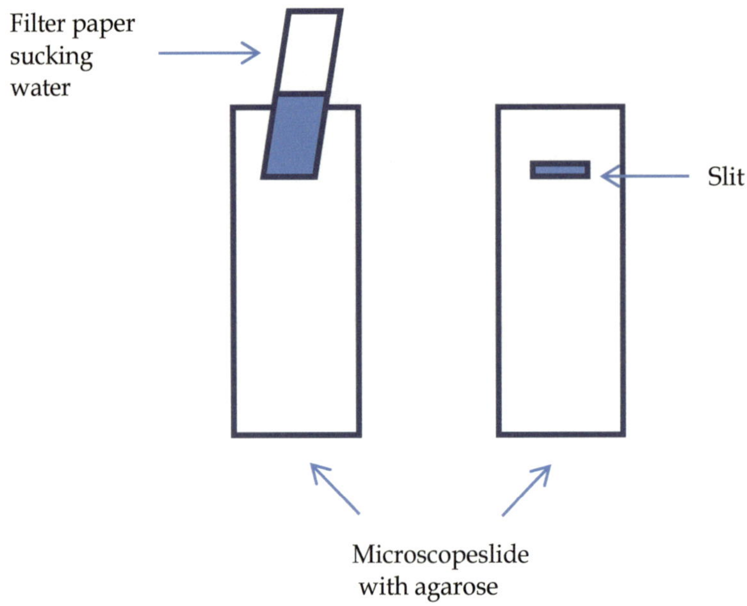

Figure 4. How to make a slit on an agar/agarose gel. The blue quadrangle at the left figure is filter paper, the blue zone shows water (buffer) migrating up in the filter paper, the blue zone on the right figure illustrates the slit

3.1.2. Electrophoresis performed with agar or agarose on glass plates

A different approach is to use glass plates 100 mm x 100 mm x 1 mm and Gel Bond film for agarose. The film is rolled on the glass plate with 10% glycerol as "glue". Strips of Dymo tape are placed on another glass plate, so that when the two glass plates are placed together by using Bulldog Clamps, the space between them is 0.2 mm. Thickness of the gel can be increased by adding more Dymo tape strips. The plate with the Dymo tapes is siliconated. The sandwich with a syringe needle between the glass plates is placed in a hot cabinet (70°C) for 30 minutes. When warm, the hot agar/agarose solution, cooked in the same way as mentioned in section 3.1.1, is added using a syringe (also hot) with a gentle pressure on the syringe. When the moulding form is nearly filled, the syringe with the needle is removed. The gels are left between the two glass plates and may be stored this way for a week in the refrigerator in a box with moist, but not soaking wet, paper towel.

Prior to applying the samples, the sandwich is opened and the gel on the Gel Bond is placed on a cooling plate. Gel is removed from the Gel Bond film, creating 1 cm rim around the gel, see Fig. 5. Application may be done either in punched holes, by a sample application strip made from plastic with slits or the sample applicator developed for the Pharmacia Phastsystem. It may be advantageous to dry the area, where the samples will be applied, with a piece of filter paper, but this operation must be done very gently. On this size of gel, 8 to 16 samples may be applied. Furthermore, as the application takes place on the top of the gel, it may be advantageous to have 10% sucrose in the sample grinding buffer. As these gels are thin, cooling from beneath is sufficient. However, there may be a risk of drying out of the gel, especially the 0.2 mm gel, however, covering the surface with a small plastic film may prevent this. The contact between the gel and the electrodes may be filter paper dipped in the buffer trays containing adequate buffer and partially covering the gel. Another possibility is to use two strips of Whatman No. 17 soaked in buffer and place the electrodes on top of these, see Fig. 5.

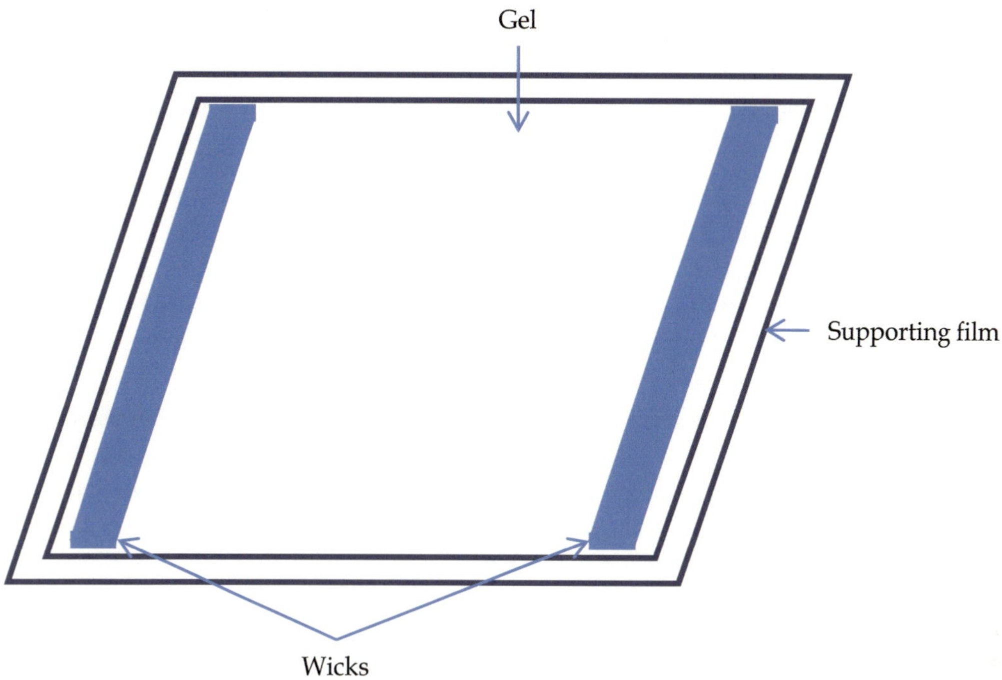

Figure 5. Set up with filter paper as buffer container

The running time depends on the enzyme studied, but as a rule a running time of about ½ - 1 hour is sufficient. Various dyes, e.g. bromphenol blue, may be applied to see how far the front is running, and the closer to the edge the better. How much voltage and current to apply depends on the buffer and the thickness of the gel. The concentration of agar/agarose will also have an impact on the electrophoretic procedure.

3.2. Starch gel electrophoresis

Horizontal starch gel electrophoresis is a well-described method that also provides buffers for gels and trays, see Ferguson (1980) and Murphy et al. (1996). This procedure uses potato starch hydrolysed, in other words, the starch molecules form a structure with pores. The pore size depends, more or less, on the starch gel concentration.

Various moulding forms can be made, and the ones used in our lab have been constructed by a plexiglass frame and Mylar film glued together with high vacuum grease. The size of the frame is 250 mm x 190 mm x 6 mm outer size, leaving a form for the gel of 210 mm x 150 mm x 6 mm, and the width of the frame is 20 mm. The size of the gel allows a horizontal slicing yielding three slices that can be stained individually, i. e. three enzymes can be detected in one gel. The concentration of starch used for this purpose was 12 %. It is possible to use thicker gels and they will allow more slices of the gel for detecting enzymes, but then the cooling has to be increased. A higher starch concentration also is possible, but the higher concentration, the slower the migration. The cooling set up is a cooling plate underneath the gel, as shown in Fig. 3, and a tray with crushed ice on the top of the gel, so you have a kind of cooling sandwich with the gel in the middle. The cooling is strongly needed, as the rather thick gel will developed a high amount of heat which will destroy the activity of the enzymes needed for detection.

The preparation of a starch gel consists of cooking the gel, evaporation of the gel and gel settling. Two thirds of the buffer used for the gel is heated to cooking point using cooking facilities as open fire, electric cooking plate or microwave oven. The starch is suspended in the rest of the buffer, the suspension is added to the hot buffer and the whole solution is swirled around for mixing. The hot solution is re-heated to boiling point two or three times, allowed to settle for a few minutes and then evaporated. It is important to avoid air in the gel solution, as it may damage the electrophoretic procedure. If evaporation is not possible, cooking the solution several times may do the job. Upon evaporation, the solution is poured into the gel frame, left for 15 minutes for settling, and then placed on the cooling plate for another 15-30 minutes. Prior to being placed on the cooling plate, the gel has to be covered with a plastic sheet to avoid loss of liquid. The gel has to be solid before applying the samples.

A slit is cut in the gel for sample application. The samples are treated as described in section 2. The supernatant of the sample is sucked up in a piece of Whatman 3MM filter paper or similar paper, size 4 mm x 6 mm, excess liquid is removed and the filter paper with the sample is applied in the slit.

The gel should run for three hours at 80 mA if possible. The contact between the electrodes in the buffer tray and the gel, see Fig. 3, is ensured with wicks made from Whatman 1. We use 3 layers of the filter paper. Other media for contact may be used, but it has to be considered whether the medium contains substances that interfere with the activity of the enzyme.

An alternative method is to use a set up like the one shown in Fig. 5, but using wider wicks also made from Whatman 17. The wicks are sucked in the tray buffer, two thirds of the paper is placed on the gel, and the electrodes are placed on the outer edge of the filter paper. If necessary, additional tray buffer is added during the electrophoresis when the power is off.

Various dyes may be used to see how far the front has moved. After the electrophoresis, the gel is cut into three slices with a string mounted in a U-frame. Guitar string E is very well suited for this purpose. The adjustment of the string is done by guitar tuning pegs attached to the frame. After slicing, the gel slices are stained, see section 4.1 on staining.

3.3. Polyacrylamide gel electrophoresis

Polyacrylamide gels are based on polymerisation of acrylamide monomers with a linking reagent, e.g. N,N'- methylenebisacrylamide, see Westermeier et al. (2001). The pore size is controlled by the concentration of acrylamide (T) and the cross-linker (C), measured in per cent. The concentration is estimated as

$$T = \frac{(a+b)\times 100}{V}$$

$$C = \frac{b\times 100}{V}$$

where a = acrylamide in g, b = N,N'- methylenebisacrylamide in g and V= the volume of the solution in mL.

If C is kept constant and T is increased, the pore size decreases.

Polyacrylamide gel electrophoresis may be performed either horizontally, as the system shown in Fig. 3, or vertically. Various equipment for electrophoresis can be purchased from professional companies. As for starch gel electrophoresis, various buffers may be applied, see e.g. Westermeier et al. (2001). Again, cooling may be important, depending on the thickness of the gel and the sensibility of the enzyme.

3.3.1. Horizontal polyacrylamide gel electrophoresis

Many companies offer ready-made gels that may be advantageous to use, as the acrylamide is toxic when unpolymerised. However, if the financial support is limited, the gels can easily be prepared in the lab, but a casting frame is needed. The casting frame can be a glass-plate covered with Gel Bond film (not needed if the glass plate is very thin) combined with a siliconated glass plate with a U-frame. The two glass plates are clamped together with Bull Dog clamps or similar equipment. A solution with the acrylamide, cross linker and catalysts, needed for the polymerisation and dissolved in the gel buffer is filled into the form with a syringe. A set up like one shown in Fig. 3 is applied for the electrophoresis. Application of the samples can be performed as described for agarose gel electrophoresis, except that the filter paper with the sample is placed on top of the gel. To improve the migration into the gel, 10% sucrose is added to the sample grinding buffer, see section 2.4. A few mA are applied for 5-15 minutes, then the filter paper is removed and a higher voltage is applied. The running time depends on the thickness of the gel and the gel buffer. If adding bromphenol blue or another dye that does not harm the sample to the grinding buffer, the migration front can be followed.

Another way to apply the samples is to construct a U-frame with slot formers. These can be made by gluing Dymo tape or other forms of tape to the glass plate, as shown in Fig. 6.

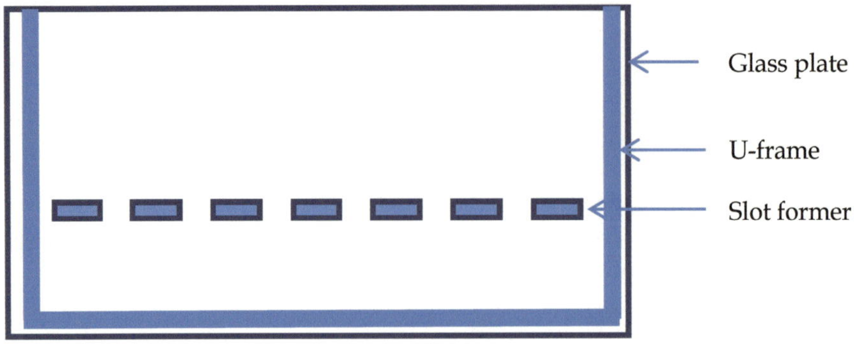

Figure 6. Template for horizontal polyacrylamide gel with slotformer

3.3.2. Vertical polyacrylamide gel electrophoresis

If a set-up for a vertical system is applied it may be an advantage first to make a running gel and then a stacking gel on top of the running gel. The stacking gel is helpful for organising the molecules according to size. By insertion of a comb in the stacking gel before the gel has solidified, holes for applying the samples can be made. The holes are rinsed with the tray buffer before the samples are applied. Then the holes are filled with the tray buffer and with a syringe the sample is applied in the bottom of the hole. It is a good help to have a dye in the sample supernatant, e.g. bromphenol blue, and also sucrose or another substance which makes the sample heavier than water. Again it may be advantageous to apply a few mA for a short while and then increase the voltage.

3.4. Isoelectric focusing in agarose

Isoelectric focusing can be performed either in agarose or polyacrylamide. Ampholytes are added to the gel before it solidifies. The ampholytes are chemical substances that have a high buffering capacity and low molecular weight. When power is applied to the gel, the ampholytes create a pH gradient across the gel, which allows proteins to be organised according to their pI value (isoelectric point (pI) where the protein is unchanged).

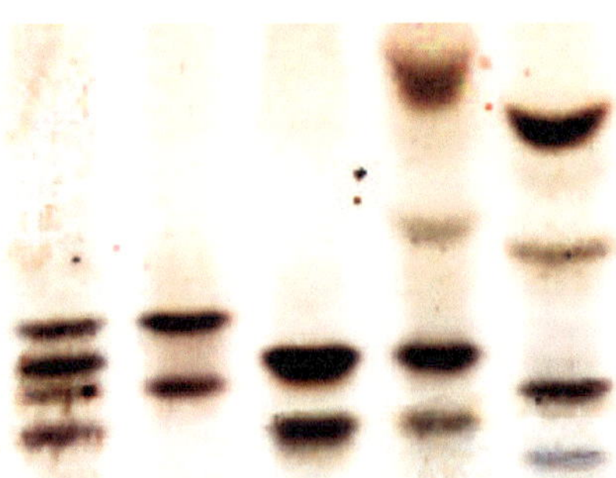

Figure 7. Zymogram of the enzyme esterase from 5 samples of plasma from minks, each allozyme consists of two bands, which means that the five individuals' four-banded phenotypes are heterozygotes. For further explanation, see Simonsen et al. (1992)

The method using agarose as gel may be performed in a very simple way similar to the method described in section 3.1.2. An outcome of the procedure is shown in Fig. 7.

4. Staining and data storage

After the electrophoresis, procedures for detecting the enzymes have to be applied, i.e. the activity of the enzyme is detected. Furthermore, it is often important to be able to return to the results for further analysis, so methods for storing the gels should be considered.

4.1. Staining

The staining procedure is based on the activity of the enzyme. In general, many proteins are separated in a single analysis, as no attempt has been made to select specific proteins from the sample supernatant, see section 2.4, and only few proteins are visible, e. g. haemoglobin, the red dye in blood. Specific methods for detecting the proteins and, hence, the enzymes are needed.

General histochemical methods for detecting proteins as staining with e.g. Coomasie Blue are often used for visualization, but the methods are unspecific and not very sensitive. A better method is to use the antigenic ability of the protein. If an antibody to the protein exists, it is possible to detect the protein by first doing electrophoresis in the usual way. Then cut the gel into pieces that are run in the second dimension in a gel with antibody. The protein can then be revealed by special histochemical methods, for further information see Axelsen et al. (1973).

The enzymes, which have catalytic abilities, are easy to detect. When electrophoresis is done, the specific substrate for the enzyme is applied to the gel and the reaction will only take place at the spot in the gel where the enzyme is located. More spots of a specific enzyme are called isozymes, i.e. proteins with the same catalytic ability, but different electric charge. If the spots can be interpreted as the result of a locus with more alleles, then the spots are identified as allozymes.

There are three main principles for visualization of enzymes:

1. Transformation of the substrate into a visible product.
2. Transformation of the substrate into a product that can react with a dye.
3. Transformation of the substrate into a product that can be used as substrate for another enzyme. This enzyme and a dye then have to be added to the solution, depending on the reaction.

An example of the first principle is the use of 4-methyl-umbelliferyl (4-MUB)-substrates, e.g. 4-MUB-acetate. The enzyme esterase splits 4-MUB-acetate into 4-MUB and acetate, 4-MUB is fluorescent and visible by UV-light.

An example of the second principle is the use of substrates that split off naphthol. By adding diazo-compounds, the visualization takes place.

An example on the third principle is dehydrogenases that use either NAD or NADP as coenzyme. This is depicted in Fig. 8.

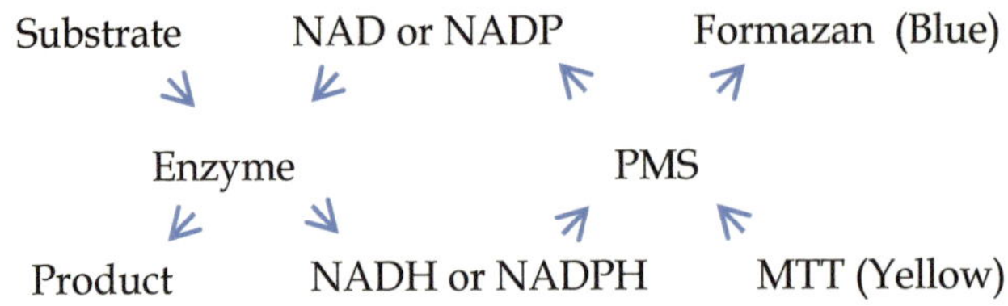

Figure 8. Visualisation of enzymes able to reduce NAD or NADP, NAD is nicotinamide dinucleotide, NADP is nicotinamide dinucleotide phosphate, NADH and NADPH are the respective reduced substances of NAD and NADP, MTT is MTT-tetrazolium salt and PMS is phenazine-methosulfate, a catalyst which transmits hydrogen ions

This method can be expanded with one step more by having an enzyme which can produce a product on which a dehydrogenase can act. A frequently used helping enzyme is glucose-6-phosphate dehydrogenase. For reactions where ammonia is released, glutamate dehydrogenase works as a linking enzyme.

Many of the reactions are light sensitive, so they have to be developed in a dark cabinet and mainly a hot cabinet will do the job.

Nowadays, it is possible to visualize about 100 different enzymes. For staining procedures, see Harris and Hopkinson (1976), Tanksley and Orton (1983), Richardson et al. (1986), Pasteur et al. (1989) and Manchenko (1994).

A couple of hints regarding staining procedure are that the reactivity of the enzyme may increase with temperature and by using the optimal buffer for the specific enzyme in the staining solution. The amount of enzyme in the sample solution may be very low and especially very thin gels have a risk of rinsing off the enzyme. Applying the stain in an agar overlay on the gel may improve the result.

Prior to performing electrophoresis, it is a good idea to check the sample solutions for enzyme activity. This can be done by using the same glass plates as described in section 3.1.2. Pour 5-10 mL staining solution in 2% agar on a glass plate, apply a droplet of the sample supernatant on the agar surface, place the glass plate in a hot cabinet for a while and if the dye is developing there a good chance that electrophoresis of that particular enzyme will work.

4.2. Gel storing

There are two methods for storing the gels; either a) dry the gels and save them or b) take a photo.

All gels can be stored, but many of the stains are sensitive to light, so the stain may fade away over time. The thin agarose gels are easy to dry. For the esterases shown in Fig. 7, once stained for enzymes, the thin agarose gel is treated with 5% acetic acid for 10 minutes and dried at room temperature. However, the 2 mm thick slides of starch gel are harder to dry. Various methods have been tested and a drying oven has been developed. An easy way is to dry the agar overlayer if the stain is applied in that way. A photo is even better, as it does not take up that much storing space. This was a tedious job 20 years ago, when you had to find the right film, develop it and to do the copying in the dark room. However, nowadays it is much easier to take a photo. The best way to do this is to have light box providing transmitted light and a stand for the camera. By using various softwares for photo editing, you can enhance the bands on the gel, see Fig. 1 and 2. From the photo, it may possible to estimate the relative amount of activity if you have a multi-banded zymogram by using e.g. the software Image J (http://rsbweb.nih.gov/ij/).

5. Data analysis

To a wide extent, electrophoresis, applied to proteins, has been used concerning genetic investigations from the sixties and to nineties. Several books and book chapters have been written about this topic, e.g. Ferguson (1980), Richardson et al. (1986), Pasteur et al. (1989), Whitmore (1990) and Murphy et al. (1990). However, now that the PCR-technique has been developed, DNA is much more suited for genetic investigations. By analysing DNA, the genes are studied directly and not after the translation of the genetic information into proteins.

Electrophoresis is a major part of the analytic methods used for both proteins and nucleic acids, the interpretation and data analysis revealed for protein electrophoresis can be applied for analysing the results achieved by studying nucleic acids and, hence, genes. A lot of software for data analysis is available on the internet (e.g. GeneAlex written by Peakall & Smouse (2006)). However, prior to carrying out advanced analysis of the data, it should checked up, whether the genetic marker is inherited according to the Mendelian laws. When dealing with proteins, it must also be considered whether the protein is codominant, i.e. whether the heterozygote can be recognised. Most studied individuals are diploids, i.e. have chromosomes in pairs, which means that an individual has two products (proteins) of each gene, two alleles. If the two products are identical in construction, then the individual is homozygous, but if the products are different, then the individual is a heterozygote. This scoring of phenotypes interpreted as genotypes is the fundament for applying protein electrophoresis for genetics.

5.1. Interpretation of zymograms

The outcome of a protein electrophoresis is called a zymogram, as mentioned in section 1. In Fig. 1, the position of three different gene products is indicated as allozyme 1, allozyme 2 and allozyme 3. Fig. 9 depicts three zymograms with allozymes designated 1 and 2.

The first zymogram in Fig. 9 is determined by a locus with two codominant alleles and the resulting protein is monomeric, which means that only one protein chain makes up the active protein. The second zymogram is the result of a locus with two codominant alleles, but the active protein is a dimer, i.e. two protein chains are needed for constructing the active protein. The third zymogram is the outcome of a locus with two codominant alleles, where four protein chains are needed for making the active protein.

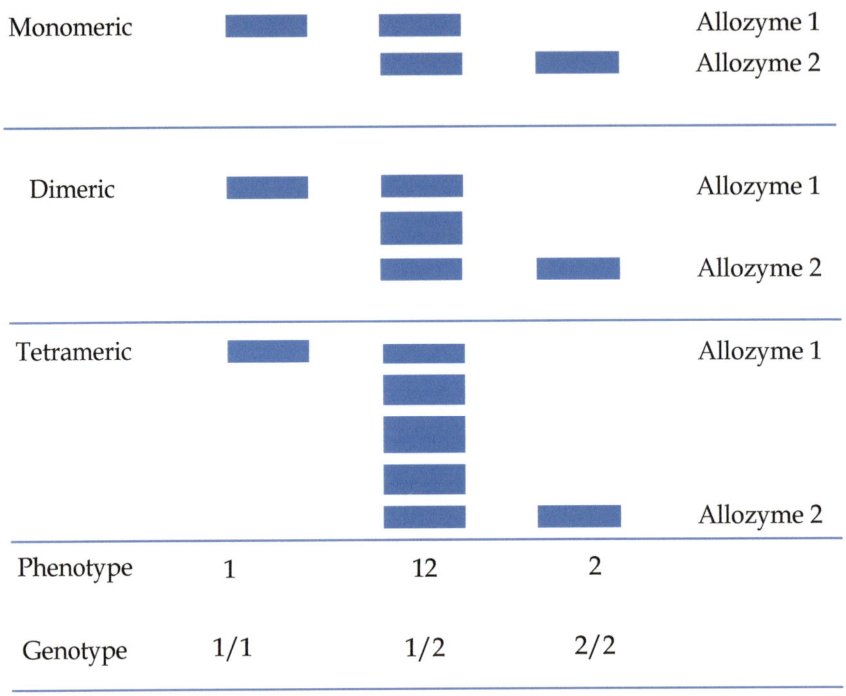

Figure 9. Depiction of three different proteins with monomeric, dimeric and tetrameric structure, each protein has two allozymes 1 and 2. Phenotype indicates the phenotypic interpretation of the zymogram and genotype the genetic interpretation.

As shown in Fig. 9, the genotypic interpretation is determined from the phenotypic interpretation. In theory, it is expected that the two bands seen for the genotype 1/2 are of equal strength for a monomeric enzyme, but that is not always the case. If the two allozymes for the dimeric enzyme combine randomly, the three banded phenotypes seen for the heterozygous individual should come out as 1:2:1. The representing allozyme 1 consists of two units of allozyme 1, the band in the middle consists of one unit of allozyme 1 and one unit of allozyme 2, and the band representing allozyme 2 consists of two units of allozyme 2. In a similar way, the five bands for the tetrameric enzyme should come out in the ratio 1:4:6:4:1, when the two units combine randomly.

Some times more zones are revealed in the zymogram. An example of this is shown in Fig. 2, where two zones are seen - one in the upper end of the gel and one in the lower end of the gel. Furthermore, there is a zone in the middle. The active protein is dimeric and the interpretation is that the enzyme is determined by two loci, one with three codominant alleles, located in the upper zone, and one with two codominant alleles, respectively. Fig. 10 depicts a more simple case for a dimeric enzyme determined by two loci with one and two codominant alleles.

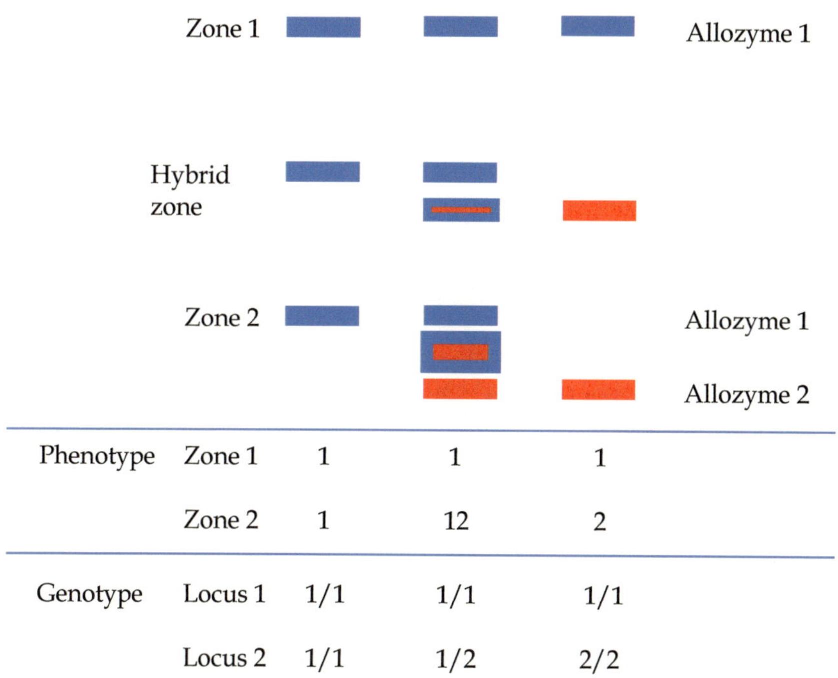

Figure 10. Zymogram of a dimeric enzyme determined by two loci with one and two codominant alleles, respectively. The hybrid zone consists of one unit from zone 1 and one from zone 2

Normally the fastest anodic migrating zone is determined by locus 1, the next by locus 2 and so on. The most common allele is designated 100 and the others are named according to the migration distance, e.g. for allele 75, 75% of the distance from application to the location of allele 100, see Fig. 11. However, the nomenclature of the alleles is not very rigid. The only feature is that a reference is needed, so the nomenclature is consistent all the way through the study.

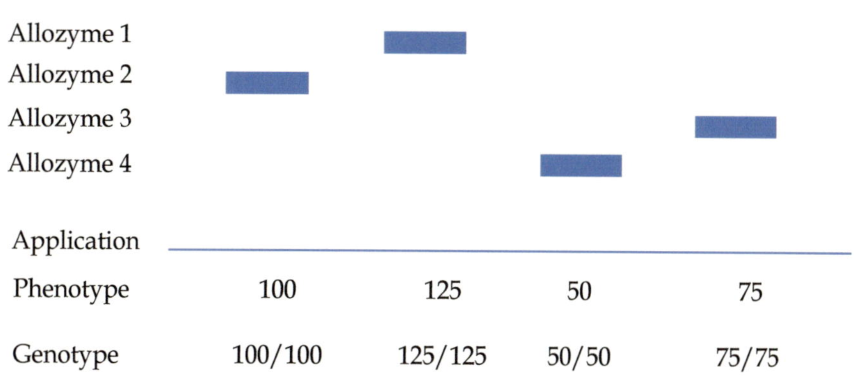

Figure 11. Designation of alleles for a monomeric enzyme

In general, an enzyme is written with capitals, the enzyme esterase as EST. If you speak about the locus, the designation is *EST* and if you speak about an allele, the name is written as *EST-1*-100 or *EST-1*^{100}, this means esterase locus No. 1 and allele 100. However, there is no strict rule about how to do this. The abbreviations of the enzymes may be found in many books on enzyme electrophoresis.

5.2. Expression

Due to the fact that enzymes are produced within an individual, the physiology of the individual as well as the environment may have an impact on expression of the enzyme. All cells have the same chromosomal set with the exception of semen and eggs, so in theory all enzymes could be expressed in all cells. However, due to cell differentiation not all enzymes are important for all cells, so a regulation of the expression in the various tissues is found. An example is shown in Fig. 12.

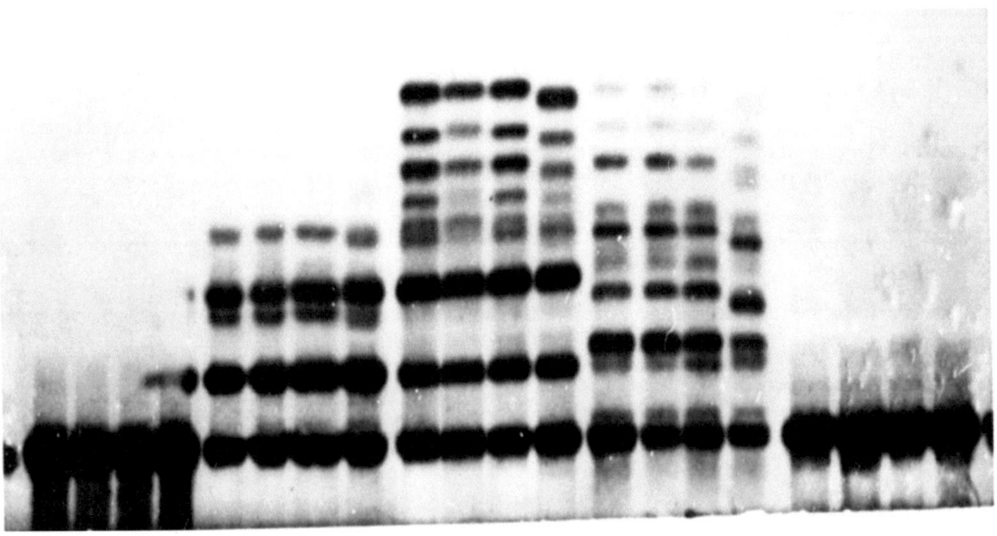

Figure 12. Zymogram of lactate dehydrogenase from four eelpouts. From left to right, 4 samples of heart tissue, 4 brains, 4 eyes without vitreous body, 4 vitreous bodies and 4 muscles. For further information, see Simonsen & Christiansen (1985)

The lactate dehydrogenase is a tetrameric enzyme determined by three loci in the eelpouts, where only one locus is expressed in all tissues investigated. The interpretation of the zymogram is difficult and is only possible when finding rare variants of the three loci (Simonsen & Christiansen 1985).

Another feature may be age, and a good example of this is haemoglobin, as described by Harris (1980). As the expression may vary due to tissue and age, it is very important to collect the samples from individuals of similar age and use the same tissue throughout an investigation.

Sometimes null alleles or silent alleles are observed. To be sure that it is a true silent allele and not just a poor sample, it is important that at least two zones are seen for the enzyme. If one zone is present and the other zone is absent in a particular individual, the conclusion regarding the presence of a silent allele is reliable. Remember that the electrophoretic procedure is a tough treatment of the proteins, and even if the allozyme is not working when developed, it may function in the individual.

Expression of an enzyme may be controlled by other genes, see e.g. Poulsen et al. (1983).

5.3. Modification

Modification of the gene product can be done in several ways. Regarding the protein structure and folding, various small molecules may have an impact on the charge of the protein. These molecules may be a part of the protein, but may not be necessary for the enzymatic function. This may result in a copy of the original product, and an example is depicted in Fig. 13.

Figure 13. Depicted zymogram of adenosine deaminase from the vitreuos body of eelpouts. Allozyme 1 and 2 are seen in many tissues, but allozymes 1′ and 2′ are only seen in the vitreous body, see Simonsen & Christiansen (1981)

The small molecules in the buffer may also have an impact on the zymogram, see Fig.14.

For the slice to the left and the one in the middle, each allozyme consists of two bands, whereas the last buffer only resolves one band for each allozyme.

Among many others, these examples merely illustrate how important it is to find the right tissue and electrophoretic procedure for the individuals studied.

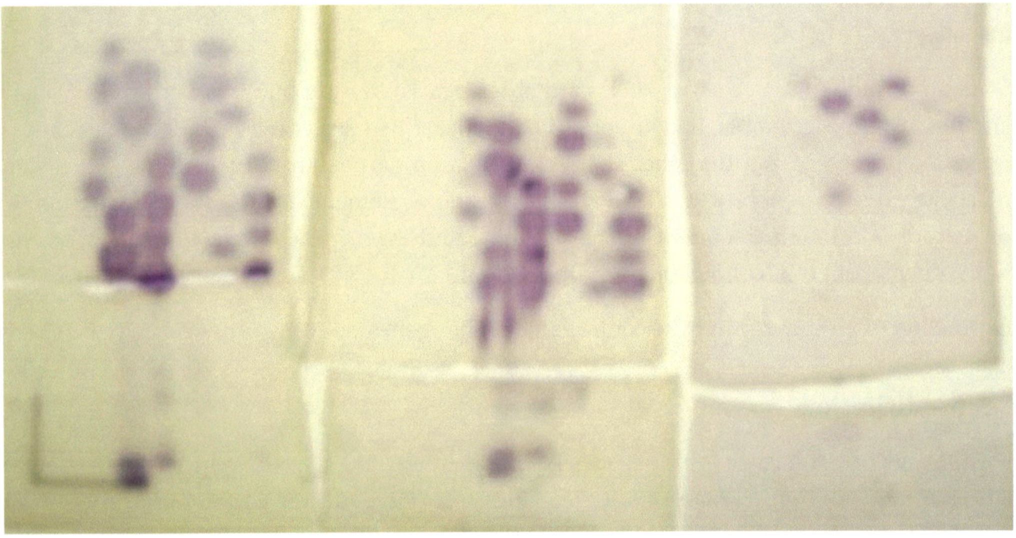

Figure 14. The enzyme phosphoglucomutase from six individuals of earthworms, analysed by using three different buffers for electrophoresis

6. Conclusion

In present times, analysis of proteins by electrophoresis is considered an old fashioned method despite the fact that it was a great improvement for studying genetic variation in populations fifty to sixty years ago. However, nowadays, the method is overruled by the DNA analytic methods. Despite this fact, the study of proteins together with the development of analytical software initiated an enormous step forward for our knowledge of population genetics and much of the software can be used for analysing the results from DNA methods.

Keep in mind that protein electrophoresis only reveals about 30% of all the genetic variation. Also keep in mind that the method is only sufficient for showing dissimilarities. When you have two allozymes with same migration distance on a gel, it cannot be concluded that they have identical amino acid composition, but at least they have the same electric charge. Finally, remember that the result of your study is not any better than the primary data, i.e. the zymogram.

Author details

Vibeke Simonsen
Aarhus University, Institute of Bioscience, Denmark

Acknowledgement

I would like to thank Aarhus University for giving me 43 years of working with allozymes. Many of the examples used in the chapter have been done in co-operation with many colleagues. It has been exiting to follow the development of the electrophoretic procedures for protein studies. Furthermore, I wish to thank Charlotte Kler for the great help with the language.

7. References

Axelsen, N.H., Krøll, J. & Weeke, B. (1973) A manual of quantitative immunoelectrophoresis: Methods and applications. Universitetsforlaget, Oslo, 169 p.

Ferguson, A. (1980) Biochemical systematics and evolution. Blackie & Son, Glasgow, 194 p.

Harris, H. (1980) The principles of human biochemical genetics. Elsevier/North-Holland, Amsterdam, 554 p.

Harris, H. & Hopkinson, D.A. (1976) Handbook of enzyme electrophoresis in human genetics. - North-Holland, Amsterdam. Pages are numbered for each section.

Hedrick, P.W. (1983) Genetics of populations. Science Books International, Boston, 629 p.

Manchenko, G.P. (1994) Handbook of detection of enzymes on electrophoretic gels. CRC Press, Boca Raton, 341 p.

Murphy, R.W., Sites Jr., J.W., Buth, D.G. & Haufler, C.H. (1996). Proteins: Isozyme electrophoresis. In D.M Hillis, C. Moritz & B.K. Mable (Eds.), Molecular systematics. Sinauer Associates, Inc., Sunderland, Massachusetts, pp. 51-120

Pasteur, N., Pasteur, G., Bonhomme, F., Catalan, J. & Britton-Davidian, J. (1989). Practical isozyme genetics. - Translated by M. Cobb, Ellis Horwood Limited, Chichester, 216 p.

Peakall, R. & Smouse, P.E. (2006) GENALEX 6: Genetic analysis in Excel. Population genetic software for teaching and research. Molecular Ecology Notes 6, 288-295.

Poulsen, H.D., Simonsen, V. & Wellendorf, H. (1983) The inheritance of six isoenzymes of Norway spruce (*Picea abies* (L.) Karst). Forest Tree Improvement 16, 12-33.

Richardson, B.J., Baverstock, P.R. & Adams, M. (1986) Allozyme electrophoresis: A handbook for animal systematics and population studies. Academic Press, London, 410 p.

Simonsen, V. & Christiansen, F.B. (1981) Genetics of *Zoarces* populations. XI. Inheritance of electrophoretic variants of the enzyme adenosine deaminase. Hereditas 95, 289-294.

Simonsen, V. & Christiansen, F.B. (1985) Genetics of *Zoarces* populations. XIV. Variation of the lactate dehydrogenase isoenzymes. Hereditas 103, 177-185.

Simonsen, V. (1992) Genetic polymorphism of esterase in plasma of American mink (*Mustela vison* L.). Animal Genetics 23, 553-555.

Soltis, D.E. & Soltis, P.E. (1990) Isozymes in plant biology, Springer, New York, 268 p.

Tanksley, S.D. & Orton, T.J. (1983). Isozymes in plant genetics and breeding. Part A - Elsevier, Amsterdam, 516 p.

Westermeier, R., Gronau, S., Beckett, P., Berkelman, T., Bülles, J., Schickle, H. & Thesseling, G. (2001) Electrophoresis in practice. A guide to methods and applications of DNA and protein separations, 3rd edition, Wiley-VCH, Weinheim, 349 p.

Whitmore, D.H. (1990). Electrophoretic and isoelectric focusing techniques in fisheries management. - CRC Press, Boca Raton, 360 p.

Chapter 7

Electrophoresis as a Useful Tool in Studying the Quality of Meat Products

Dario G. Pighin

Additional information is available at the end of the chapter

http://dx.doi.org/10.5772/45761

1. Introduction

The study of meat quality usually involves a great number of methodologies at the laboratory. Thus, new technology and automatic equipments are used in combination to well-known methodologies in order to improve the study of the constituents of a given sample. Several methodologies are usually carried out in combination, especially when composition and functionality of food compounds are studied.

Meat and meat products are mainly composed of water, proteins, lipids, minerals and carbohydrates (Bender, 1992). Most sensory and functional properties depend on the quantity and quality of these compounds. In the last years, several studies aimed its efforts on the improvement of sensory quality of meat products by means of the incorporation of additives (Szerman et al., 2012).

At present, ready-to-cook and pre-cooked foods are installed in life. Regarding meat market, *sous vide* processing appears to be an interesting option for offering different kinds of beef products for consumption. It consists of the cooking/pasteurization process to a raw material packaged in a heat-stable vacuum container. The products are stored at 0-3 °C and are able to be re-heated and consumed even after 5 weeks (Vaudagna et al., 2002). Besides the enhancement of shelf life, several other advantages have been associated to *sous vide* processing, such as increased flavour profile, increased tenderness and nutritional loss reduction (Church & Parsons, 2000; Vaudagna et al., 2002).

The *sous vide* system has been extensively studied in meat and meat products (Church & Parsons, 2000; García-Segovia et al., 2007). Nevertheless, there is still a lack of knowledge regarding the application of this mild cooking system to whole-muscle beef. Moreover, this pre-cooked system presents important technological inconvenients like juice retention inside the packaging, which would affect profitability and sensorial and nutritional characteristics of the final product (Church & Parsons, 2000; Vaudagna et al., 2002). On this regard, additives such as sodium chloride and alkaline phosphates have being frequently used for meat products manufacturing in order to increase water holding capacity (WHC) of meat and, consequently, reducing cooking weight loss (Baublits et al., 2006; Pietrasik & Shand, 2004, 2005; Vaudagna et al., 2008). Results obtained showed that the incorporation of NaCl (0.70% or 1.20% w/w) plus TPP (0.25% w/w), and posterior cooking at mild temperatures (60-65 °C) increase the WHC of *Semitendinosus* muscles and reduce cooking weight losses. The increased WHC would be responsible for the tenderness increment found, probably as a consequence of the swelling of myofibrils. Nevertheless, the exact physicochemical events involved have not been completely elucidated. Thus, the study of the structure and functionality of meat proteins emerges as an important issue to take into account in order to elucidate the effect of additives incorporated into meat products and the mechanisms involved in the improvement of sensorial and physical properties.

Regarding meat proteins, it is known that they have a widely range of size between approximately 20 kDa to 3,000 kDa (Warris, 2000). Among them, high molecular weight proteins display a key role in conferring functional properties to meat (Warris, 2000). Its approach usually represents a challenge at the laboratory. Myosin, actin, titin and nebulin are four of the major proteins of the skeletal sarcomere, and possibly the most important myofibrillar proteins having interactions with each other in the muscular tissue. Myosin (aprox. 540 kDa) and actin (aprox. 43 kDa) are the most noticeable contractile proteins of the thick and thin filaments, respectively (Clark et al., 2002). Titin (aprox. 3,000 kDa) and nebulin (aprox. 800 kDa) are two of the giant myofibrillar proteins, acting as protein rulers by regulating the assembly of myosin and actin filaments in the sarcomere (Wang, 1996). At present, several methods from different sources have been reported for their isolation. However, most of them are complex, tedious and mainly focused on the isolation of one protein at a time.

Currently, electrophoresis represents one of the most and reliable methods used in studying meat proteins in the laboratory. It does not require expensive equipment and its outcomes allow to be analyzed by means of several techniques like densitometry and differential scanning calorimetry (DSC), among others. In general terms, electrophoresis refers to the migration of charged particles in a particular medium under the influence of an electric field. It constitutes a powerful and broadly used method for the analysis of complex mixtures of analytes from different sources. Separated molecules can be visualized by means of specific staining. One remarkable advantage is that the dried support medium can be finally kept as a permanent record. It has been stated that the migration rate (cm/s) of a given particle submitted to electrophoresis relies on several factors related to the shape, charge and size of the particle and to the electrical field strength and the temperature of the system as well. Thus, the mobility of the particle –migration rate per unit of field strength (V/cm)- results directly proportional to net charge of the particle and inversely proportional to its molecular size at a given viscosity of the electrophoretic support (Karcher & Landers, 2006). Several support mediums had been described. Among them, polyacrylamide gels (PAGE) offer several advantages when compared to others matrixes, especially when proteins mixtures are been analyzed (Karcher & Landers, 2006). In this case, protein separation takes place on the basis of charge-to-mass-ratio and molecular size. The addition of sodium dodecyl sulphate to the system (SDS) denatures proteins conferring to them a constant charge-to-mass ratio. Thus, under described conditions, proteins separation only depends on its relative molecular weights (MW). Gradient SDS-PAGE is usually recommended when proteins mixture spans wide MW range. In this case, the decreasing pore size also contributes to sharpen the proteins bands.

2. Practical applications

A great deal of attention has been paid to the most important proteins of the myofibrillar system because of their potential role in meat processing. Several authors have succeeded in the isolation and/or enrichment of some of these proteins (Ho et al., 1994; Pan & Damodaran, 1994; Wang, 1982; Wang & Greaser, 1985). Nevertheless, most of the described methodologies use to be difficult to carry out because of their complexity, associated in part with the exerted interactions among these proteins throughout the myofibrils (Houmeida et al., 1995; Linke et al., 2002).

The purpose of the present chapter is to compile own published data which illustrate the usefulness of electrophoresis analysis in meat research, either used alone or in combination to other techniques. The first part describes a simple methodology to simultaneously co-enrich myosin, actin, titin and nebulin from beef muscles. It is believed that the described method will provide enriched myofibrillar proteins, which could be used for more specific analysis, like microscopy studies, immunological and/or differential scanning calorimetric analysis. The second component of this chapter is focused on the study of the effect of saline additives on muscle proteins structure to improve meat quality. Since DSC has become one of the most useful techniques for studying the thermal behavior of components of biological systems, its approach in combination to SDS-PAGE is also described. The methodologies used provide useful information in order to explain the improvement of sensorial properties observed in meat treated with saline additives and submitted to *sous vide* system.

3. Methodology

3.1. Materials

Beef *Semitendinosus* muscles were dissected from British breed steer carcasses 48 h post slaughter, following the Guidelines of National Sanitary Authority (SENASA, 1969). Slaughter procedure was carried out in a commercial meatpacking and processing plant with good manufacturing practices. High molecular weight (HMW) standard was purchased from Amersham-Pharmacia (Buckinghamshire, England). Bovine serum albumin (BSA, fraction V), enzyme inhibitors and mouse monoclonal anti-nebulin antibody were obtained from SIGMA (St. Louis, MO, USA). Goat anti-mouse IgG antibody conjugated to alkaline phosphatase was purchased from Bio-Rad (Cambridge, MA). All other reagents were of the purest grade available (analytical grade) and obtained from SIGMA or Bio-Rad.

Semitendinosus muscles were trimmed free of fat and connective tissue before measuring pH with a portable puncture pHmeter (Metrohm AG CH-9101, Herisau, Switzerland). Muscles with pH 24 h values ranging between 5.5 and 5.7 were selected for the studies carried out.

3.2. Methods

3.2.1. *Isolation of titin, nebulin, MHC and actin*

3.2.1.1. *Separation of myofibrillar proteins*

The separation of myofibrillar proteins of *Semitendinosus* muscles was conducted according to Boehm et al. (1998), and as detailed in Pighin & Gonzalez (2008). Protein concentration was assayed according to Lowry et al. (1951), using BSA as standard. Discontinuos SDS-PAGE (see details below) was performed to evaluate the protein profile. The rest of the extracted fraction was diluted as described in Pighin & Gonzalez (2008) to reach a final concentration of 5 g protein/L.

3.2.1.2. *SDS-PAGE analysis and enrichment of titin, nebulin, MHC and actin*

Partial isolation of proteins was achieved by means of ammonium sulfate salt precipitation, using four different saturation ranges: 0–20, 20–40, 40–60 and 60–100 g/L. After each precipitation step, protein solution was centrifuged at 23,000 x g for 20 min at 4 °C. The resulting precipitate was resuspended in SDS buffer (3% SDS, 10% glycerol, 1 mM EDTA, 5 mM DTT, 10 mM Tris-HCl, pH 8.0), and the saline extract was dialyzed overnight against buffer SDS (nitrocellulose membrane, MWCO: 12–14 kDa). Discontinuous SDS-PAGE (see details below) was used to monitor the protein composition of each precipitate.

SDS-PAGE analysis conducted according to Laemmli (1970) with some modifications was performed in order to monitor the protein profile of extracts and to continue the isolation of proteins. Electrophoresis was run in a lineal gradient of acrylamide concentration (3–12%) in a Mini Protean III system (Bio-Rad), at constant voltage (130 V) during 90 min. For comparison purposes, stacking and resolving gel solutions were prepared both with and without glycerol (50 mL/L). A HMW calibration stained kit was used to estimate the molecular weight (MW) of the different proteins. In order to monitor protein composition, each line was loaded with 50 mg of proteins, and with 120 mg for isolation purposes. Gels were stained with 1 g/L Coomassie Brilliant Blue R-250 solution. Images were captured with a Bio-Rad GS-800 Imaging Calibrated Densitometer and processed by Quantity One 1-D Analysis software (Bio-Rad, version 4.4.1) to determine relative front (Rf), MW and relative amount of each protein. When a protein was not accurately identified in the electrophoretic gel by staining, the Western blot technique was utilized as a tool to reveal the presence of this protein. This analysis was applied to nebulin. For this purpose, a non-stained SDS-PAGE gel was transferred overnight onto a nitrocellulose membrane (0.45-mm pore size) at 90 mA using Mini Trans-Blot Cell (Bio-Rad). Then, the presence of nebulin was detected with the commercial mouse monoclonal anti-nebulin clone NB2 antibody (diluted 1:400), followed by the secondary antibody anti-mouse IgG raised in goat and conjugated to alkaline phosphatase (dilution 1:1,000). Once the protein of interest was located on the nitrocellulose membrane, its position on an electrophoretic gel could be established by comparison.

Once MHC, actin, nebulin and titin were identified, they were excised from a non-stained gel, and individually eluted in an Electro-Eluter model 422 (Bio-Rad, MWCO: 12–15 kDa) as detailed in Pighin & Gonzalez (2008). Finally, each protein was collected in a volume of approximately 500 mL, dialyzed overnight, lyophilized and stored at -80 °C. The purity of the isolated extracts was finally monitored by discontinuous SDS-PAGE as previously described.

3.2.2. *Study of protein contribution to the effect of saline addition to meat*

3.2.2.1. *Incorporation of saline additives*

Brines containing NaCl and/or TPP were incorporated into the selected muscles (10% w/w) with an automatic multi-needle injector (36 needles, Fricor, Buenos Aires, Argentina) to give the final concentrations (g/100 g injected muscle) of 0.70 NaCl; 0.25 TPP; 0.70 NaCl + 0.25 TPP, and 1.20 NaCl + 0.25 TPP. Non injected (NI) muscles were used as control. Injected and NI muscles were vacuum packaged (Cryovac CN510, Sealed Air Co., Buenos Aires, Argentina) and submitted to continuous tumbling at 5.0 rpm, 8 h, 2-4 °C (Lance Industries tumbler, model LT-15, Allenton, USA) in order to improve brine sorption and protein extraction. Two slices (2.0 cm wide) of each muscle were frozen at -80 °C until thermal analysis and the rest of the muscles were kept at 2-4 °C until protein extraction.

3.2.2.2. *Myofibrillar and sarcoplasmic proteins extraction and myofibrils isolation*

Myofibrillar proteins extraction was conducted as mentioned in **3.2.1.1.** After centrifugation, the supernatant containing sarcoplasmic proteins was separated and stored at -80 °C until electrophoretic analysis. The precipitate was suspended in SDS buffer and filtered through a home-made nylon strainer to remove connective tissue. An aliquot of the filtrate, which included proteins of the myofibrillar system, was subjected to SDS-PAGE analysis. Protein concentration of both sarcoplasmic and myofibrillar extracts was measured according to Lowry et al. (1951).

Myofibrils isolation from minced meat of the different saline treatments was conducted according to Culler et al. (1978), and as detailed in Pighin et al. (2008). The filtered myofibrils collected were used for DSC analysis. Protein concentration of isolated myofibrils and of whole muscles was determined by a macro-Kjeldahl method (Foss Tecator equipment, Sweden), with a conversion factor of N = 6.25.

3.2.2.3. *SDS-PAGE analysis and thermal analysis*

Electrophoretic profile of protein extracts was monitored by discontinuous SDS-PAGE (Laemmli, 1970), carried out under the same conditions previously described in **3.2.1.2.**

A Perkin-Elmer Pyris-1 differential scanning calorimeter (DSC) was employed to study the thermal properties of proteins of the whole muscle and of the isolated myofibrils. For temperature and heat flow calibration, Indium was used as standard (melting point 156.6 °C, enthalpy 28.46 J/g). Five to fifteen milligrams of each sample, accurately weighed, were placed into an aluminum pan, which was hermetically sealed and equilibrated for 2 min at the initial scanning temperature. During the study, a heating rate of 10 °C /min was applied over a 20-90 °C temperature range. All samples were re-scanned (submission to a second thermal analysis) to check for the irreversibility of the denaturation process. Endotherm (mW vs. temperature), denaturation enthalpy (DH) and endothermic peak (Td) were obtained using the software Pyris-7 (Perkin-Elmer). The total enthalpy change (DHT) was estimated by the algebraic sum of the individual denaturation enthalpies.

3.2.2.4. Statistical analysis

The experiments described were conducted in duplicate. The study of saline incorporation included 5 treatments (NI; 0.70% NaCl; 0.25% TPP; 0.70% NaCl + 0.25% TPP; 1.20% NaCl + 0.25% TPP). A set of five muscles were assigned at random to each treatment. SDS-PAGE analysis and DSC measurements were performed in all muscles included in the treatments. Analysis of variance (ANOVA) for DSC measurements was conducted using the statistical package SAS (version 8, SAS Institute Inc., 2004, Cary, NC). Duncan´s Multiple Range Test (with a significance of a = 0.05) was used to determine differences among treatments.

4. Results and discussion

4.1. Isolation of titin, nebulin, MHC and actin

4.1.1. Separation of myofibrillar proteins

Figure 1 shows the electrophoretic pattern of supernatant and precipitate fractions obtained after the extraction of myofibrillar proteins. It can be seen (lane 2) that most of the soluble proteins (28–70 kDa) remained in the supernatant fraction. Instead, proteins of the precipitate (lane 3) have a wider MW range (25–2,500 kDa). The precipitate included almost exclusively myofibrillar proteins, yielding 64 g of extracted protein per kilogram of muscular tissue. It can also be observed that the most intense band is MHC (MW: 220 kDa, Rf: 0.47), which represents approximately 35 % of total myofibrillar proteins extracted. Actin (MW: 43 kDa) is the second most intense (20 %) having one of the largest Rf (0.92). Titin band represents approximately 4 % of the proteins extracted. It has the lowest Rf (0.06) in the SDS-PAGE analysis with an estimated MW of 2,500 kDa. A similar MW for titin was found by Kellermayer & Grama (2002) and Kurzban & Wang (1988). A comparable electrophoretic pattern of myofibrillar proteins has been reported previously by Chen et al. (2005).

Nebulin could not be precisely identified on the present SDS-PAGE due to the presence of a set of several proteins in the range of 600–800 kDa. Therefore, the use of monoclonal antibodies appeared to be a useful tool to corroborate nebulin localization in the gel. Figure 2 shows the Western blot analysis of protein extract (precipitate fraction). Only one band was detected by the commercial kit against nebulin, with an Rf of 0.20 and an estimated MW of 710 kDa. It is important to denote that the MW of titin and nebulin were calculated in approximation because of the absence of suitable HMW standards for SDS-PAGE. Nevertheless, the MW found for nebulin is consistent with previous published data reporting a value of 600– 900 kDa for this protein (Locker & Wild, 1986; Wang, 1982).

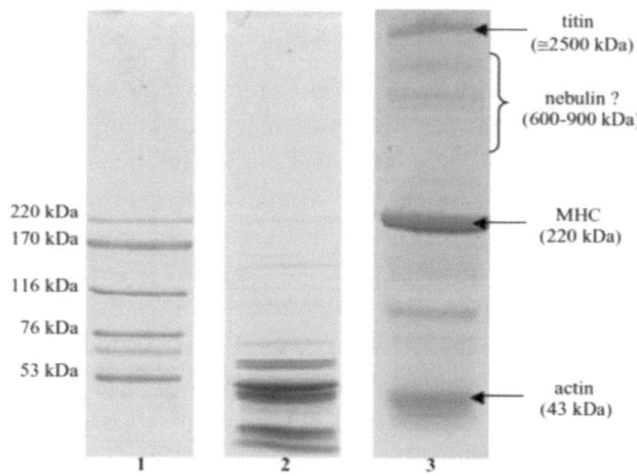

Figure 1. Myofibrillar protein isolation and SDS-PAGE (3–12%) analysis of protein fractions Lane 1 = HMW standard. Lane 2 = supernatant. Lane 3 = precipitate. (Pighin & Gonzalez, 2008)

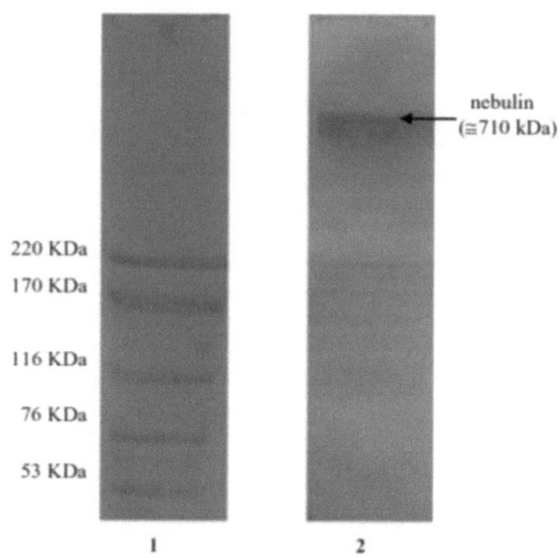

Figure 2. Western blot analysis of myofibrillar protein extract Lane 1 = HMW standard. Lane 2 = myofibrillar protein extract. (Pighin & Gonzalez, 2008)

It is important to remark that the use of gradients is usually recommended when SDS-PAGE analysis is applied to high-MW proteins, in order to improve separation and resolution of bands. The addition of glycerol has been reported to facilitate gradient formation and also to improve resolution of the SDS-PAGE technique (Sobieszek, 1994). However, Chen et al.

(2005) did not find a great improvement of the gel system with the use of this reagent. Present research supports the first statement, given the fact that the addition of glycerol both to the resolving and the stacking gel solutions has successfully improved band resolution (comparison data not shown).

4.1.2. Enrichment of titin, nebulin, MHC and actin

The isolation of myofibrillar proteins requires time-consuming procedures that usually separate them individually. Thus, myosin isolation generally involves the removal of sarcoplasmic proteins by means of washings with diluted phosphate buffer and subsequent protein extraction with specific solutions containing elevated amounts of KCl (Dudziak & Foegeding, 1988; Hermanson et al., 1986). Actin isolation requires a prior separation of G-actin by breaking intermolecular linkages of F-actin. One of the most commonly used agents is acetone, which can break the linkages without denaturing the protein (Syrovy, 1984). After that, a second isolation stage involves several steps including washing, filtration, centrifugation, polymerization/depolymerization cycles, dialysis and centrifugation (Pardee & Spudich, 1982; Spudich & Watt, 1971). Gel filtration chromatography can also be employed for further actin purification (Kuroda, 1982). More recently, Perez-Juan et al. (2007) have succeeded in the isolation of actomyosin and actin in an interesting attempt to amalgamate myofibrillar protein isolations in a single extraction process.

High-molecular weight proteins (i.e. titin and nebulin) extraction specifically involves the addition of protease inhibitors to avoid degradation during processing. The isolation methodology initiates with myofibrils extraction and posterior protein–SDS solubilization, followed by the precipitation of titin/nebulin–SDS complex by salt fractionation. For further purification, the use of gel filtration is recommended because of the great size of these proteins (Wang, 1982). A more exhaustive extraction of titin was proposed by Pospiech et al. (2002), which demanded several centrifugation steps followed by suspension/ homogenization of the sediments in specific solutions, dialysis, and hydroxyapatite and ion-exchange chromatography. Previous description shows the requirement of time and effort to achieve the isolation of one or another myofibrillar protein. For that reason, the present research was focused on looking for a more simple methodological strategy.

Salting out fractionation is believed to be a useful step for protein isolation because the procedure is inexpensive and relatively easy to apply. Sodium chloride had been proposed for the precipitation of the giant proteins, titin and nebulin (Wang, 1982; Wang & Wright, 1988). However, in the present research, the use of ammonium sulfate as precipitating agent produced a proper isolation of these two proteins and also of myosin and actin. In order to select the appropriate saturation range, different ammonium sulfate ranges were assayed: 0–20, 20–40, 40–60 and 60–100 g/L. After each centrifugation step, the protein profile of the precipitated fractions was monitored by SDS-PAGE. The different saturation ranges showed the following protein pattern (Fig. 3): the band corresponding to MHC (MW: 220 kDa) was the most noticeable one, showing similar concentrations in all saline ranges (see numbers between brackets). Actin (MW: 43 kDa) was also detected in all fractions, being its intensity higher in the saline ranges of 40–60 g/L and 60–100 g/L. Protein profiles of the ranges 0–20 g/L and 20–40 g/L were similar, except for the amount of the titin band (2,500 kDa), which is about four times higher in the second saline range. A saturation range of 40–60 g/L showed that the amount of titin and nebulin extracted is the highest. A higher saturation range (60–100 g/L) did not include the giant proteins in the precipitated fraction.

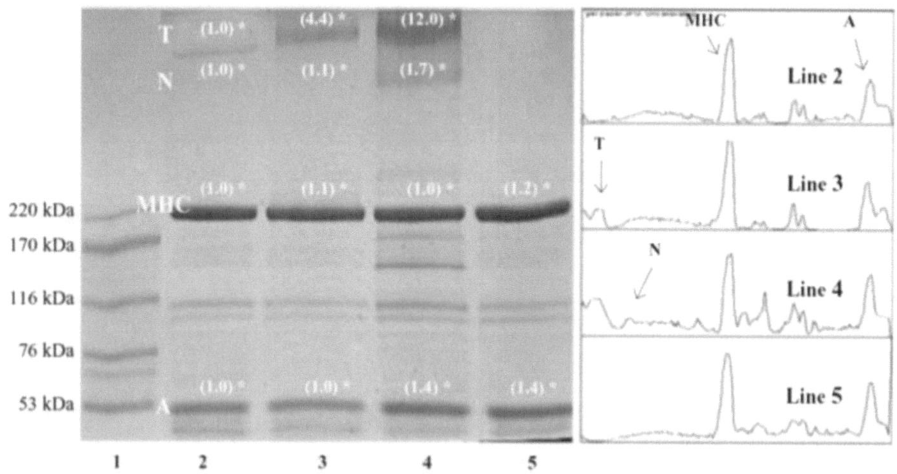

Figure 3. SDS-PAGE analysis of the different ammonium sulfate precipitates
Lane 1 = HMW standard. Lane 2 = range 0–20 g/L. Lane 3 = range 20–40 g/L. Lane 4 = range 40–60 g/L. Lane 5 = range 60–100 g/L. T, titin; N, nebulin; MHC, myosin heavy chain; A, actin. *Numbers between brackets mean times increment. (Pighin & Gonzalez, 2008)

If this protein profile (Fig. 3) is densitometrically compared to the myofibrillar protein extract (Fig. 1), it becomes evident that titin and nebulin have been greatly enriched in the saline range of 40–60 g/L, having 3.6 and 4.0 times increment, respectively.

Even though the analysis of the gel in Fig. 3 indicates that the saline ranges of 20–40 g/L and 40–60 g/L were appropriate for isolation and enrichment of titin, and 40–60 g/L was so for the enrichment of nebulin, the unified range of 20–60 g/L appeared to be an interesting option to increase the amount of these proteins in subsequent experiments. Instead, the saline range of 60–100 g/L was suggested to isolate actin and MHC. Once the saline precipitation ranges were selected, SDS-PAGE was chosen to continue the purification protocol. For this purpose, each ammonium sulfate precipitate was resuspended and electrophoretically developed in an SDS-PAGE gel (3–12%). After the identification of the proteins on the gel by means of its Rf's, they were first excised, and then submitted to electroelution.

Figure 4 shows the SDS-PAGE of each electroeluted protein (lanes 1–4) and the Western blot analysis of nebulin (lane 5). It can be seen that MHC (lane 2) was obtained in an almost pure state. Actin and titin (lanes 3 and 4) were obtained enriched but not in a pure state. The reason for this result is that actin and titin concentrations obtained after the salt precipitation and the electrophoresis steps were low and made difficult the removal of a thinner band from the electrophoretic gel to improve their isolation by electroelution. Nebulin band - stained with Coomassie Brilliant Blue dye- was not accompanied by any other protein band (data not shown). However, the band was vaguely stained and it required the identification and confirmation by the previously mentioned immunoblotting technique (lane 5).

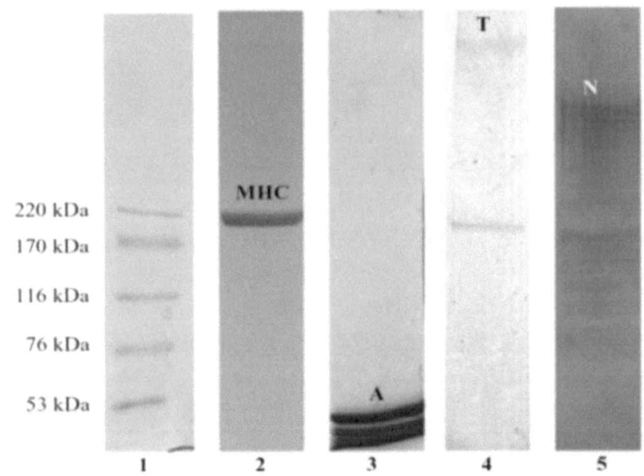

Figure 4. SDS-PAGE and Western blot analysis of the electroeluated proteins
Lane 1 = HMW standard. Lane 2 = myosin heavy chain extract (SDS-PAGE analysis). Lane 3 = actin (A) extract (SDS-PAGE analysis). Lane 4 = titin (T) extract (SDS-PAGE analysis). Lane 5 = nebulin (N) extract (Western blot analysis).
(Pighin & Gonzalez, 2008)

4.2. Study of protein contribution to the effect of saline addition to meat

4.2.1. Electrophoretic analysis of proteins extracts

The electrophoretic analysis of sarcoplasmic proteins extracts belonging to saline treated muscles are shown in Fig. 5. As can be seen, proteins bands ranged between 30 and 75 kDa. It is worth noting that the intensity of the band with an estimated MW of 70 kDa was increased about 2.5 times in samples of muscles containing NaCl alone or combined with TPP (see arrows, lanes 3-A, 3-B and 4-B) with respect to NI. Even more, an additional band with an estimated MW of 65 kDa was revealed in these saline treated muscles (see dashed arrows, lanes 3-A, 3-B and 4-B). On the contrary, the presence of TPP alone did not modify the protein pattern found in NI muscles (lanes 2-A and 2-B). These findings suggest that the presence of NaCl collaborates in the solubilization of myofibrillar proteins, releasing them to the soluble fraction.

Myofibrillar proteins were separated by SDS-PAGE into a broad range of molecular weight (Fig. 6). Bands corresponding to MHC (MW 220 kDa, see arrow) and actin (MW 46 kDa, see dashed arrow) were the most noticeable ones. No differences among treatments were verified by means of densitometric analysis. This result does not agree with the one previously shown in Fig. 5, which had suggested an augmented solubilization of myofibrillar proteins as a consequence of the application of NaCl into muscles. Hence, an attenuated intensity of any of the bands in Fig. 6 was expected. It is probable that the increment of 70 kDa or the appearance of 65 kDa band found in Fig. 5 is a result of the degradation of myofibrillar proteins.

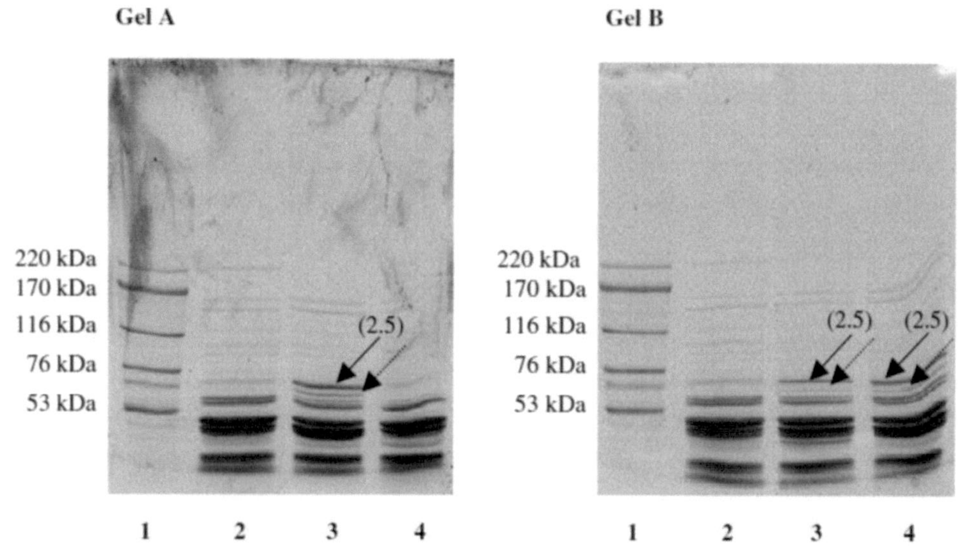

Figure 5. SDS-PAGE analysis of sarcoplasmic proteins obtained from muscles treated with NaCl and/or TPP
Gel A: Lane 1, HMW-SDS; Lane 2, NI; Lane 3, 0.70 % NaCl; Lane 4, 0.25 % TPP.
Gel B: Lane 1, HMW-SDS; Lane 2, NI; Lane 3, 0.70 % NaCl + 0.25 % TPP; Lane 4, 1.20 % NaCl + 0.25 % TPP.
(Pighin et al., 2008)

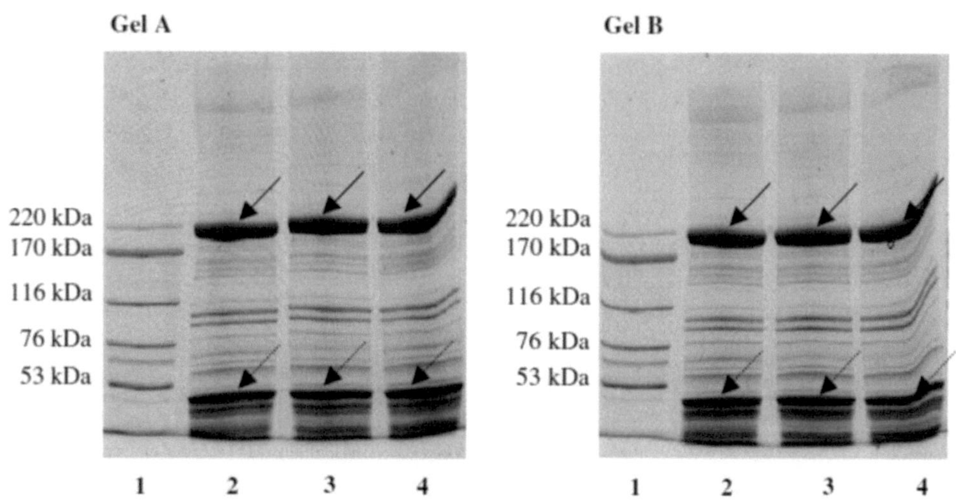

Figure 6. SDS-PAGE analysis of myofibrillar proteins extracted from muscles treated with NaCl and/or TPP
Gel A: Lane 1, HMW-SDS; Lane 2, NI; Lane 3, 0.70 % NaCl; Lane 4, 0.25 % TPP.
Gel B: Lane 1, HMW-SDS; Lane 2, NI; Lane 3, 0.70 % NaCl + 0.25 % TPP; Lane 4, 1.20 % NaCl + 0.25 % TPP.
(Pighin et al., 2008)

4.2.2. Thermal analysis of whole muscle

Thermal analysis of NI whole muscles (Fig. 7) showed three characteristic endothermic peaks. Published data allowed to relate them to myosin and its subunits denaturation (I; 57.0 °C), sarcoplasmic proteins and collagen denaturation (II; 67.4 °C) and actin denaturation (III; 80.3 °C) (Graiver et al., 2006; Kijowski & Mast, 1988; Stabursvik & Martens, 1980). Enthalpies involved in each thermal transition were 1.32, 0.65 y 1.74 J/g protein, respectively. Re-scanned samples did not show any thermal response, corroborating the irreversibility of the transitions. The calculated DHT of NI muscles was 3.71 J/g protein. It can be seen in Table 1 that NaCl treatment (0.70 %) significantly reduced myosin DH and actin Td. Only slight decreases were found in myosin Td and actin DH. These results make evident that actin was destabilized by NaCl, becoming the more susceptible protein to thermal denaturation (decreased Td). Instead, the effect of NaCl upon myosin could be seen on the DH, which reduction could be related to the decrease of hydrogen bonds and/or to the increment of protein aggregation (exothermic event) exerted by the salt. Total enthalpy (DHT) for NaCl-treated muscles (calculated by adding DH of individual values related to transitions I, II, and III) decreased from 3.71 J/g protein (NI muscle) to 3.04 J/g protein, demonstrating system instability.

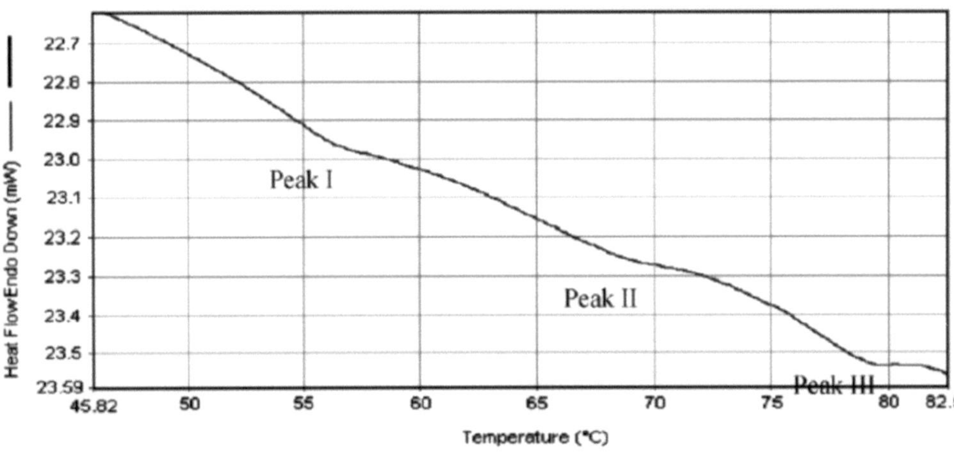

Figure 7. Thermal analysis of NI whole muscles
(Pighin et al., 2008)

Treatment	Peak I (myosin)		Peak II (sarcoplasmic proteins/collagen)		Peak III (actin)		
	T_d (°C)	ΔH (J/g protein)	T_d (°C)	ΔH (J/g protein)	T_d (°C)	ΔH (J/g protein)	ΔH_T (J/g protein)
NI	57.0 ab	1.32 a	67.4 c	0.65 ab	80.3 a	1.74 a	3.71
0.70% NaCl	56.4 b	0.70 b	67.1 c	0.73 a	77.2 c	1.61 a	3.04
0.25% TPP	57.6 a	1.75 a	70.9 a	0.50 bc	80.8 a	2.03 a	4.28
0.70% NaCl + 0.25% TPP	58.1 a	1.79 a	69.7 ab	0.38 c	78.5 b	1.66 a	3.83
1.20% NaCl + 0.25% TPP	56.3 b	1.40 a	68.6 bc	0.51 bc	76.8 c	0.84 b	2.75

Table 1. Endothermic peaks (Td) and denaturation enthalpies (DH) of whole muscles treated with NaCl and/or TPP
a-c Means within a column having different letters are significantly different ($p < 0.05$).
(Pighin et al., 2008)

The described findings reveal an important effect of the salt in decreasing the thermal stability of proteins. In order to support this fact, it was proposed that addition of neutral salt causes anions to compete with water molecules for specific sites of proteins, altering its hydration properties, and consequently requiring lower denaturation energies (Von Hippel & Wong, 1965). Treatment with TTP (0.25 %) significantly increased sarcoplasmic proteins/collagen Td. Also, Td of myosin and DH of actin were slightly increased. Whole muscle treated with 0.25 % TPP increased the DHT from 3.71 (NI muscles) to 4.28 J/g protein. These findings show an important stabilizing effect exerted by TPP, probably by altering hydrophobic interactions rather than affecting pH or ionic strength (Trout & Schmidt, 1986). With regard to the effect of phosphates on meat, Kijowski & Mast (1988) studied the effect of different phosphates (sodium pyrophosphate or sodium tripolyphosphate) on chicken breast muscles and its isolated myofibrils. They stated that low levels of sodium pyrophosphate or sodium tripolyphosphate (0.25 % or 0.50 %) caused the maximum stabilization (DHT) of chicken muscle and isolated myofibrils by means of myosin stabilization (increased DH). These results agree with present ones, even they were obtained in beef muscles.

When both additives were used together (0.70 % NaCl + 0.25 % TPP), Td corresponding to sarcoplasmic proteins/collagen was significantly increased in comparison to NI muscles, suggesting stabilization of the related proteins. This stabilization went along with a significant decreased of the DH, probably associated to an increment of the exothermic protein aggregation. On the contrary, actin Td showed a significant reduction in comparison to NI. Destabilization of actin in the presence of NaCl and phosphates was also described by Kijowski & Mast (1988) in chicken breast muscles treated with 2 % NaCl and 0.25/0.50 % pyrophosphate. When the effect of this combination of salts is compared to the one previously described, 0.70 % NaCl effect, it can be speculated that two effects appear to coexist: the same stabilizing effect of TPP on sarcoplasmic proteins/collagen and the destabilizing effect of NaCl on actin. Even though, the last effect is smaller in magnitude than the one produced by NaCl alone. It is important to denote that DHT of the muscle treated with the combination of salts (0.70 % NaCl + 0.25 % TPP) was slightly higher than the NI one (3.83 vs. 3.71 J/g protein, respectively). Comparing the DHT obtained individually for NaCl and TPP treatments, it can be seen that the effect of the addition of both salts is contrary to the effect of NaCl alone and in the same direction as TPP. Hence, the final increase could be assigned to the increase (even non significant) of myosin DH induced by TPP. Even more, when NaCl was raised to 1.20%, a maximal drop in actin stability was achieved (higher reduction of Td and DH). This event was accompanied by a maximal drop in DHT (from 3.71 to 2.75 J/g protein). It becomes evident that the increased amount of NaCl in the brine produced a destabilizing effect similar to the one found for NaCl alone, which also turned to be predominant to expenses of the effect of TPP.

Regarding this issue, Findlay & Barbut (1992) had stated that the addition of NaCl (0.5-2.0 %) reduced myosin Td and conversely TPP (0.2-0.6 %) increased its stability. Additionally, DHT varied depending on salts concentration. At low NaCl levels (less than 1.00 %) the effect of TTP on the increment of DHT was predominant. Instead, as NaCl increased the phosphate effect was minimized. In accordance with such observations, present results indicate that NaCl exerts a destabilizing effect in the presence of TPP (0.25 %) that becomes stronger as NaCl level raises.

4.2.3. Thermal analysis of isolated myofibrils

Thermal analysis of isolated myofibrils from NI muscle showed only two endothermic peaks (Fig. 8), corresponding to denaturation enthalpies of myosin and its subunits (I: 58.1 °C) and actin (II: 73.5 °C). Data obtained from Table 2 showed a significant decrease of myosin and actin DH in myofibrils isolated from 0.70 % NaCl-treated muscles and washed in the same brine solution. No changes in Td of these proteins were found. The destabilizing effect of NaCl was in agreement with data obtained from the whole muscle. TPP (0.25 %) treatment produced a significant increase of both, myosin and actin Td. In contrast, DH of myosin was strongly reduced ($p < 0.05$) by the treatment. These findings indicate that the salt is capable of stabilizing actin and myosin structures. However, the smaller energy required to denature myosin (lower DH) suggests a small amount of protein available in the myofibril fraction. Supporting this comment, it had been proposed that phosphates can induce myosin solubilization by promoting both, thick filament depolymerisation and actomyosin dissociation (Granicher & Portzehl, 1964; Xiong, 2005). Even though, an increased protein aggregation (exothermic event) taking place immediately after myosin denaturation, might not be discarded. In the case of whole muscle treated with TPP, myosin and actin Td did not suffer any change and their DH were only slightly increased. Therefore, it can be affirmed that the effect of TPP on the whole muscle and myofibrils is almost alike.

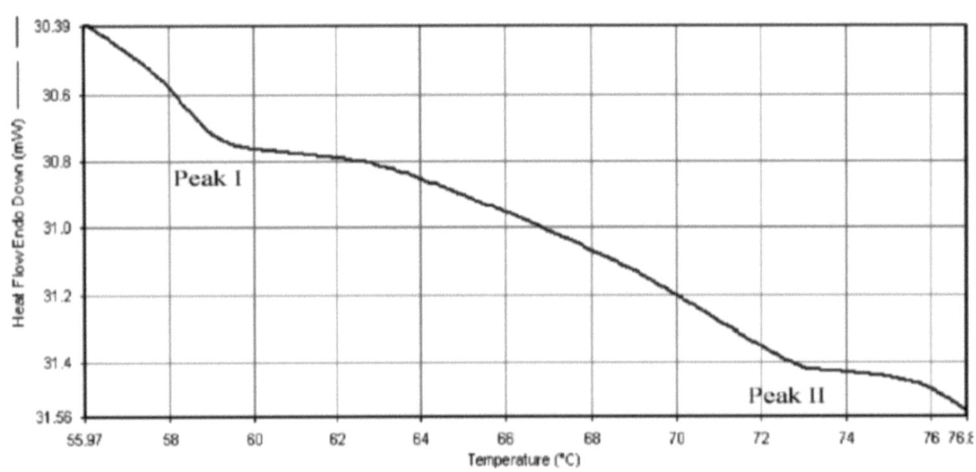

Figure 8. Thermal analysis of myofibrils isolated from NI muscles (Pighin et al., 2008)

When NaCl and TPP were used together (0.70 % NaCl + 0.25 % TPP) both, myosin and actin Td of isolated myofibrils significantly decreased ($p < 0.05$). Enthalpy of the proteins was also decreased, being the myosin DH reduced in almost 2.60 J/g protein with respect to NI muscles (NI: 3.87 vs. 1.27 J/g protein; $p < 0.05$). These findings confirm the destabilizing effect of NaCl previously described for actin in the whole muscle. Even more, this effect seems to be stronger in the myofibril fraction since myosin transition was also affected. Further increments of NaCl (1.20 %) -in the presence of TPP- produced similar effects on both proteins, with the exception of an intensified effect on actin DH. It is important to remark that the effect of both additives together was definitively higher than the effect of any of the additives individually. Offer & Knight (1998) proposed that polyphosphates -at the concentrations used in the industry- may assist NaCl in causing depolymerisation of thick filaments and dissociation of actomyosin, facilitating in consequence the myosin extraction. The action of NaCl as a mild structure-breaker aided by the presence of TPP could be the reason of the significant reduction of protein stability found.

Treatment	Peak I (myosin)		Peak II (actin)	
	T_d (°C)	ΔH (J/g protein)	T_d (°C)	ΔH (J/g protein)
NI	58.1 b	3.87 a	73.5 b	2.10 a
0.70% NaCl	57.9 b	2.16 b	74.9 b	1.44 c
0.25% TPP	63.2 a	0.73 d	83.4 a	1.97 ab
0.70% NaCl + 0.25% TPP	51.5 c	1.27 c	68.7 c	1.53 bc
1.20% NaCl + 0.25% TPP	51.1 c	1.56 c	67.6 c	0.80 d

Table 2. Endothermic peaks (Td) and denaturation enthalpies (DH) of myofibrils isolated from muscles treated with NaCl and/or TPP
a-d Means within a column having different letters are significantly different ($p < 0.05$).
(Pighin et al., 2008)

As it was mentioned, the physicochemical mechanisms involved in water binding and retention in the protein matrix of cooked meat were not completely elucidated. These results showed that the combined presence of NaCl (0.70 % or 1.20 %) and TPP (0.25 %) changes the heat susceptibility of muscle proteins (especially actin) so that they denature and coagulate at a lower cooking temperature. These modifications suggest the occurrence of protein conformational changes due to the salts, probably by altering hydrophobic and electrostatic interactions that stabilize the protein structure (Franks & Eagland, 1975). Evidently, these changes collaborate in the binding and posterior retention of water inside the tissue by any of the mechanisms proposed for the action of NaCl and polyphosphates (Offer & Knight, 1998). Increased solubilization of proteins found in whole muscles treated with NaCl (alone or in combination with TPP) could be related to the removal of transverse myofibrillar proteins which may act as structural constrains to myosin extraction. Also, some other low molecular weight myofibrillar proteins could be removed. The softening of the structure together with conformational modifications of myofibrillar proteins would allow entrapping water inside the tissue. This event would be the reason of the increased WHC of *Semitendinosus* muscle found by Vaudagna et al. (2008).

5. Conclusion

The isolation protocol previously described in this chapter has been demonstrated to be a successful methodology for individually isolating four of the main myofibrillar proteins.

Despite the proteins were not isolated in a complete pure state, this approach seems to be useful for using them for further analysis. The estimation of proteins' Rf by means of an immunological method looks like an interesting tool to complement the identification of proteins in unclear electrophoretic patterns.

The thermal behavior of the myofibrillar proteins is quite involved in the thermal behavior of whole muscle. NaCl incorporation into meat leads to an important protein destabilizing effect upon myofibrillar proteins, even at relative low doses when compared to industrial usage. The use of NaCl in combination with TPP produces a destabilizing global effect, suggesting that TPP may assist NaCl in acting as structure-breaker. The increase of WHC of *Semitendinosus* muscles injected with saline additives and submitted to *sous vide* cooking would be mainly associated to the conformational modifications of myofibrillar proteins and to the weakening of myofibrillar structures due to protein removal.

Taken together, the use of gel electrophoresis demonstrated to be a useful tool not only in isolating major muscle proteins but also in contributing to study the effect of saline additives in meat, in order to improve sensorial properties. Its combination with advanced technologies like DSC represents an interesting issue to look forward in the meat science approach.

Author details

Dario G. Pighin
Institute of Food Technology, National Institute of Agricultural Technology – INTA, Argentina

Acknowledgement

Special thanks are given to the Instituto Nacional de Tecnología Agropecuaria (INTA), Consejo Nacional de Investigaciones Científicas y Tecnológicas (CONICET) and Universidad de Morón (UM) for their financial support in the studies carried out and compiled in the present chapter.

6. References

Baublits, R.T.; Pohlman, F.W. Jr.; Brown, A.H.; Yancey, E.J. & Johnson, Z.B. (2006). Impact of muscle type and sodium chloride concentration on the quality, sensory and instrumental colour characteristics of solution enhanced whole-muscle beef. *Meat Science*, Vol.72, No.4, pp. 704–712, ISSN: 0309-1740.

Bender, A. (1992). *Meat and meat products in human nutrition in developing countries*, FAO, http://www.fao.org/docrep/T0562E/T0562E00.htm.

Boehm, M.L.; Kendall, T.L.; Thompson, V.F. & Goll, D.E. (1998). Changes in the calpains and calpastatin during postmortem storage of bovine muscle. *Journal of Animal Science*, Vol.76, pp 2415–2434. ISSN: 0021-8812.

Culler, R.D.; Parrish, F.C. Jr.; Smith, G.C. & Cross, H.R. (1978). Relationship of myofibril fragmentation index to certain chemical, physical and sensory characteristics of bovine longissimus muscle. *Journal of Food Science*, Vol.43, pp. 1177-1182. ISSN: 1750-3841.

Church, I.J. & Parsons, A.L. (2000). The sensory quality of chicken and potato products prepared using cook-chill and sous vide methods. *International Journal of Food Science and Technology*, Vol.35, pp. 361–368. ISSN: 0950-5423.

Chen, S.P.; Sheu, J.R.; Lai, C.Y.; Lin, T.Y.; Hsiao, G. & Fong, T.H. (2005). Detection of myofibrillar proteins using a step gradient minigel with an ambiguous interface. *Analytical Biochemistry*, Vol.338, pp. 270–277. ISSN: 0003-2697.

Clark, K.A.; Mcelhinny, A.S.; Beckerle, M.C. & Gregorio, C.C. (2002). Striated muscle cytoarchitecture: An intricate web of form and function. *Annual Review of Cell and Developmental Biolology*. Vol.18, pp. 637–706. ISSN: 1081-0706.

Dudziak, J.A. & Foegeding, E.A. (1988). Isolation of actomyosin and myosin from post-rigor turkey breast ant thigh. *Journal of Food Science*, Vol.53(S), pp. 1287– 1290. ISSN: 1750-3841.

Findlay, C.J. & Barbut, S. (1992). A response surface investigation of the effects of sodium chloride and tripolyphosphate on the thermal properties of beef muscle. *Meat Science*, Vol.31, pp. 155–164. ISSN: 0309-1740.

Franks, F. & Eagland, D. (1975). The role of solvent interactions in protein conformation. *CRC Critical Reviews in Biochemistry*, Vol.3 No.2, pp. 165–219. ISSN: 0045-6411.

García-Segovia, P.; Andrés-Bello, A.; & Martínez-Monzó , J. (2007). Effect of cooking method on mechanical properties, color and structure of beef muscle (*M. pectoralis*). *Journal of Food Engineering*, Vol.80 No.3, pp. 813–821. ISSN: 0260-8774.

Graiver, N.; Pinotti, A.; Califano, A. & Zaritzki, N. (2006). Diffusion of sodium chloride in pork tissue. *Journal of Food Engineering*, Vol.77, pp. 910–918. ISSN: 0260-8774.

Granicher, D. & Portzehl, H. (1964). The influence of magnesium and calcium pyrophosphate chelates, of free magnesium ions, free calcium ions, and free pyrophosphate ions on the dissociation of actomyosin in solution. *Biochimica et Biophysica Acta*, Vol.86, pp. 567–578. ISSN: 0304-4165.

Hermanson, A.M.; Harbitz, O. & Langton, M. (1986). Formation of two types of gels from bovine myosin. *Journal of the Science of Food and Agriculture*, Vol.37, pp. 69–84. ISSN: 1097-0010.

Ho, C.Y., Stromer, M.H. & Robson, R.M. (1994). Identification of the 30-kDa polypeptide in post mortem skeletal muscle as a degradation product of troponin-T. *Biochimie* Vol.76, pp. 369–375. ISSN: 0300-9084.

Houmeida, A.; Holat, J.; Tskhovrebova, L. & Trinick, J. (1995). Studies of the interaction between titin and myosin. *The Journal of Cell Biology*, Vol.131, pp. 1471–1481. ISSN: 0021-9525.

Karcher, R.E. & Landers, J.P. (2006). Electrophoresis, In *TIETZ Textbook of Clinical Chemistry and Molecular Diagnostics*. Carl A. Burtis, Ph.D. Edward R. Ashwood, M.D. David E. Bruns, M.D., Eds. Fourth Edition. pp. 102-111. Elsevier Inc. ISBN-13: 978-0-7216-0189-2. ISBN-I0: 0-7216-0189-8. Washington, USA.

Kellermayer, M.S.Z. & Grama, L. (2002). Stretching and visualizing titin molecules: Combining structure, dynamics and mechanics. *Journal of Muscle Research and Cell Motility,* Vol.23, pp. 499–511. ISSN: 0142-4319.

Kijowski, J.M. & Mast, M.G. (1988). Effect of sodium chloride and phosphates on the thermal properties of chicken meat proteins. *Journal of Food Science,* Vol.53 No.2, pp. 367–370. ISSN: 1750-3841.

Kuroda, M. (1982). Direct extraction of G-actin from the myosin-removed myofibrils under the conditions of low ionic strength. The *Journal of Biochemistry (Tokyo),* Vol.92, pp. 1863–1872. ISSN: 1756-2651.

Kurzban, G.P. & Wang, K. (1988). Giant polypeptides of skeletal muscle titin: Sedimentation equilibrium in guanidine hydrochloride. *Biochemical and Biophysical Research Communications,* Vol.150, pp. 1155–1161. ISSN: 0006-291X.

Laemmli, U.K. (1970). Cleavage of structural proteins during the assembly of the head of Bacteriophage T4. *Nature,* Vol.227, pp. 680–685. ISSN: 0028-0836.

Linke, W.A.; Kulke, M.; Li, H.; Fujita-Becker, S.; Neagoe, C.; Manstein, D.I.; Gautel, M. & Fernandez, J.M. (2002). PEVK domain of titin: An entropic spring with actin-binding properties. *Journal of Structural Biology,* Vol.137, pp. 194–205. ISSN: 1047-8477.

Locker, R.H. & Wild, D.J.C. (1986). Acomparative study of high molecular weight proteins in various types of muscle across the animal kingdom. *The Journal of Biochemistry (Tokyo),* Vol.99, pp. 1473–1484. ISSN: 1756-2651.

Lowry, O.H.; Rosebrough, M.J.; Farr, M.L. & Randall, R.L. (1951). Protein measurement with the Folin phenol reagent. *The Journal of Biological Chemistry,* Vol.193, pp. 265–275. ISSN: 0021-9258.

Offer, G. & Knight, P. (1998). The structural basis of water-holding in meat. Part I: General principles and water uptake in meat processing. In R. Lawrie (Ed.), *Developments in Meat Science 4,* pp. 63-171, ISBN-13: 978-0853349860, London, Elsevier Applied Science.

Pan, K.M. & Damodaran, S. (1994). Isolation and characterization of titin T1 from bovine cardiac muscle. *Biochemistry,* Vol.33, pp. 8255–8261. ISSN: 0006-2960.

Pardee, J.D. & Spudich, J.A. (1982). Purification of muscle actin. *Methods in Enzymology,* Vol.85, pp. 164–179. ISSN: 0076-6879.

Perez-Juan, M.; Flores, M. & Toldrá, F. (2007). Simultaneous process to isolate actomyosin and actin from post-rigor porcine skeletal muscle. *Food Chemistry,* Vol.101 No.3, pp. 1005–1011. ISSN: 0308-8146.

Pietrasik, Z. & Shand, P.J. (2004). Effect of blade tenderization and tumbling time on the processing characteristics and tenderness of injected cooked roast beef. *Meat Science,* Vol.66 No.4, pp. 871–879. ISSN: 0309-1740.

Pietrasik, Z. & Shand, P.J. (2005). Effects of mechanical treatments and moisture enhancement on the processing characteristics and tenderness of beef semimembranosus roast. *Meat Science,* Vol.71 No.3, pp. 498–505. ISSN: 0309-1740.

Pighin, D.G. & Gonzalez, C.B. (2008). Enrichment of myofibrillar proteinsfrom beef muscle by a simple sequential method. *Journal of Muscle Foods,* Vol.19, pp. 362-373. ISSN: 1046-0756.

Pighin, D.G.; Sancho, A.M. & Gonzalez, C.B. (2008). Effect of salt addition on the thermal behavior of proteins of bovine meat from Argentina. *Meat Science*, Vol.79, pp. 549-556. ISSN: 0309-1740.

Pospiech, E.; Greaser, M.L.; Mikolajczack, B.; Chiang, W. & Krzymdzinska, M. (2002). Thermal properties of titin from porcine and bovine muscles. *Meat Science*, Vol.62, pp. 187–192. ISSN: 0309-1740.

SENASA (National Service of Animal Health) Decreto #4238/68 (1969). *Despostadero*. In Reglamento de Inspección de Productos, Subproductos y Derivados de Origen Animal (G. Say, ed.) pp. 115–116, Buenos Aires, Argentina.

Sobieszek, A. (1994). Gradient polyacrylamide gel electrophoresis in presence of sodium dodecyl sulfate: A practical approach to muscle contractile and regulatory proteins. *Electrophoresis*, Vol.15, pp. 1014–1020. ISSN: 0173-0835.

Spudich, J.A. & Watt, S. (1971). The regulation of rabbit skeletal muscle contraction. *The Journal of Biological Chemistry*, Vol.246 No.15, pp. 4866–4871. ISSN: 0021-9258.

Stabursvik, E. & Martens, H. (1980). Thermal denaturation of proteins in post rigor muscle tissue as studied by differential scanning calorimetry. *Journal of the Science of Food and Agriculture*, Vol.31, pp. 1034–1042. ISSN: 1097-0010.

Syrovy, I. (1984). Separation of muscle proteins. *Journal of Chromatography A*, Vol.300, pp. 225– 247. ISSN: 0021-9673.

Szerman, N.; Gonzalez, C.B.; Sancho, A.M.; Grigioni, G.; Carduza, F. and Vaudagna, S.R. (2012). Effect of the addition of conventional additives and whey proteins concentrates on technological parameters, physicochemical properties, microstructure and sensory attributes of *sous vide* cooked beef muscles. *Meat Science*, Vol.90, pp. 701–710. ISSN: 0309-1740.

Trout, G.R. & Schmidt, G. (1986). Effect on the functional properties of restructural beef rolls: The role of pH, ionic strength, and phosphate type. *Journal of Food Science*, Vol.51, pp. 1416–1423. ISSN: 1750-3841.

Vaudagna, S. R.; Sánchez, G.; Neira, M. S.; Insani, E. M.; Picallo, A. B.; Gallinger, M. M. & Lasta, J.A. (2002). *Sous vide* cooked beef muscles: Effects of low temperature - long time treatments on their quality characteristics and storage stability of product. *International Journal of Food Science and Technology*, Vol.37, pp. 425-441. ISSN: 0950-5423.

Vaudagna, S.R.; Pazos, A.A.; Guidi, S.M.; Sanchez, G.; Carp, D.J. & Gonzalez, C.B. (2008). Effects of salt addition on *sous vide* cooked whole beef muscles from Argentina. *Meat Science*, Vol.79 No.3, pp. 470–482. ISSN: 0309-1740.

Von Hippel, P.H. & Wong, K.Y. (1965). On the conformational stability of globular proteins. The effects of various electrolytes and nonelectrolytes on the thermal ribonuclease transition. *The Journal of Biological Chemistry*, Vol.240 No.10, pp. 3909–3923. ISSN: 0021-9258.

Wang, K. (1982). Purification of titin and nebulin. *Methods in Enzymology*, Vol.85, pp. 264–274. ISSN: 0076-6879.

Wang, K. (1996). Titin/Connectin and nebulin: Giant proteins rulers of muscle structure and function. *Advances in Biophysics*, Vol.33, pp. 123–134. ISSN: 0065-227X.

Wang, K. & Wright, J. (1988). Architecture of the sarcomere matrix of skeletal muscle: Immunoelectron microscopic evidence that suggests a set of parallel inextensible nebulin filaments anchored at the Z line. *The Journal of Cell Biology,* Vol.107, pp. 2199–2212. ISSN: 0021-9525.

Wang, S.M. & Greaser, M.L. (1985). Immunocytochemical studies using a monoclonal antibody to bovine cardiac titin on intact and extracted myofibrils. *Journal of Muscle Research and Cell Motility,* Vol.6, pp. 293–312. ISSN: 0142-4319.

Warris, P.D. (2000) The Chemical Composition and Structure of Meat, In *Meat Science. An Introductory Text*. pp. 37-67. CABI Publishing, ISBN: 0-85199-424-5, Oxfordshire, UK.

Xiong, Y.L. (2005). Role of myofibrillar proteins in water-binding in brine-enhance meats. *Food Research International,* Vol.38, pp. 281–287. ISSN: 0963-9969.

Chapter 8

Application of the Different Electrophoresis Techniques to the Detection and Identification of Lactic Acid Bacteria in Wines

Lucía González-Arenzana, Rosa López,
Pilar Santamaría and Isabel López-Alfaro

Additional information is available at the end of the chapter

http://dx.doi.org/10.5772/45759

1. Introduction

The microorganisms play an essential role in winemaking since a mixed culture of numerous microorganisms including fungal, yeast, and bacteria species are involved in this process and are the responsible for the final quality of the wine (Bisson et al., 1993). Therefore, in order to control the fermentation processes knowing and understanding the complex microbiota involved in them is necessary.

Yeast are able to convert sugar from grapes into ethanol and many other changes that lead to wine. Lactic acid bacteria (LAB) that are often present on the surface of the grapes and can represent significant populations in musts (Lonvaud-Funel, 1999) play dual roles in wine fermentations: as wine spoilage agents and as the main effectors of malolactic fermentation (MLF). Numerous studies have been conducted on the LAB that occur on grapes, grape musts and wines and it is generally agreed that a succession of species happens during the different stages of winemaking and conservation of wines (Ribéreau-Gayon et al., 2006). Most bacterial species present in wine fermentations have been identified by traditional microbiological techniques involving cultivation. However, as it was observed with microbial ecology studies of other environments, cultivation-dependent methods often exhibit biases resulting in an incomplete representation of the true present bacterial diversity (Amann et al., 1995; Hugenholtz et al., 1998). Applications of culture-independent molecular techniques to monitor the microbial successions of various food and beverage fermentations have revealed microbial constituents and microbial interactions not witnessed by previous plating analyses (Giraffa & Neviani, 2001).

Denaturing gradient gel electrophoresis (DGGE) and temperature gradient gel electrophoresis to separate bacterial 16S ribosomal DNA (rDNA) amplicons are common culture independent methods employed to characterize microbial communities from specific environmental niches (Lopez et al., 2003; Muyzer & Smalla, 1998). These approaches are attractive since they enable to detect individual species as well as to get overall profiling of community structure changes with time.

Otherwise, ecology, interactions and development of the different bacterial strains during alcoholic fermentation (AF) and MLF are still a field of active research. Efficient and precise methods of strain identification and discrimination have been developed during the last years, either to prepare well-defined starters of biotechnological interest in winemaking, or to quickly assess the presence of certain strains in a wine, or to gain insight of such a complex ecosystem as wine.

Pulsed field gel electrophoresis (PFGE) has proved to be an useful tool for the identification of a wide LAB strains variety and especially for species belonging to the genus *Lactobacillus* (Charteris et al., 1997). Several restriction enzymes have been used for obtaining profiles of *Oenococcus oeni* strains: *Not*I (Kelly et al., 1993; Prevost et al., 1995; Sato et al., 2001), *Apa*I and *Sfi*I (Daniel et al., 1993; Kelly et al., 1993; Lopez et al., 2007). Other used methods for LAB strain identification include ribotyping or restriction fragment length polymorphism (RFLP) of genes encoding rRNAs (Charteris et al., 1997; Zavaleta et al., 1997), random amplified polymorphic DNA (RAPD-PCR) analysis with arbitrary primers (Cocconcelli et al., 1997; Lopez et al., 2008; Spano et al., 2002), rep-polymerase chain reaction (rep-PCR) analysis (Parry et al., 2002) and multilocus sequence typing (MLST) system (de las Rivas et al., 2004; de las Rivas et al., 2006). RAPD-PCR and PFGE of macrorestriction fragments are the most frequently used (Guerrini et al., 2003; Tenreiro et al., 1994; Viti et al., 1996; Zapparoli et al., 2000; Zavaleta et al., 1997), and more recently Ruiz et al. (Ruiz et al., 2008) have obtained a better discrimination in the study of bacterial diversity by the combination of results from those two techniques.

For all these reasons, the aims of this work were: (a) applying the DGGE, PFGE and RAPD-PCR techniques to the analysis of the LAB species diversity and the intraspecific diversity of *Oenococcus oeni* in a winery of La Rioja region during three consecutive vintages, (b) getting a better knowledge of bacterial ecology throughout both AF and MLF in wine elaborated with Tempranillo, the classic red grape variety of Spain and native of Appellation of Origin Rioja, (c) evaluating the occurrence of genotypes from commercial *Oenococcus oeni* strains between the autochthonous *Oenococcus oeni* strains isolated from non inoculated fermenting wines, and (d) contributing to the maintenance of the LAB biodiversity in Rioja red wines.

2. Methodology

2.1. Wine production and wine samples

Traditional red wine fermentations from c.v. Tempranillo local grapes of 2006, 2007 and 2008 vintages at one winery of the Spanish northern region of Rioja were studied. Winemaking practices were the typical of this wine-producing area: AFs were conducted in the presence of grape skins, seeds and stalks, after the addition of sulphur dioxide and until the residual reducing sugar content was under 2 g/L. At this final point of AF, wines were drawn off into tanks and were allowed to undergo spontaneous MLF with the endogenous microbiota. The sampled winery had never used commercial starters for MLF. One fermentation tank was sampled in each vintage. Wine samples were collected aseptically for chemical and microbiological analysis at different times: must (stage 1), tumultuous AF (density around 1,025; stage 2), at final AF (< 2 g/L glucose + fructose; stage 3), initial MLF (consumption of 10% of the initial malic acid; stage 4), tumultuous MLF (consumption of 60% of the initial malic acid; stage 5) and at final MLF (L-malic acid concentration < 0.5 g/L; stage 6).

2.2. Commercial *Oenococcus oeni* starter samples

Sixteen commercial starter cultures employed to induce MLF derived from six different companies were analyzed. These commercial cultures were selected between the most frequently used in Spain.

2.3. Chemical analysis of the musts and wines

Alcohol degree, pH, total acidity, volatile acidity, reducing sugars, free and total sulphur dioxide and L-malic and L-lactic acid content were measured according to the European Community Official Methods (European Community, 1990).

2.4. Culture dependent methods

2.4.1. Bacterial enumeration and isolation

Must or wine samples were diluted in sterile saline (0.9% NaCl) solution and plated on modified MRS agar (Scharlau Chemie S.A., Barcelona, Spain) plates supplemented with tomato juice (10% v/v), fructose (6 g/L), cysteine-HCl (0.5 g/L), D,L-malic acid (5 g/L) and pymaricine (50 mg/L) (Acofarma, S. Coop., Terrassa, Spain). Samples were incubated at 30 ºC under strict anaerobic conditions (Gas Pak System, Oxoid Ltd., Basingstoke, England) for at least ten days, and viable counts were reported as the number of CFU/mL. Fifteen colonies from each wine sample were selected for reisolation and identification. Isolates were stored in 20% sterile skim milk (Difco) at –20 ºC.

Every commercial lyophilized starter culture was hydrated in saline solution (0.9% NaCl) and then 100 µL aliquot from the appropriate dilution was plated at the surface of modified MRS agar without pymaricine. Because of their low viability in laboratory conditions (Maicas et al., 1999a; Maicas et al., 1999b) glycine (40 mM) and ethanol (10% v/v) were added to this medium. The plates were incubated for at least 10 days at 30 ºC under anaerobic atmosphere (Gas Pak System, Oxoid Ltd.) and five colonies were isolated from each one.

2.4.2. Species identification

Species identification was carried out by previously recommended methods, which included bacteria morphology, Gram staining, and catalase (Holt et al., 1994). *Oenococcus oeni*, *Lactobacillus plantarum* and *Lactobacillus brevis* species were confirmed by the species-specific PCR method (Beneduce et al., 2004; Zapparoli et al., 1998). In case of the identification of other species, PCR amplification of partial 16S rRNA genes was performed with WLAB1 and WLAB2 as previously described (Lopez et al., 2007). PCR products were sequenced by Macrogen Inc. (Seoul, South Korea) and sequences were used for comparison to the data in GenBank using the Basic Local Alignment Search Tool (BLAST) (Altschul et al., 1990).

2.4.3. Oenococcus oeni typification by PFGE

PFGE was carried out according to the method described by Birren et al. (1993) with some modifications (Lopez et al., 2007) for agarose block preparation. Because of the difficulty of typing commercial strains first of all these cells underwent to fifteen minutes of ultrasounds, moreover a higher quantity of lysozime (100 µL/block) was added and incubated for 2 h. Macrorestriction analysis was performed with two endonucleases: *Sfi*I, following the method reported by López et al. (2007), and *Apa*I by the method reported by Larisika (2008) with the following modifications for optimal separation of fragments: 1.2% (w/v) agarose gels were submitted to 24 h with a pulse ramping between 0.5 and 20 s at 14 ºC and 6 V/cm in a CHEF DRII apparatus (Bio-Rad).

2.4.4. Oenococcus oeni typification by RAPD-PCR

RAPD-PCR was carried out following the procedure described by Ruiz et al. (2010b) with some modifications: MgCl 100 mM, dNTP 50 mM and primer M13 100 mM. RAPD-PCR reaction was developed in a total volume of 50 µL and it was carried out with a Perkin Elmer, GeneAmp PCR System 2400 thermocycler. 20 µL of amplified products were resolved by electrophoresis in a 1.4% agarose gel in 0.5x TBE (45 mM Tris base, 89 mM, boric acid, 2.5 mM EDTA pH 8) for 3 h at 70 V.

2.4.5. Numerical analysis of PFGE and RAPD-PCR images

The conversion, normalization and further processing of images were carried out by InfoQuest™ FP software version 5.10 (Bio-Rad, USA). Comparison of the obtained PFGE patterns was performed with Pearson's product-moment correlation coefficient and the Unweighted Pair Group Method using Arithmetic averages (UPGMA). Comparison of the pulse types from the PFGE and RAPD was made by composite data set comparison with average molecular analysis by Unweighted Pair Group Method using Arithmetic averages (UPGMA) (Ruiz et al., 2008).

2.5. Culture independent methods

2.5.1. Direct DNA extraction from wines samples

A volume of 10 mL of each must or wine sample was centrifuged (30 min, 10000xg, 4 ºC). The supernatant was discarded and 1.2 mL of saline solution (NaCl 0.9%) and 2.4 mL of zirconium hydroxide (7 g/L) were added to the pellet to facilitate pelleting of the bacteria in wine (Lucore et al., 2000). After horizontal shaking during 10 min at room temperature, the suspension was again centrifuged (10 min, 500xg, 7 ºC) and finally DNA was purified from the cell pellet by using a PowerSoil® DNA isolation kit (MO BIO Laboratories, Inc., Carlsbad, CA USA) as per the manufacturer's instructions.

2.5.2. PCR conditions

PCR was performed using an Applied Biosystem, GeneAmp® PCR System 2700 thermocycler at a final volume of 50 µL. To amplify the region V4 to V5 of 16S rDNA gene, primers WLAB1 and WLAB2^{GC} were used as López et al. described (2003). Moreover, primers *rpo*B1, *rpo*B1o, and *rpo*B2^{GC} were employed to amplify the region of the *rpoB* gene as it was described by Renouf et al. (2006) with the next modifications: 0.5 µM of each primer, 1 mM dNTP mix and 0.5 µL of *PfuUltra* II Fusion HS DNA Polymerase (Stratagene).

2.5.3. PCR-DGGE analysis

The separation of the respective PCR products was performed with the D-CODE™ universal mutation detection system (Bio-Rad, Hercules, Calif.). PCR products obtained from WLAB1-WLAB2^{GC} primers were run on 8% (wt/V) polyacrilamide gels in a running buffer containing 2 M Tris base, 1 M Glacial acetic acid and 50 mM EDTA pH 8 (TAE), and a denaturing gradient from 35 to 55% of urea and formamide. The electrophoresis was performed at 20 V for 10 min, and 80 V for 18 h at a constant temperature of 60 °C. PCR products generated with the *rpo*B1, *rpo*B1o, and *rpo*B2^{GC} primers were separated with 8% polyacrylamide gels containing a 32 to 50% urea-formamide gradient. Electrophoresis was performed for 10 min at 20 V, and 16 h at 60 V at a constant temperature of 60 °C. The DGGE gels were stained in ethidium bromide after the electrophoresis and then were visualized with UV transiluminattion (GelDoc, Bio-Rad). Blocks of polyacrylamide gels which contained selected DGGE bands were excised and later incubated overnight in 20 µL of sterile and pure water at 4 ºC to make DNA bands diffuse to the liquid. One microliter of this solution was used to reamplify the PCR product.

2.5.4. DNA sequencing and phylogenetic analysis

PCR products were sequenced by Macrogen Inc. (Seoul, South Korea). The quality and characteristics of the obtained sequences were analyzed with the software InfoQuest™ FP 5.10, only those ones considered as appropriate were used for comparison to the GenBank database with the Basic Local Alignment Search Tool (BLAST) (Altschul et al., 1990). After this preliminary study, our sequences and their homologous ones (obtained from the Nucleotide Database: http://www.ncbi.nlm.nih.gov/nuccore) were assembled and submitted to phylogenetic and evolutionary analysis with MEGA version 4.0.2 (Tamura et al., 2007). The Neighbor-Joining analysis (Saitou & Nei, 1987) allowed to get information about the relations between the gotten sequences and the reliability of the identifications provided by the Nucleotide Database. The bootstrap test was based on 1000 replicates (Felsenstein, 1985).

The evolutionary distances were computed using the Maximum Composite Likelihood method (Tamura et al., 2007) that allowed to calculate the equivalent units to the base substitutions per site.

3. Results and discussion

3.1. Oenological parameters of wine samples and fermentation development

Results for analytical composition of wines during three vintages are displayed in Table 1. Data were within the usual range of Tempranillo wines from this Spanish region (González-Arenzana et al., 2012b). After completion AF, alcohol content ranged between 13.0% and 14.0%, pH was between 3.32 and 3.64 and free SO_2 level was between 4.24 and 18.1 mg/L. During MLF a decrease in total acidity and a subsequent increase in pH were observed. In addition, an increase in volatile acidity was noted as it was expected. The wine from 2006 vintage showed less restrictive parameters for microbial growth, so it presented higher pH and the lowest values of alcohol content and SO_2.

Year	2006		2007		2008	
Stage	3	6	3	6	3	6
Alcohol content (% v/v)	13.0	-	13.8	-	14.0	-
pH	3.64	3.83	3.41	3.57	3.32	3.50
Total acidity (g/L tartaric acid)	7.98	5.91	7.63	6.71	9.00	7.20
Volatile acidity (g/L acetic acid)	0.25	0.46	0.37	0.49	0.26	0.37
Total SO_2 (mg/L)	28.4	-	38.1	-	31.6	-
Free SO_2 (mg/L)	4.24	-	18.1	-	13.2	-
L-malic acid (g/L)	2.60	0.04	1.48	0.16	2.61	0.21
L-lactic acid (g/L)	-	1.81	-	1.10	-	1.72

-: not analyzed

Table 1. Analytical composition of wines at final AF (stage 3) and final MLF (stage 6) at each vintage.

AF completion lasted for six, sixteen and eleven days in 2006, 2007 and 2008 vintages, respectively. Viable LAB counts during AF were in the range of 10^2 - 10^3 CFU/mL, increasing to 10^7 – 10^8 CFU/mL during MLF, similar to spontaneous MLF results reported by other authors (European Community, 1990; Lopez et al., 2008). The development of the MLF was related to the viable population of LAB and there was a relation between bacterial population and decrease in L-malic acid (data not shown). Important differences in MLF duration were observed between vintages and MLF completion lasted for 21, 239 and 136 days in 2006, 2007 and 2008 vintages, respectively. Different temperatures at each vintage (wine temperature below 12 ºC after AF in 2007) and the lack of temperature control in the winery were the determinant factors in these differences, but factors such as pH, composition of the wine and the interaction with other microorganisms implicated in the fermentation could also influence, as it has been reported by other authors (du Plessis et al., 2004; Lonvaud-Funel, 1999; Reguant et al., 2005a; Reguant et al., 2005b).

3.2. Species identification

3.2.1. Culture dependent microbiological analysis

Figure 1 shows the number of isolates of the viable LAB species identified at each stage and year of vinification. A total of 251 LAB isolates were recovered and identified as belonging to eight different species. The greatest diversity of LAB species was detected during the AF. *Oenococcus oeni* was present in all studied stages of the fermentation process except in 2007. It was isolated in must and tumultuous AF in 2006 and 2008 vintages, and it was the only species isolated at MLF in the three years, being therefore the predominant species, followed by *Lactobacillus plantarum, Leuconostoc mesenteroides* and *Lactobacillus mali*. The other non-*Oenococcus oeni* species appeared at stages 1-2 in variable rates in the three vintages. A similar distribution of species has been also reported by other authors (Fugelsang & Edwards, 2007; Ruiz et al., 2010) and they also concluded that *Oenococcus oeni* was the main responsible species for MLF. The diversity of species found at each year was different, being the number of species isolated in 2007 almost double that in 2006 and 2008, and missing *Oenococcus oeni* until the end of the 2007 AF, a fact that could also have influence in the MLF duration, as it was indicated above.

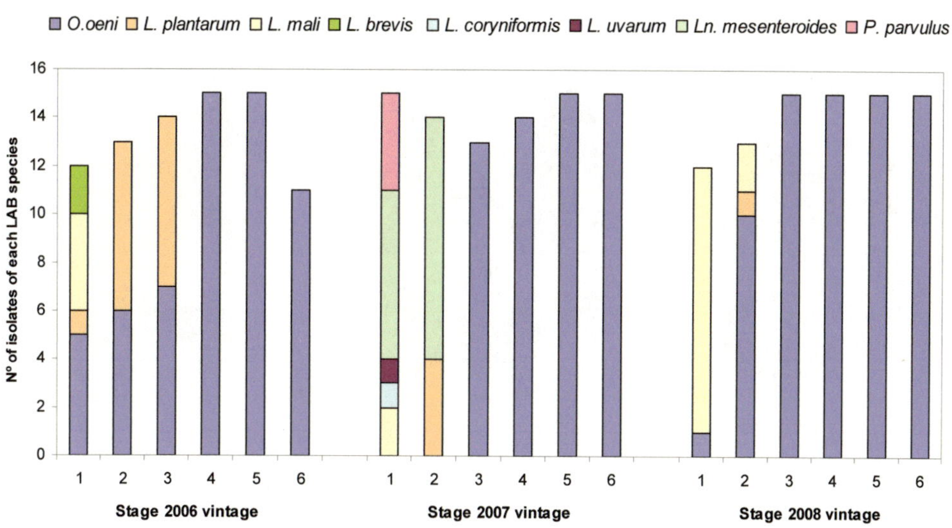

Figure 1. Number of isolates of the viable LAB species identified at each stage of vinification (1, must; 2, tumultuous AF; 3, final AF; 4, initial MLF; 5, tumultuous MLF; 6, final MLF) and in each vintage.

3.2.2. Culture independent microbiological analysis and comparison with culture dependent method

PCR-DGGE analysis of the sampled wine fermentations in the three studied vintages using primers WLAB1/2 (16S rDNA-based primer sets) and primers *rpo*B1/1o/2 (β subunit of the

RNA polymerase-based primer sets) (Figure 2a and 2b, respectively) revealed different species and a profile of bacterial community structure changes during AF and MLF.

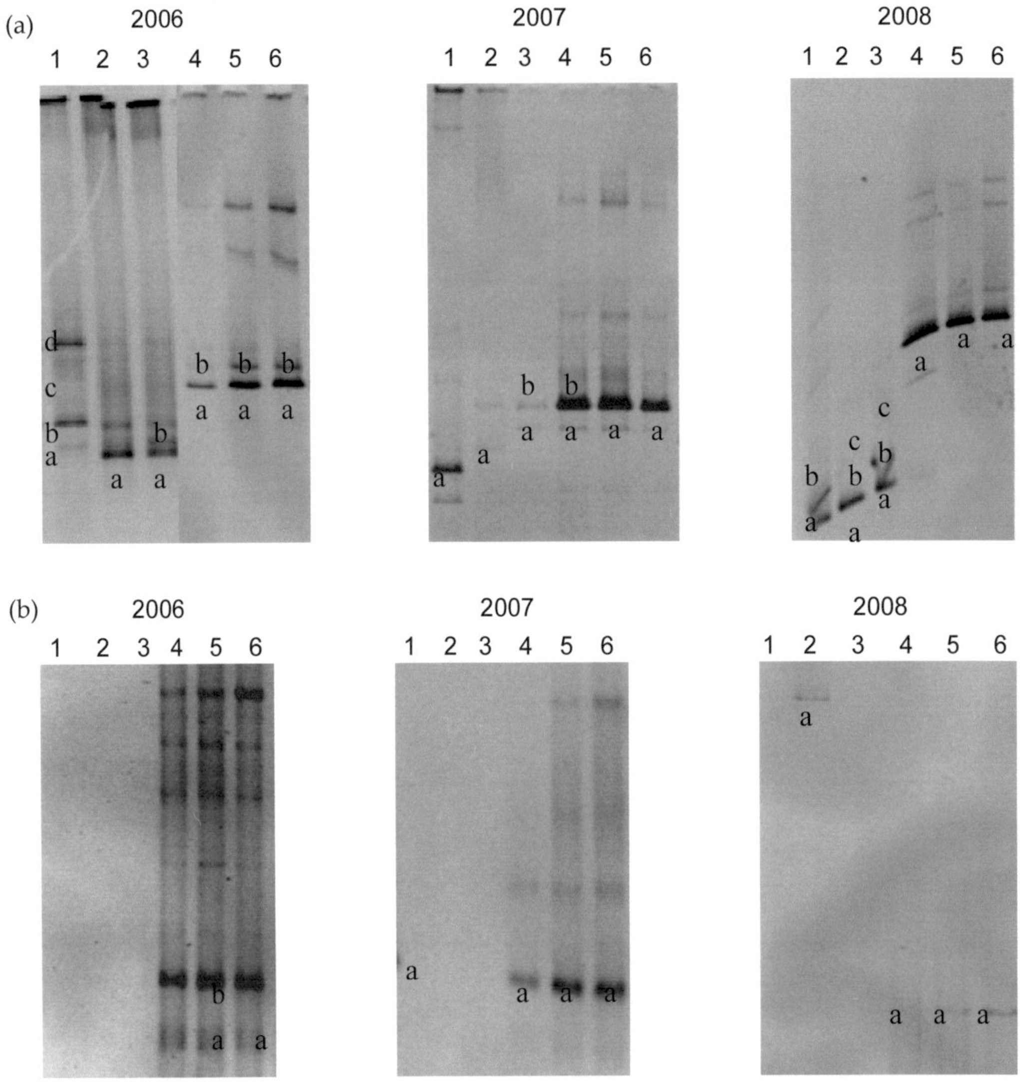

Figure 2. DGGE gels of the sampled wines at each vintage and stage of vinification (1, must; 2, tumultuous AF; 3, final AF; 4, initial MLF; 5, tumultuous MLF; 6, final MLF). Letters indicate bands excised for each gene, 16S rRNA (a) and *rpo*B (b).

The sequences obtained from the DNA excised DGGE-bands of each sample and their homologous ones from Nucleotide Database (Altschul et al., 1990) constituted a tree for each studied gene (Figure 3). Figure 3a shows a tree based on 16S rDNA gene composed by four ramifications or branches belonging to the genus *Oenococcus, Lactobacillus, Weissella* and *Leuconostoc*; and Figure 3b shows a tree based on *rpo*B gene composed by three branches belonging to *Oenococcus, Leuconostoc* and *Lactobacillus.*

(a)

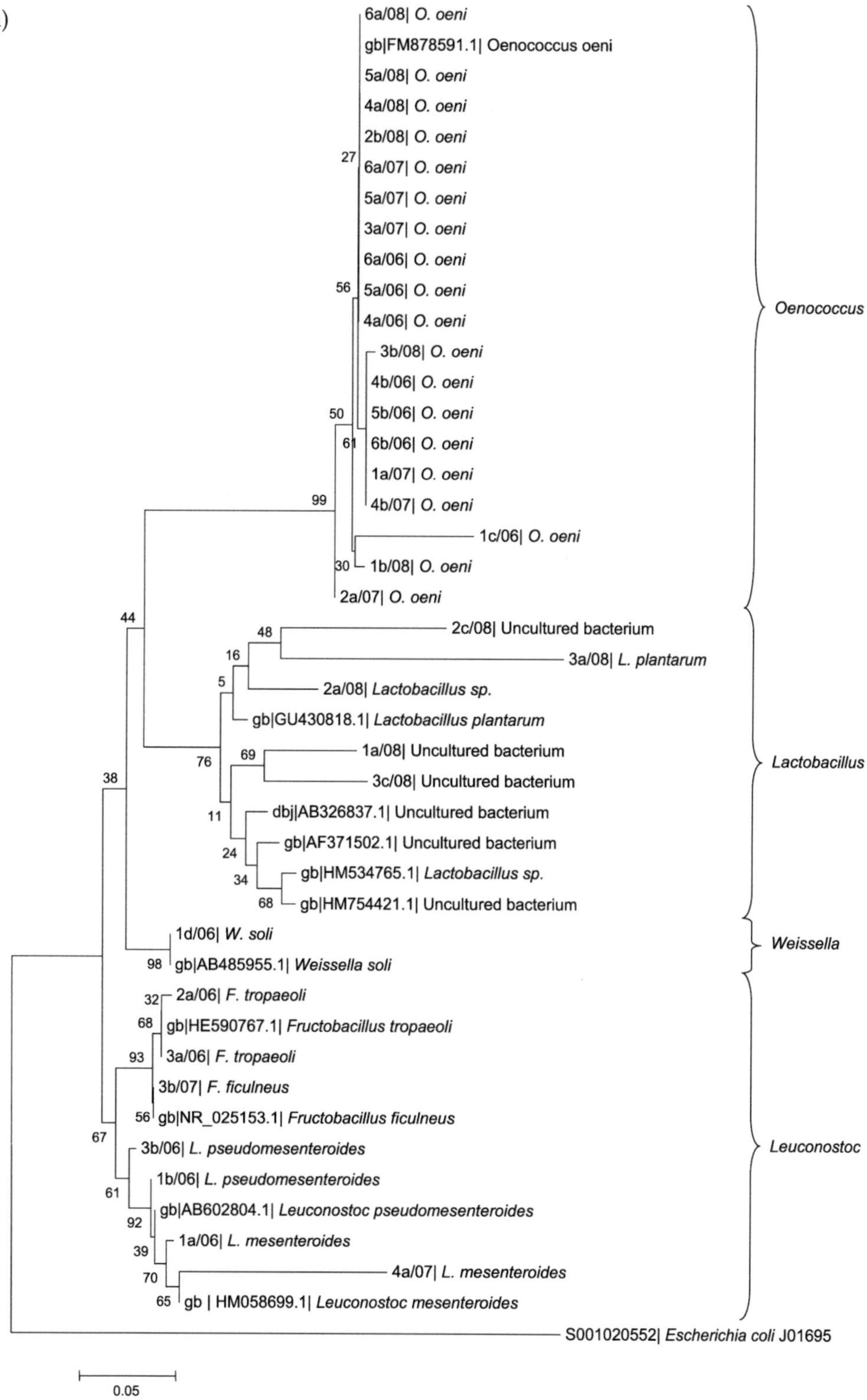
6a/08| O. oeni
gb|FM878591.1| Oenococcus oeni
5a/08| O. oeni
4a/08| O. oeni
2b/08| O. oeni
6a/07| O. oeni
5a/07| O. oeni
3a/07| O. oeni
6a/06| O. oeni
5a/06| O. oeni
4a/06| O. oeni
3b/08| O. oeni
4b/06| O. oeni
5b/06| O. oeni
6b/06| O. oeni
1a/07| O. oeni
4b/07| O. oeni
1c/06| O. oeni
1b/08| O. oeni
2a/07| O. oeni
2c/08| Uncultured bacterium
3a/08| L. plantarum
2a/08| Lactobacillus sp.
gb|GU430818.1| Lactobacillus plantarum
1a/08| Uncultured bacterium
3c/08| Uncultured bacterium
dbj|AB326837.1| Uncultured bacterium
gb|AF371502.1| Uncultured bacterium
gb|HM534765.1| Lactobacillus sp.
gb|HM754421.1| Uncultured bacterium
1d/06| W. soli
gb|AB485955.1| Weissella soli
2a/06| F. tropaeoli
gb|HE590767.1| Fructobacillus tropaeoli
3a/06| F. tropaeoli
3b/07| F. ficulneus
gb|NR_025153.1| Fructobacillus ficulneus
3b/06| L. pseudomesenteroides
1b/06| L. pseudomesenteroides
gb|AB602804.1| Leuconostoc pseudomesenteroides
1a/06| L. mesenteroides
4a/07| L. mesenteroides
gb | HM058699.1| Leuconostoc mesenteroides
S001020552| Escherichia coli J01695
Oenococcus
Lactobacillus
Weissella
Leuconostoc
27
56
50
61
99
30
44
48
16
5
76
69
11
24
34
68
38
98
32
68
93
56
67
61
92
39
70
65
0.05

(b)

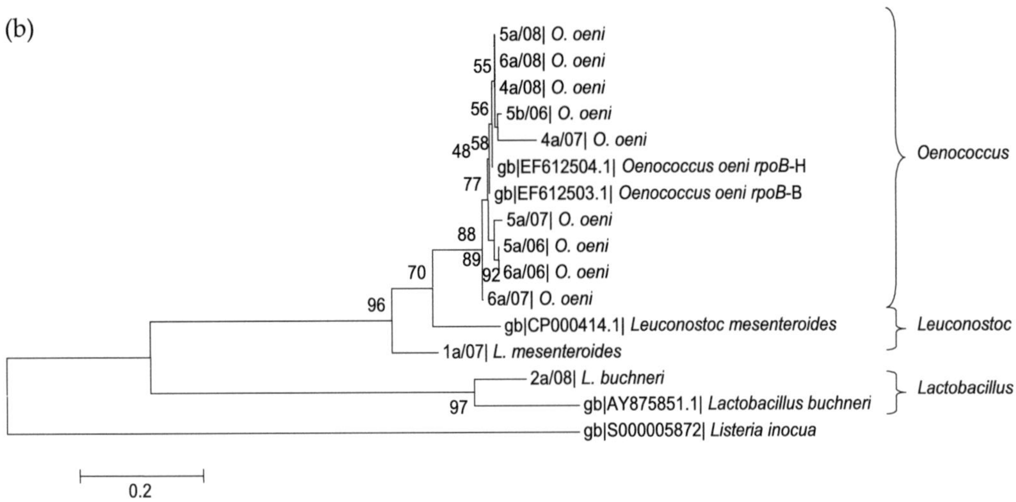

Figure 3. Phylogenetic trees generated from the method of 16S rDNA (a) and *rpo*B (b) sequences of recovered bands were inferred using the Maximun Parsimony method (Felsenstein, 1985). Numbers over the branches are bootstrap values (1000 repetitions). *Escherichia coli* and *Listeria inocua* were used as outgroups. Scale bar represents (calculated) distance. Reference strains are closely related to the sequenced bands and accession number of each gene is indicated. Band number isolated from fermenting wines are indicated by the sampled stage of vinification (1, must; 2, tumultuous AF; 3, final AF; 4, initial MLF; 5, tumultuous MLF; 6, final MLF) the letter of the position in the gel and the isolation year.

The species identification at each stage of vinification in the three studied years with culture independent techniques (16S rDNA/PCR-DGGE and *rpo*B/PCR-DGGE) and culture dependent method (plating on modified MRS) are recovered in Table 2.

A total of fourteen different LAB species were identified in the three studied vintages by traditional and culture independent methods. PCR-DGGE analysis allowed to identify nine species in comparison to the eight ones detected by culture in plate of the sampled wine. Thus, *Fructobacillus ficulneus, Fructobacillus tropaeoli, Lactobacillus buchneri, Leuconostoc pseudomesenteroides* and *Weissella* were not detected in the employed culture medium, while *Lactobacillus coryniformis, Lactobacillus brevis, Lactobacillus mali, Lactobacillus uvarum* and *Pediococcus parvulus* were not detected by PCR-DGGE. Therefore, results obtained by both methods were complementary and demonstrated the importance of using a combined analytical approach to explore microbial communities as other authors have concluded in different ecological niches (Iacumin et al., 2009). In relation with DGGE analysis, PCR-amplified bands from 16S rDNA gene gave better results than the *rpo*B bands amplification, with eight and three identified species, respectively. Nevertheless, the results obtained with the two genes were again complementary so *Lactobacillus buchneri* was only detected by *rpo*B/PCR-DGGE.

Results about diversity of LAB species found at each year and stage of vinification were very similar to those described above for culture dependent method. The greatest diversity was detected again during AF, opposite to MLF were only two species were present, *Oenococcus oeni* and *Leuconostoc mesenteroides*. The presence of *Leuconostoc mesenteroides* in

2007 at stage 4 confirmed again the possible competition of other LAB with *Oenococcus oeni* and their influence in MLF duration.

Detected LAB species	2006						2007						2008					
	1*	2	3	4	5	6	1	2	3	4	5	6	1	2	3	4	5	6
Fructobacillus ficulneus	-	-	-	-	-	-	-	-	w	-	-	-	-	-	-	-	-	-
Fructobacillus tropaeoli	-	w	w	-	-	-	-	-	-	-	-	-	-	-	-	-	-	-
Lactobacillus coryniformis	-	-	-	-	-	-	c	-	-	-	-	-	-	-	-	-	-	-
Lactobacillus brevis	c	-	-	-	-	-	-	-	-	-	-	-	-	-	-	-	-	-
Lactobacillus buchneri	-	-	-	-	-	-	-	-	-	-	-	-	-	b	-	-	-	-
Lactobacillus mali	c	-	-	-	-	-	c	-	-	-	-	-	c	c	-	-	-	-
Lactobacillus plantarum	c	c	c	-	-	-	-	c	-	-	-	-	-	c	w	-	-	-
Lactobacillus uvarum	-	-	-	-	-	-	c	-	-	-	-	-	-	-	-	-	-	-
Lactobacillus sp.	-	-	-	-	-	-	-	-	-	-	-	-	-	w	-	-	-	-
Leuconostoc mesenteroides	w	-	-	-	-	-	bc	c	-	w	-	-	-	-	-	-	-	-
Leuconostoc pseudomesenteroides	w	-	w	-	-	-	-	-	-	-	-	-	-	-	-	-	-	-
Oenococcus oeni	wc	c	c	wc	◊	◊	w	w	wc	◊	◊	◊	wc	wc	wc	◊	◊	◊
Pediococcus parvulus	-	-	-	-	-	-	c	-	-	-	-	-	-	-	-	-	-	-
Weissella sp.	w	-	-	-	-	-	-	-	-	-	-	-	-	-	-	-	-	-
UB.1	-	-	-	-	-	-	-	-	-	-	-	-	w	-	-	-	-	-
UB.2	-	-	-	-	-	-	-	-	-	-	-	-	-	w	-	-	-	-
UB.3	-	-	-	-	-	-	-	-	-	-	-	-	-	-	w	-	-	-
Total species nº	7	3	4	1	1	1	6	3	2	2	1	1	3	6	3	1	1	1
Plate species nº	4	2	2	1	1	1	5	2	1	1	1	1	2	3	1	1	1	1
Total DGGE species nº	4	1	2	1	1	1	2	1	2	2	1	1	2	4	3	1	1	1
16S rDNA/DGGE species nº	4	1	2	1	1	1	1	1	2	2	1	1	2	3	3	1	1	1
*rpo*B/DGGE species nº	-	-	-	-	1	1	1	-	-	1	1	1	-	1	-	1	1	1

*stage of vinification (1, must; 2, tumultuous AF; 3, final AF; 4, initial MLF; 5, tumultuous MLF; 6, final MLF)
-; not detected
UB; uncultured bacterium
w; detected with 16S rDNA/DGGE
b; detected with *rpo*B/DGGE
c; detected in plate after culture
◊; LAB species detected with all the employed methods

Table 2. LAB species detected with culture independent techniques (16S rDNA/PCR-DGGE and *rpo*B/PCR-DGGE) and culture dependent methods at each stage of vinification (1, must; 2, tumultuous AF; 3, final AF; 4, initial MLF; 5, tumultuous MLF; 6, final MLF) during three vintages.

3.3. Strain typing of *Oenococcus oeni*

3.3.1. PFGE analysis of the strains of this study

Identification of the *Oenococcus oeni* strains of this study was successfully achieved by PFGE of DNA digested with *Sfi*I and each strain presented a characteristic PFGE pattern. Digestions with *Apa*I enzyme were not more discriminating than *Sfi*I restriction (data not shown). Cluster analysis and visual inspection of the PFGE patterns from the 187 *Oenococcus oeni* isolates recovered from wine fermentations in the three years gave a total number of 37 distinct genotypes (Figure 4). Twenty-four of them were detected in 2006, five in 2007 and fifteen in 2008 vintages (Figure 5). The lower clonal diversity found in 2007 vintage would be related with the low temperatures during MLF (below 12 ºC) what could make few genotypes be able to get adapted to the difficult conditions.

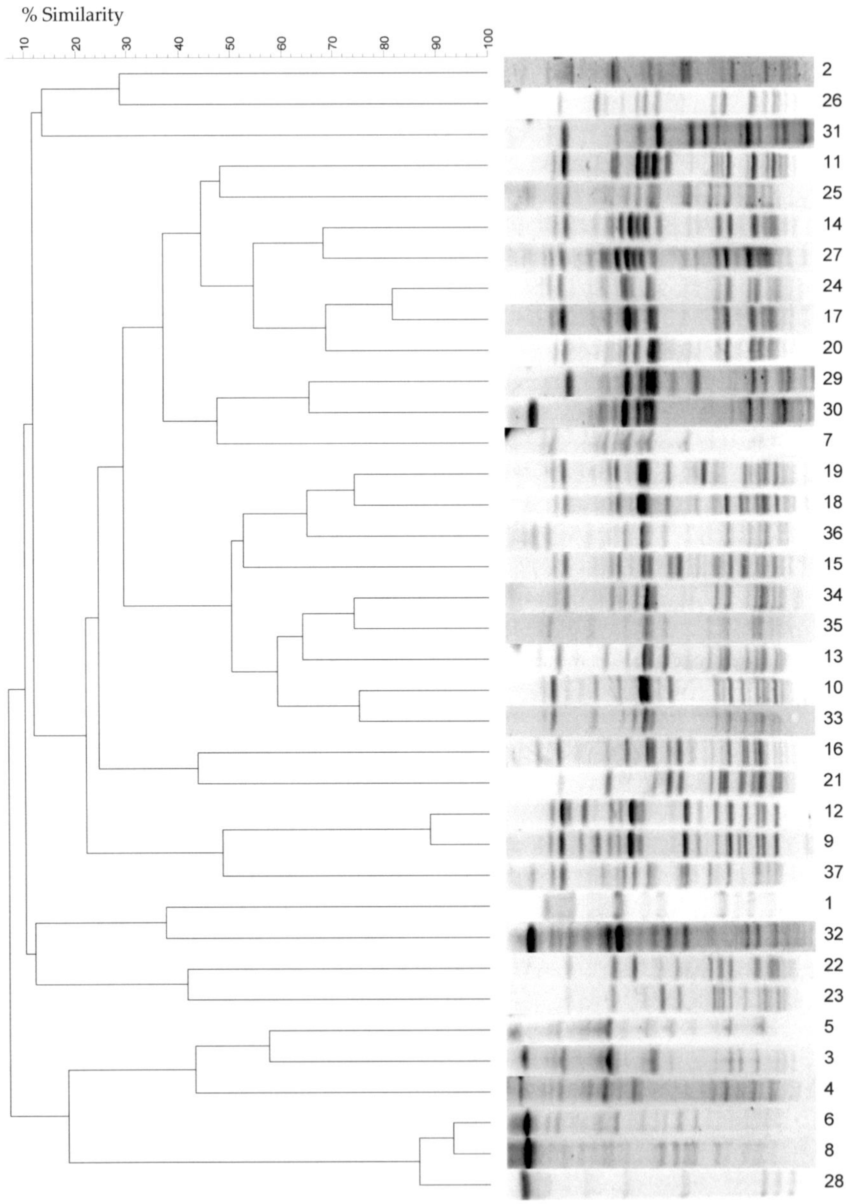

Figure 4. UPGMA dendrogram based on the *Sfi*I PFGE patterns of the 37 *Oenococcus oeni* genotypes. Figure from González-Arenzana et al. (2012a).

Comparing coincident genotypes for the three vintages, it was observed that between the genotypes isolated in 2006 vintage two were found in 2007 (genotypes 18 and 20) and four in 2008 (genotypes 3, 13, 17 and 18). Moreover, two genotypes isolated in 2007 (genotypes 18 and 25) were also detected in 2008 vintage. Only one genotype (18) was identified in the three studied years. The frequency of participation of each genotype varied from year to year, thus dominant genotypes one year were minority or not present at other one which suggested the adaptation of *Oenococcus oeni* strains to the winery conditions every year. Similar results were reported by other authors in studies about bacteria and yeast populations (Gutierrez et al., 1999; Reguant et al., 2005a; Ruiz et al., 2010; Santamaría, 2009).

Interestingly, no genotype was isolated in all fermentation stages so fourteen genotypes appeared only at AF (stages 1-3), six were present at all MLF stages (4, 5 and 6) and three of them were also detected at the end of AF. Most fermentation stages showed mixed *Oenococcus oeni* strains populations, which confirmed that several *Oenococcus oeni* strains occurred in a single spontaneous MLF (Lopez et al., 2007; Renouf et al., 2009; Ruiz et al., 2010). The number of different genotypes identified at each stage ranged from 0 to 5, and from 3 to 9, at stages 1-3 and 4-6, respectively. Genotypes 9 and 18 in 2006 vintage, 20 and 26 in 2007, and 13 in 2008 were the predominant ones during MLF. Three out of these genotypes (13, 18 and 20) could be considered as interesting *Oenococcus oeni* strains for the selection of new malolactic starter cultures as individual or mixed strains, because in addition to be dominant in most of the MLF stages at each vintage, they were isolated at more than one year in quality wines.

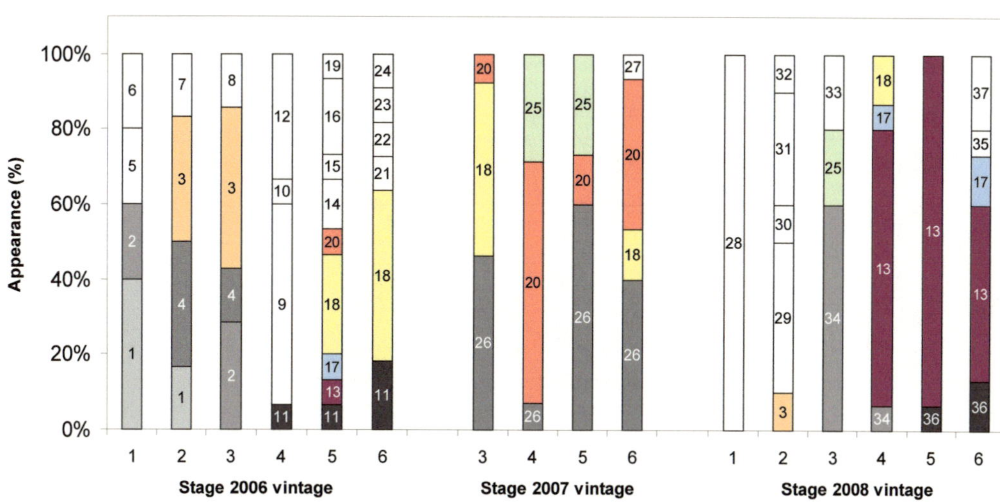

Figure 5. Frequency of appearance (%) of *Oenococcus oeni* genotypes at each stage (1, must; 2, tumultuous AF; 3, final AF; 4, initial MLF; 5, tumultuous MLF; 6, final MLF) and vintage. Genotypes without color only appeared once, genotypes with grey color appeared only in one year at more than one stage and genotypes with color appeared in more than one vintage.

3.3.2. Comparison of the PFGE and RAPD-PCR profiles from the wine fermentation strains and commercial strains

The thirty-seven genotypes of the indigenous *Oenococcus oeni* strains from wine fermentations and fourteen PFGE and RAPD-PCR patterns obtained of strains from several commercial cultures were submitted to comparison by bioinformatics and visual analysis (data not shown). Figure 6 shows that genotypes I, II, III and IV from commercial starter cultures resulted indistinguishable from four indigenous genotypes (23, 13, 21 and 18, respectively), despite commercial malolactic cultures had never been employed in the sampled cellar.

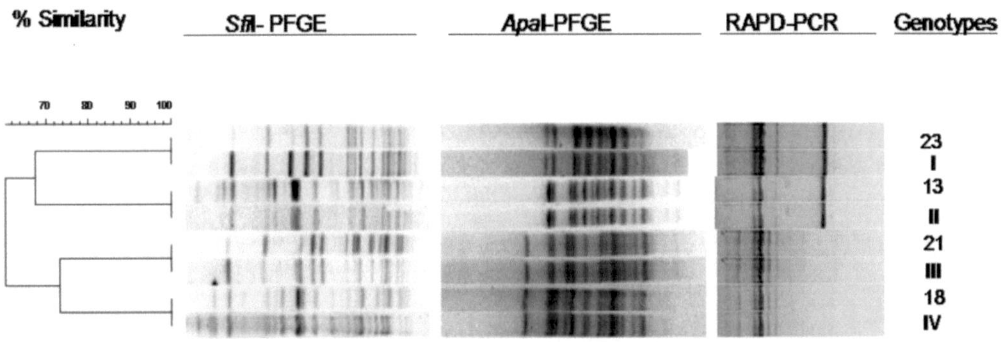

Figure 6. Consensus dendrogram obtained by combination *Sfi*I-PFGE, *Apa*I-PFGE and M13 RAPD-PCR patterns corresponding to the commercial strains (I to IV) and their respective indistinguishable genotypes from indigenous *Oenococcus oeni* strains from this winery.

These all four autochthonous genotypes were detected in 2006 vintage, year which showed the greatest strain diversity, two of them occurred in 2008 and one in 2007, with an important frequency of appearance in some cases (Table 3).

Genotype	Isolation stage	Frequency of appearance at each vintage (%)		
		2006	2007	2008
13	4-5-6	2	-	45
18	3-4-5-6	15	14	3
21	6	2	-	-
23	6	2	-	-

Table 3. *Oenococcus oeni* genotypes indistinguishable to commercial patterns; isolation stage (1, must; 2, tumultuous AF; 3, final AF; 4, initial MLF; 5, tumultuous MLF; 6, final MLF) and frequency (%) of their appearance (frequency of appearance (%) = nº of isolates that presented a specific PFGE pattern × 100/total nº of isolates per vintage) at each vintage.

Therefore, despite two of these strains had been previously considered as interesting for the selection of new malolactic starter cultures, the possible identical strain identification with already marketed strains suggested reject these two indigenous *Oenococcus oeni* isolates from a future selection process regardless of their oenological properties as malolactic starter cultures.

4. Conclusion

This study has been a contribution to a better description of the LAB ecology along the process of Tempranillo wines winemaking.

The study about the microbial diversity of viable LAB populations showed that the species diversity was higher at the AF stage where eight different species were identified. *Oenococcus oeni* was detected during AF in variable proportions and it became the majority species during spontaneous MLF.

This work allowed to increase the endogenous strain collection of LAB isolated from fermenting wines of the Appellation of Origin Rioja what meant a contribution to the preservation of biodiversity and wine peculiarity of this region and a starting point for future research.

The analysis of the total LAB populations by culture independent techniques (PCR-DGGE) showed that the species diversity detected along the winemaking process was higher than the one found by the study of viable LAB, identifying up to nine different LAB species. The LAB species variability was also higher at the previous stages to the MLF. Once spontaneous MLF started this variability was greatly reduced, with *Oenococcus oeni* and *Leuconostoc mesenteroides* as the only detected species.

The results obtained with culture dependent and independent techniques were complementary so in studies conducted in microbial ecology they both should be used to achieve a broader view of the studied ecosystem.

PFGE has shown to be a suitable method for strain differentiation, for monitoring individual strains and determining which strains actually survive and carry out MLF. The results of *Oenococcus oeni* typification indicated the high diversity of indigenous *Oenococcus oeni* strains responsible for MLF of the wines of this study and the complexity of the ecology involved in a fermentating wine. The frequency of participation of each genotype varied from year to year, thus dominant genotypes one year were minority or not present at other one, which suggested the adaptation of *Oenococcus oeni* strains to the winery conditions every year.

Several genotypes could be considered as interesting *Oenococcus oeni* strains for the selection of new malolactic starter cultures as individual or mixed strains because, in addition to be isolated at more than one year in quality wines, they were dominant in most of the MLF stages at each vintage. The comparison of the patterns from commercial cultures and the genotypes from indigenous *Oenococcus oeni* strains showed four indistinguishable genotypes. The presence of these four genotypes for one to three years, and in some cases with a high frequency of appearance, demonstrated the significance of this study in order to exclude these genotypes from a future selection process.

Author details

Lucía González-Arenzana, Rosa López, Pilar Santamaría and Isabel López-Alfaro*
ICVV, Instituto de Ciencias de la Vid y del Vino (Gobierno de La Rioja, Universidad de La Rioja and CSIC) C/ Madre de Dios 51, Logroño (La Rioja), Spain

Acknowledgement

This work was supported by funding and predoctoral grant (B.O.R. 6th March, 2009) of the Government of La Rioja, the I.N.I.A. project RTA2007-00104-00-00 and FEDER of the European Community and was made possible by the collaborating winery.

5. References

Altschul, S.F.; Gish, W.; Miller, W.; Myers, E.W. & Lipman, D.J. (1990). Basic local alignment Search Tool. *Journal of Molecular Biology,* Vol.215, No.3, (October 1990) pp. 403-410, ISSN 1041-4347.

Amann, R.; Ludwig, W. & Scheleifer, K. (1995). Phylogenetic Identification and In-situ detection of Individual Microbial-cells without Cultivation. *Microbiological Reviews,* Vol.59, No.1, (March 1995), pp.143-169, ISSN 0146-0749.

Beneduce, L.; Spano, G.; Vernile, A.; Tarantino, D. & Massa, S. (2004). Molecular Characterization of Lactic Acid Populations associated with Wine Spoilage. *Journal of Basic Microbiology,* Vol.44, No.1, pp. 10-16, ISSN 0233111X.

Birren, B. & Lai, E. (1993). *Pulsed field gel electrophoresis. A practical guide.* Academic Press, ISBN 0-12-101290-5, San Diego, United States.

Bisson, L.F. & Kunkee, R.E. (1993). Microbial Interactions during Wine Production, *In: Mixed cultures in biotechnology,* McGraw Hill, Zeikus, J.G. & Johnson (Eds.), 37-68, ISBN 0-07-072844-5pp., New York, USA.

Charteris, W.; Kelly, P.; Morelli, L. & Collins, J. (1997). Selective Detection, Enumeration and Identification of Potentially Probiotic *Lactobacillus* and *Bifidobacterium* Species in Mixed Bacterial Populations. *International Journal of Food Microbiology,* Vol.35, No.1, (March 1997), pp. 1-27, ISSN 0168-1605.

Cocconcelli, P.; Parisi, M.; Senini, L. & Bottazzi, V. (1997). Use of RAPD and 16S rDNA Sequencing for the Study of Lactobacillus Population Dynamics in Natural Whey Culture. *Letters in Applied Microbiology,* Vol.25, No.1, (July 1997), pp. 8-12, ISSN 0266-8254.

DanielL, P.; Dewhaele, E. & Hallet, J. (1993). Optimization of Transverse Alternating-field Electrophoresis for Strain Identification of *Leuconostoc oenos. Applied Microbiology and Biotechnology,* Vol.38, No.5, (February 1993), pp. 638-641, ISSN 0175-7598.

de las Rivas, B.; Marcobal, A. & Muñoz, R. (2006). Development of a Multilocus Sequence Typing method for analysis of *Lactobacillus plantarum* strains. *Microbiology-Sgm,* Vol.152, (January 2006), pp. 85-93, ISSN 1350-0872.

* Corresponding Author

de las Rivas, B.; Marcobal, A. & Muñoz, R. (2004). Allelic Diversity and Population Structure in *Oenococcus oeni* as Determined from Sequence Analysis of Housekeeping Genes. *Applied and Environmental Microbiology,* Vol.70, No.12, (December 2004), pp. 7210-7219, ISSN 0099-2240.

du Plessis, H.W.; Dicks, L.M.T.; Pretorius, I.S.; Lambrechts, M.G. & du Toit, M. (2004). Identification of Lactic Acid Bacteria Isolated from South African Brandy Base Wines. *International Journal of Food Microbiology,* Vol.91, No.1, (February 2004), pp. 19-29, ISSN 0168-1605.

European Community. (1990). Community Methods for the Analysis of Wine. Commission regulation no. 26/76/90.No. *Journal of European Community,* pp. 1-191.

Felsenstein, J. (1985). Confidence-limits on Phylogenies - an Approach using the Bootstrap. *Evolution,* Vol.39, No.4, (December 2004), pp. 783-791, ISSN 0014-3820.

Fugelsang, K. & Edwards, C. (2007). Microbial Ecology during Vinification. In: *Wine Microbiolgy: Practical Applications and Procedures,* Springer (Ed.), pp. 82-100, ISBN 978-0-387-33341-0, New York, USA.

Giraffa, G. & Neviani, E. (2001). DNA-based, Culture-independent Strategies for Evaluating Microbial Communities in Food-associated Ecosystems. *International Journal of Food Microbiology,* Vol.67, No.1-2, (July 2001), pp. 19-34, ISSN 0168-1605.

González-Arenzana, L.; Santamaría, P.; López, R.; Tenorio, C. & López-Alfaro, I. (2012a). Ecology of Indigenous Lactic Acid Bacteria along DifferentWinemaking Processes of Tempranillo Red Wine from La Rioja (Spain). *The Scientific World JOURNAL,* Vol.2012, (March 2012), pp. (7), ISSN 1537-744X.

González-Arenzana, L.; López, R.; Santamaría, P.; Tenorio, C. & López-Alfaro, I. (2012b). Dynamics of Indigenous Lactic Acid Bacteria Populations in Wine Fermentations from La Rioja (Spain) during Three Vintages. *Microbial Ecology,* Vol.63, (January 2012) pp. 1-8, ISSN 00953628.

Guerrini, S.; Bastianini, A.; Blaiotta, G.; Granchi, L.; Moschetti, G.; Coppola, S.; Romano, P. & Vincenzini, M. (2003). Phenotypic and Genotypic Characterization of *Oenococcus oeni* Strains Isolated from Italian Wines. *International Journal of Food Microbiology,* Vol.83, No.1, (May 2003), pp. 1-14, ISSN 0168-1605.

Gutiérrez, A.R.; Santamaría, P.; Epifanio, S.; Garijo, P. & López, R. (1999). Ecology of Spontaneous Fermentation in One Winery during 5 Consecutive Years. *Letters in Applied Microbiology,* Vol.29, No.6, (December 1999), pp. 411-415, ISSN 0266-8254.

Holt, J. G., Krieg, N. R., Sneath, P. H. A., Staley, J. Y., & Williams, S. T. (1994). *Bergey´s Manual of Determinative Bacteriology,* Lippincott Williams & Wilkins, ISBN 0-683-00603-7, Baltimore, USA.

Hugenholtz, P.; Goebel, B. & Pace, N. (1998). Impact of Culture-independent Studies on the Emerging Phylogenetic View of Bacterial Diversity. *Journal of Bacteriology,* Vol.180, No.18, (September 1998), pp. 4765-4774, ISSN 0021-9193.

Iacumin, L.; Cecchini, F.; Manzano, M.; Osualdini, M.; Boscolo, D.; Orlic, S. & Comi, G. (2009). Description of the Microflora of Sourdoughs by Culture-dependent and Culture-independent Methods. *Food Microbiology,* Vol.26, No.2, (April 2009), pp. 128-135, ISSN 0740-0020.

Kelly, W.; Huang, C. & Asmundson, R. (1993). Comparison of *Leuconostoc oenos* Strains by Pulsed-Field Gel-Electrophoresis. *Applied and Environmental Microbiology,* Vol.59, No.11, (November 1993), pp. 3969-3972, ISSN 0099-2240.

Lonvaud-Funel, A. (1999). Lactic Acid Bacteria in the Quality Improvement and Depreciation of Wine. *Antonie Van Leeuwenhoek International Journal of General and Molecular Microbiology,* Vol.76, No.1-4, (November 1999), pp. 317-331, ISSN 0003-6072.

López, I.; López, R.; Santamaría, P.; Torres, C. & Ruiz-Larrea, F. (2008). Performance of Malolactic Fermentation by Inoculation of Selected Lactobacillus plantarum and *Oenococcus oeni* Strains Isolated from Rioja Red Wines. *Vitis,* Vol.47, No.2, (February 2008), pp. 123-129, ISSN 0042-7500.

López, I.; Ruiz-Larrea, F.; Cocolin, L.; Orr, E.; Phister, T.; Marshall, M.; VanderGheynst, J. & Mills, D.A. (2003). Design and Evaluation of PCR Primers for Analysis of Bacterial Populations in Wine by Denaturing Gradient Gel Electrophoresis. *Applied and Environmental Microbiology,* Vol.69, No.11, (November 2003), pp. 6801-6807, ISSN 0099-2240.

López, I.; Torres, C. & Ruiz-Larrea, F. (2008). Genetic Typification by Pulsed-Field Gel Electrophoresis (PFGE) and Randomly Amplified Polymorphic DNA (RAPD) of Wild *Lactobacillus plantarum* and *Oenococcus oeni* Wine Strains. *European Food Research and Technology,* Vol.227, No.2, (June 2008), pp. 547-555, ISSN 1438-2377.

López, I.; Tenorio, C.; Zarazaga, M.; Dizy, M.; Torres, C. & Ruiz-Larrea, F. (2007). Evidence of Mixed Wild Populations of *Oenococcus oeni* Strains during Wine Spontaneous Malolactic Fermentations. *European Food Research and Technology,* Vol.226, No.1-2, (November 2007), pp. 215-223, ISSN 1438-2377.

Lucore, L.; Cullison, M. & Jaykus, L. (2000). Immobilization with Metal Hydroxides as a Means to Concentrate Food-borne Bacteria for Detection by Cultural and Molecular ;ethods. *Applied and Environmental Microbiology,* Vol.66, No.5, (May 2000), pp. 1769-1776, ISSN 0099-2240.

Maicas, S.; González-Cabo, P.; Ferrer, S. & Pardo, I. (1999a). Production of *Oenococcus oeni* Biomass to Induce Malolactic Fermentation in Wine by Control of pH and Substrate Addition. *Biotechnology Letters,* Vol.21, No.4, (April 1999), pp. 349-353, ISSN 0141-5492.

Maicas, S.; Pardo, I. & Ferrer, S. (1999b). Continuous Malolactic Fermentation in Red Wine using Free *Oenococcus oeni. World Journal of Microbiology & Biotechnology,* Vol.15, No.6, (December 1999), pp. 737-739, ISSN 0959-3993.

Muyzer, G. & Smalla, K. (1998). Application of Denaturing Gradient Gel Electrophoresis (DGGE) and Temperature Gradient Gel Electrophoresis (TGGE) in Microbial Ecology. *Antonie Van Leeuwenhoek International Journal of General and Molecular Microbiology,* Vol.73, No.1, (January 1998), pp. 127-141, ISSN 0003-6072.

Parry, C. E., Zink, R. & Mills, D. A. (2002). *Oenococcus oeni* Strain Differentiation by rep-PCR. *American Journal of Enology and Viticulture,* Vol.53, No. 3, pp.253A, ISSN 0002-9254.

Prevost, H.; Cavin, J.; Lamoreux, M. & Divies, C. (1995). Plasmid and Chromosome Characterization of *Leuconostoc oenos* Strains. *American Journal of Enology and Viticulture,* Vol.46, No.1, (August 1993), pp. 43-48, ISSN 0002-9254.

Reguant, C.; Carreté, R.; Constanti, M. & Bordons, A. (2005a). Population Dynamics of Oenococcus oeni Strains in a New Winery and the Effect of SO_2 and Yeast Strain. *FEMS Microbiology Letters,* Vol.246, No.1, (May 2005), pp. 111-117, ISSN 0378-1097.

Reguant, C.; Carreté, R.; Ferrer, N. & Bordons, A. (2005b). Molecular Analysis of *Oenococcus oeni* Population Dynamics and the Effect of Aeration and Temperature during Alcoholic Fermentation on Malolactic Fermentation. *International Journal of Food Science and Technology,* Vol.40, No.4, (April 2005), pp. 451-459, ISSN 0950-5423.

Renouf, V.; Claisse, O.; Miot-Sertier, C. & Lonvaud-Funel, A. (2006). Lactic Acid Bacteria Evolution during Winemaking: Use of *rpo*B Gene as a Target for PCR-DGGE Analysis. *Food Microbiology,* Vol.23, No.2, (April 2006), pp. 136-145, ISSN 0740-0020.

Renouf, V.; Vayssieres, L.C.; Claisse, O. & Lonvaud-Funel, A. (2009). Genetic and Phenotypic Evidence for Two groups of *Oenococcus oeni* Strains and their Prevalence during Winemaking. *Applied Microbiology and Biotechnology,* Vol.83, No.1, (May 2009), pp. 85-97, ISSN 0175-7598.

Ribéreau-Gayon, P. Dubourdieu, D. Donèche & B. Lonvaud, A. (Ed.). (2006). Handbook of enology, John Wiley & Sons, ISBN-13: 978-0955706905, New York, USA.

Ruiz, P.; Izquierdo, P.M.; Seseña, S. & Palop, M.Ll. (2010). Selection of Autochthonous *Oenococcus oeni* Strains according to their Oenological Properties and Vinification Results. *International Journal of Food Microbiology,* Vol.137, No.2-3, (February 2010), pp. 230-235, ISSN 0168-1605.

Ruiz, P.; Izquierdo, P.M.; Seseña, S. & Palop, M. Ll. (2008). Intraspecific Genetic Diversity of Lactic Acid Bacteria from Malolactic Fermentation of Cencibel Wines as derived from Combined Analysis of RAPD-PCR and PFGE Patterns. *Food Microbiology,* Vol.25, No.7, (October 2008), pp. 942-948, ISSN 0740-0020.

Saitou, N. & Nei, M. (1987). The Neighbor-Joining Method - a New Method for Reconstructing Phylogenetic trees. *Molecular Biology and Evolution,* Vol.4, No.4, (July 1987), pp. 406-425, ISSN 0737-4038.

Santamaría Aquilúe, P. (2009). *Ecología de la Fermentación Alcohólica en la D.O. C.A. Rioja: Selección de Levaduras para la Elaboración de Vinos Tintos.* (Unpublished. Universidad de La Rioja,

Sato, H.; Yanagida, F.; Shinohara, T.; Suzuki, M.; Suzuki, K. & Yokotsuka, K. (2001). Intraspecific Diversity of *Oenococcus oeni* Isolated during Red Wine-making in Japan. *FEMS Microbiology Letters,* Vol.202, No.1, (August 2001), pp. 109-114, ISSN 0378-1097.

Spano, G.; Beneduce, L.; Tarantino, D.; Zapparoli, G. & Massa, S. (2002). Characterization of *Lactobacillus plantarum* from Wine Must by PCR Species-specific and RAPD-PCR. *Letters in Applied Microbiology,* Vol.35, No.5, (June 2002) pp. 370-374, ISSN 0266-8254.

Tamura, K.; Dudley, J.; Nei, M. & Kumar, S. (2007). MEGA4: Molecular Evolutionary Genetics Analysis (MEGA) Software Version 4.0. *Molecular Biology and Evolution,* Vol.24, No.8, (August 2007), pp. 1596-1599, ISSN 0737-4038.

Tenreiro, R.; Santos, M.; Paveia, H. & Vieira, G. (1994). Inter-strain Relationships Among Wine Leuconostocs and their Divergence from other Leuconostoc Species, as Revealed by Low-Frequency Restriction Fragment Analysis of Genomic DNA. *Journal of Applied Bacteriology,* Vol.77, No.3, (September 1994), pp. 271-280, ISSN 0021-8847.

Viti, C.; Giovannetti, L.; Granchi, L. & Ventura, S. (1996). Species Attribution and Strain Typing of *Oenococcus oeni* (formerly *Leuconostoc oenos*) with Restriction Endonuclease Fingerprints. *Research in Microbiology,* Vol.147, No.8, (October 1996), pp. 651-660, ISSN 0923-2508.

Zapparoli, G.; Reguant, C.; Bordons, A.; Torriani, S. & Dellaglio, F. (2000). Genomic DNA Fingerprinting of *Oenococcus oeni* Strains by Pulsed-Field Gel Electrophoresis and Randomly Amplified Polymorphic DNA-PCR. *Current Microbiology,* Vol.40, No.6, (June 2000), pp. 351-355, ISSN 0343-8651.

Zapparoli, G.; Torriani, S.; Pesente, P. & Dellaglio, F. (1998). Design and Evaluation of Malolactic Enzyme Gene Targeted Primers for Rapid Identification and Detection of *Oenococcus oeni* in Wine. *Letters in Applied Microbiology,* Vol.27, No.5, (November 1998), pp. 243-246, ISSN 0266-8254.

Zavaleta, A.; Martínez-Murcia, A. & Rodríguez-Valera, F. (1997). Intraspecific Genetic Diversity of *Oenococcus oeni* as derived from DNA Fingerprinting and Sequence Analyses. *Applied and Environmental Microbiology,* Vol.63, No.4, (April 1997), pp. 1261-1267, ISSN 0099-2240.

Chapter 9

Isoenzyme Analyses Tools Used Long Time in Forest Science

Malgorzata K. Sulkowska

Additional information is available at the end of the chapter

http://dx.doi.org/10.5772/45756

1. Introduction

1.1. What kind of tool are isoenzymes

A specific group of proteins – enzymes, characterized by similar chemical properties are isoenzymes (Hunter & Markert, 1957, Tanskley & Orton, 1983, Goncharenko, 1988, Mejnartowicz, 1990, Sułkowska, 1995). They are characterized by similar chemical, structural and catalytic properties almost identical within their functional group. They are characterized the same function in cell reactions, however they are products of different genes (described as allozymes).

Their origin is explained by gene duplication of polyploidisation mutation, which can create kinds of pseudogene – or nucleic acid hybrydisation. Different forms of an enzyme that are coded by variant alleles at the same locus are called allozymes, while isozymes are products coded by genes located at different loci. Even then, isozymes and allozymes are variants of the same genes and they possess the same catalytic functions they can be easily identified in biochemical way. First of all the substitutions of different amino acids are responsible for their variant electric charge and this can be the way how to identify them by electrophoresis process. This charge characteristics of isozymes and allozymes is the basis to use them as molecular markers.

An important feature of this group of proteins is that the presence of both alleles of a gene is disclosed in a manner independent of each other so cold co-dominant character, so it is possible to request a degree of heterozygosity within the study population characteristics (Rider & Taylor, 1980). In practice, this means that there is a simple way to distinguish heterozygotes from homozygotes. The isoenzyme molecules are proteins. It is important to define them as the first products of gene translation, so they possess information about their coding region of DNA. Their heredity of responsible genes is explained by Mendelian character. A large number of genes encoding enzymes shows a significant degree of polymorphism. The process of natural selection is still important factor, because genes are specialised to different functions.

Long time in forest sciences, the isoenzyme markers were the best tools to analyse genetic variation of populations despite of the different limits and restrictions of this method. Nowadays we have more informative tools based on DNA markers such as sequencing, microsatellites, PCR-RFLP and single genes analysis like SNPs. Most of this DNA based markers is still developed to make them more suitable in analysis of plant organisms. The level of complication of plant genome is higher comparing to animal. Average of size of *Pinus* genome is about 30 billion base pair (Grotkopp et al., 2004). We have up to now many possibilities to analyse the polymorphism of DNA fragments of plant material but not the particular genes. So, the isoenzyme markers represent still one of the best markers close to DNA level. It is possible to assess the variation of individuals at different level: within species, within population, and among populations within species. It is worth to add they are quick and cheap marker systems and good alternative to assay and identify level of genetic variation as pilot study of populations (Bakshi & Könnert, 2011) as well as conservation biology activities e.g. gene bank – enables choosing of proper sample for long time conservation (Bednorz et al., 2006) or as well as in quantifying mating system analysis (Mrazikowa & Paule, 1990).

The proteins of enzyme are consisted of polypeptides and are molecules characterised the specific conformation. The quaternary structure of particular enzyme forms may be different in relation to combination of peptides involved to build their molecules. We can observed (Figure 1) monomeric enzymes with two the same allozymes, dimeric forms with three allozymes and tetrameric proteins with fives allozymes. We can sometimes observe on gel after electrophoresis bands of some intermediate mobility reaction.

The isoenzyme analysis method enables the assessment of variability of isoenzymes in different types of tissues: young leaves, buds, pollen and seeds. There is possibilities to analyse diploid as well as haploid tissue. Especially it is important, when we study coniferous species genetic variation, then it is great tool to analyse mating system of trees, because of possibilities of using haploid tissue of mother trees and diploid tissue of embryos. The analysis requires very small amounts of plant material and is a very sensitive method. Simultaneously, it is possible to identify many samples/or individuals. The great advantage of this method is low cost of chemicals used to performed the studies.

1.2. Advantages of isoenzymes and week points of the utilisation

The using of isoenzyme electrophoresis method is useful tool especially in assessment of gene frequency of specific genes, determining of genetic similarities and genetic distances between the two objects e.g.:

- Good tools to support conservation and management of forest trees genetic resources
    - To characterize forest tree stands genetic structure

  - To asses initial gene pole
  - To present selection processes of forest stands and to maintain rich of natural diversity of stands
- Genetic characteristics of different forest tree species reproductive material:
  - Mother stands and progeny stands gene flow analysis
  - Seed orchards and progeny plantations mating system
  - The representation of populations selected for preservation in gene banks or in situ or ex situ measures
- Solving of problems from seed stands management point of view:
  - clonal/pedigree identification/selection process
  - pollen contamination especially important in management of artificial tree stands like forest seed plantations
  - patterns of gene flow and mating system in natural and artificial stands

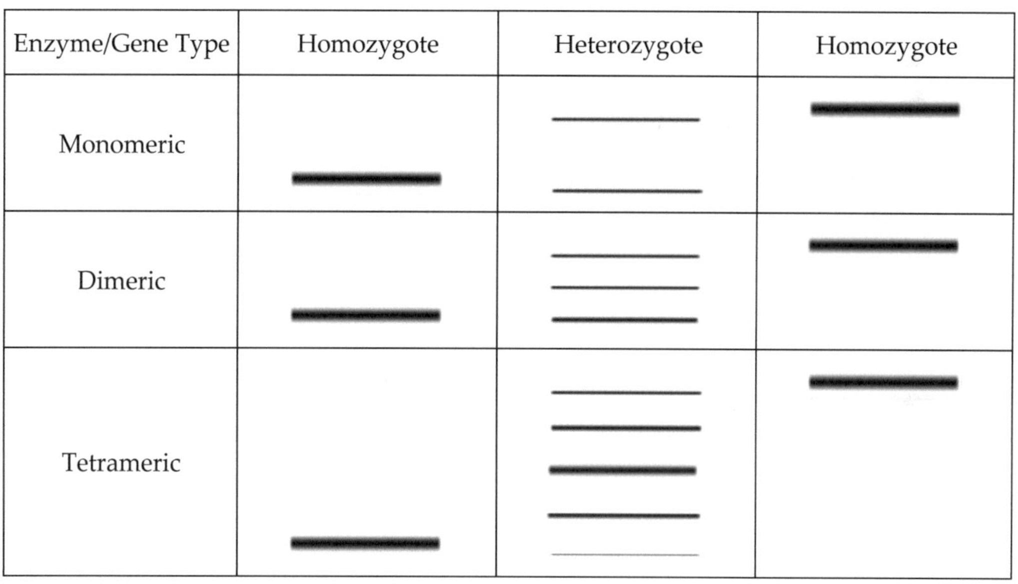

Figure 1. The graphic illustration of bands of proteins on the gel after electrophoresis process when we study various types of isoenzyme proteins.

1.3. Methodical problems

The isoenzyme studies are suitable only when their heritability is explained by Mendelian character – often the same enzyme systems are not suitable when we study different species. The enzyme loci are not randomly distributed over the genome, so they are not representative for total genome variation (Hubby & Lewontin, 1966). We have problems to compare analysed data with reference studies when the number of analysed enzyme system is different. At the first studies some enzyme systems were described by different numbers enzyme gene loci systems e.g. peroxidases or number of loci controlling analysed enzyme. For example GOT (transaminase glutamine oxylo-acetate) was assumed to be coded by 2 or

3 or 5 gene loci for Scots pine (Müller-Starck et al., 1992). Some alleles are not identify as the bands on the gel (null alleles). It is not possible to use particular enzyme markers when linkage disequilibrium is occurred for analysed loci, so it obvious to verify the genetic control system and inheritance of chosen enzyme markers.

Still seems to opened the question are allozyme neutral or adaptive?

Some researchers showed the results studies only for polymorphic enzyme systems without any information about monomorphic ones, so the genetic variation was described as higher than it was real.

The obtained information about isoenzyme variation is not a representative sample of the total variability of the analyzed species, which illustrates the genetic variability within the ranges of their occurrence, due to too small amount of tested samples. Evaluation of variation by this method applies only to a very small part of the genome.

2. Methods

2.1. The idea of isoenzyme electrophoresis

The charge characteristics of isoenzymes enables to use them as genetic markers, which can be distingished. First of all the substitutions of different amino acids are responsible for their variant electric charge and this can be the way how to identify them by electrophoresis process. The identification of proteins molecule characterised a specific electric charge can be dissolved in a buffering solution, after homogenisation of plant or animal tissue. The non-denatured proteins are separated on the carrier gel and they migrate under the influence of an applied electric field. The molecules of enzyme proteins can change conformation shape and it will be followed by modification of net charge. This changes of molecules charge can be detect using electrophoresis method.

The rate of migration of molecules in the gel and segregation of bands of enzymes in the gel is the result of interaction of: the applied electric field, pH of gel and electrode buffers (Concle et al., 1982). After completion of electrophoresis on gel plate, the proteins are stained by using the appropriate staining reactions.

2.2. Migration process of proteins in different media and their visualization possibilities

The carrier of proteins in electrophoresis may be different types of substances: polyacrylamide gels and potato starch gels. The second one were historically used, but the great advantage of them still is the possibility of cutting them to obtain slices, which can be analise to identify different protein markers using the same gel and tissue material, because all the proteins are active in the gel.

Proteins separated in the gel during electrophoresis can be detected by staining or UV analysis e.g. fluorescent esterase enzyme. Methods of detection of enzymes are relatively well developed and reported in many publications (Concle et al., 1982, Goncharenko, 1988, Wang & Szmidt, 1989, Liengsiri et al., 1990). The visualisation of enzymes is possible by precipitation of soluble indicators like tetrazolium salts using cofactors of NAD or NADP to transfer them into reduced insoluble forms (Figure 2). It is important to put attention that the staining reactions require the active forms of enzymes and they simulate work of proteins in functional tissues of organisms. Most of reactions are performed in the temperature 37 °C. During relatively short time one day of laboratory work we have result and evaluation of the analysis for about 50 samples e.g. individuals using 10 or more enzyme systems represented by one or more loci each of them. It is important as well that the staining process does not requires complicated laboratory equipment. The resolving of proteins during electrophoresis can be performed in horizontal or vertical chambers, but the crucial is to protect the activity of enzymes by providing of cooling system. The power of electric field applied for migration of proteins is often up to 280V, so the heating of gel is observed and have to be reduced.

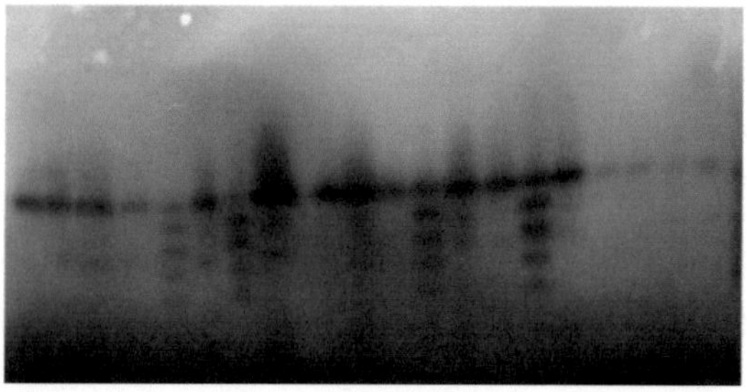

Figure 2. Menadione reductase electrophoretic pattern in European beech. We can observed The monomeric enzyme form with two the same allozymes (homozygote form), dimeric form with three allozymes (heterozygote form) and tetrameric proteins with fives allozymes (heterozygote form). Photo M. Sulkowska

2.3. Genetic differentiation evaluation process

2.3.1. Analysed parameters to describe population genetic variation and differentiation

The basic parameters evaluated to measure population genetic variation and differentiation are usually as it follows: number of alleles per locus, heterozygosity (*H*), percentage of polymorphic loci, genetic distance (G_{ST}). The terms of heterozygosity, diversity and differentiation are explained in different way by various researchers. Most of people used to work with genetic parameters elaborated by Nei (1973, 1978), when we consider heterozygosity observed or expected, what is estimated as well as a measure of genetic diversity heterozygosity observed (*Ho*), heterozygosity expected (*He* – or *D* – gene diversity). The value of this parameter is calculated as values from zero (no heterozygosity) up to nearly 1.0 (when we observe a large number of nearly equally frequent alleles). Instead of average number of alleles per locus more precise used to be measure of effective number of alleles per locus (n_e) – Crow & Kimura (1970).

$$He = 1 - \sum_{i=1}^{n} pi^2 \tag{1}$$

$$n_e = \frac{1}{\sum_{i=1}^{n} pi^2} \tag{2}$$

p_i – frequency of n allele occurrence in population

Heterozygosity is often one of the most important "parameters" when we describe the genetic data. Using this measure we explain the general trends in the structure of analysed populations – even is their history and future genetic structure is concerned. Low values of heterozygosity is influenced by small population size and processes of genetic drift e.g. bottlenecks effect. A lot of heterozygotes in population is equal high genetic variability and the opposite. When we compare the level of the observed and expected heterozygosity in balanced populations concerning random and open mating system it means under Hardy-Weinberg equilibrium and heterozygosity is higher than expected we can expect flow of alien pollen outside of population. If the observed heterozygosity is lower than expected we can assume that some inbreeding processes may occur in the population.

The interpopulational variation is described very often as G_{ST} (Nei, 1973) and it is used as equivalent of F_{ST} statistics (1969, 1978) and it enables to asses for each population the distance from other populations.

$$G_{ST} = \frac{H_T - H_S}{H_T} \tag{3}$$

H_T - heterozygosity interpopulational
H_S - heterozygosity intrapopulational

$$F_{IT} = (H_T - H_I)/H_T \tag{4}$$

F_{ST}, cold as fixation index is the measure of proportion of the total genetic variance within subpopulations in relative to the total genetic variance. The values of this parameter can range from 0 to 1. High F_{ST} implies a considerable degree of differentiation among populations.

F_{IS} (inbreeding coefficient) is the proportion of the variance in the subpopulation contained in an individual. High F_{IS} implies a considerable degree of inbreeding. Values can range from -1 (outbred) to +1 (inbred).

$$F_{ST} = (H_T - H_S)/H_T \tag{5}$$

$$F_{IS} = (H_S - H_I)/H_S \tag{6}$$

H_T - heterozygosity total for population
H_S - heterozygosity within subpopulation
H_I - heterozygosity of individual

2.3.2. *Software*

One of the oldest programs enabling computing of isoenzyme analysis data is BIOSYS-1. This program was elaborated to help biochemical population geneticists to describe the analysis of electrophoretically detectable allelic variation. It can be utilised to study allele frequencies and genetic variability measures, to test for deviation of genotype frequencies from Hardy-Weinberg, expectations, to calculate F-statistics, to perform heterogeneity of chi-square analysis, to calculate a variety of similarity and distance coefficients, and to construct dendrograms using among others cluster analysis procedures. The program, documentation, and test data are available from the authors (Swofford & Selander, 1981).

Another one interesting software enabling as well analysing of DNA data markers is POPGENE (Yeh & Boyle, 1997). The current version of POPGENE is designed specifically for the analysis of co-dominant and dominant markers using haploid and diploid data. It performs most types of data analysis encountered in population genetics and related fields. It can be used to compute summary statistics, including: allele frequency: estimates gene frequencies at each locus from raw data, effective number of alleles per locus, percentage of all loci that are polymorphic, observed homozygosity, expected homozygosity, Shannon Index, gene diversity Nei's (1973), F-Statistics, Gene Flow: estimates gene flow from the estimate of G_{ST} or F_{ST} and many others parameters.

POPGENE is a good tool in analysing and simulations studies of population genetics, including: Hardy-Weinberg Equilibrium, multiple allele and loci inheritance, natural selection, genetic drift, migration, mutation and inbreeding.

Applied procedure of isoenzyme analysis is consisted of the following steps:

- Sample preparation (collecting of samples, homogenization and extraction of proteins from the tissue)
- Preparation of gels and running buffers requirements regarding analysed enzyme systems
- Development of isoenzyme electrophoresis
- Detection and staining of proteins
- Computing of the obtained data base with utilisation of proper software

3. Results

3.1. Genetic variation characteristics

The investigations of beech variation and differentiation in Europe showed (Gömöry et al., 2003, Sułkowska et al., 2012). The aim of this study was the assessment of genetic diversity and differentiation patterns of European beech (*Fagus sylvatica* L.) populations within its natural range in Poland and to compare them to those in other neighbouring European countries including Slovakia, the Czech Republic, Ukraine, and even Romania, which was reported previously (Paule et al. 1995). These stands cover 5.2% of the forest area in Poland, and form the predominant forest tree communities throughout the Carpathians and Sudety Mountains, and the moraine landscape of the Pomeranian Lake District. Varying environmental conditions have resulted in a great number of ecotypes and populations which are characterised by various ecological requirements. Poland represents the northeastern limit of the beech's natural distribution. Genetic diversity and differentiation was assessed using allozyme gene markers employing 9 enzyme systems: glutamate-oxaloacetate transaminase (GOT - EC 2.6.1.1), leucine aminopeptidase (LAP- EC 3.4.11.1), isocitrate dehydrogenase (IDH - EC 1.1.1.42), malate dehydrogenase (MDH - EC 1.1.1.37), menadione reductase (MNR - EC 1.6.99.2), phosphoglucomutase (PGM - EC 2.7.5.1), phosphoglucose isomerase (PGI - EC 5.3.1.9), peroxydase (PX - EC 1.11.17) and shikimate dehydrogenase (SKDH - EC 1.1.1.25). The data revealed: high genetic diversity of beech, similar like in other neighboring European populations, slight decrease of average number of alleles per locus and level of differentiation towards the North of the natural range limit, which generally confirm the migration paths after glaciations but it is not the basis to distinguish geographic regions.

The population differentiation of beech provenances of selected seed stands and their progenies for chosen genetic parameters and on the basis of soil characteristics of the habitats were studied (Sulkowska et al., 2008, Sulkowska, 2010). Beech populations occurring toward the northeast of the natural range were characterised by a decreasing the average number of alleles per locus and percentage of polymorphic loci. The highest genetic differentiation was found in the East Carpathians.

3.2. Ecotype variation characteristics

3.2.1. Geographic trends with an example of coniferous species - Scots Pine (Pinus sylvestris L.)

Allozyme differentiation in chosen European populations of Scots Pine (*Pinus sylvestris* L.) were studied in 17 populations from North and East-Central Europe (Prus-Głowacki et al., 1993). The samples were collected from provenance trial in Lubień (Poland). The provenances from Scandinavia, northern Poland, Netherlands and Belgium were more heterozygotic, more polymorphic and characterised higher number of alleles per locus. The source of seeds used to establish the provenance trial was unknown as far their autochthonous or introduced origin is concerned, but the results indicate a degree of coincident agreement with geographical distribution of stands. This effect seems to be blubbered by human activity (uncontrolled seed transfer).

The studies of 11 enzyme systems concerned on the genetic variation of *Pinus sylvestris* from Spain in relation to other European populations revealed genetic dissimilarity of populations from this region. The differences were observed as far as slightly higher number of alleles per locus, but lower heterozygosity level in populations from Spain (Prus-Głowacki & Stephan, 1994).

In Poland, the isoenzyme studies of 5 systems variability of 8 populations revealed existence of two groups of populations – North and South groups (Krzakowa, 1979). There was a high variability within all analysed loci. The importance of existence of this groups was undertaken as well by field studies and low regulations (Dz.U. 04, nr 67, poz. 621, 2004, Matras, 2005).

3.2.2. Site plasticity with an example of deciduous species – European beech (Fagus sylvatica L.

Present genetic structure of beech population in Europe was formed many different factors not only environmental and genetic but also anthropogenic. Different environmental condition resulted in great number of ecotypes and populations, that characterized various ecological requirements (Dzwonko, 1990, Giertych, 1990). Very important factors affected the gene pool were glacial epoch and the location of beech refugia, for postglacial migration paths of species (Szafer, 1935, Huntley & Birks, 1983, Ralska-Jasiewiczowa, 1983, Hazler et al., 1997). One of the first studies on genetic variability of isoenzymes of European beech - *Fagus sylvatica* L. were conducted in France. The study of genes encoding peroxidase system. There was a relationship between the frequencies of particular alleles encoding these enzymes, and analyzed population, geographical and environmental factors (Cugen et al., 1985). Genetic variation of beech - *Fagus sylvatica* L., was also analyzed on a wider scale within the six enzyme loci, 130 population for southern and western Europe. Observed correlation between the frequency of alleles of genes encoding peroxidase and climate (Comps, et al., 1990).

Some genotypes are eliminated during natural selection process, when they are not efficient in the environment, it is shown using isoenzyme markers (Müller-Starck, 1985). In most case populations characterised higher level of genetic variation seems to be more tolerant to harmful environmental factors (Starke et al., 1996).

In a continuous process of verifying the adaptation of individuals, which occurs in nature, it is contributed to both human and natural selection. This was proven among other by differences in the genetic structure of two provenances of beech German and Romanian, grown in a greenhouse and natural conditions (Kim, 1985). It was found that beech seedlings with *Lap-A2* allele always were characterised a higher survival. Homozygous for the allele *Lap-A2* survived better in a greenhouse, while heterozygotes were characterized by higher vitality in the natural environment.

3.3. Using of isoenzyme markers in coniferous seed orchard research

Isoenzyme markers are very important tool concerning genetic parental or progeny identity (Wheeler & Jech, 1992). The estimation of gene flow in mating system of seed orchards is crucial for proper use of seeds to asses and if possible avoid pollen contamination outside of the stands and seed orchard genetic efficiency. It plays important role for gene conservation of the stands and the improvement in forest tree selection (Concle, 1972, Rudin & Lingren, 1977).

The investigations of 122 trees in seed stand in Sweden were conducted for adult trees, embryos of seeds and progeny of the stand (Yazdani et al., 1985). The analyses revealed significant variation among different groups of studied objects at 5 enzyme loci. Genetic frequencies of alleles were close to Hardy-Weinberg equilibrium, but the deviations were found for embryos.

In Poland, also featured on the study of inheritance of a certain enzyme - GOT (transaminase glutamino-oxylo-acetate) for selected two homozygous plus trees, under the terms of this gene, and the pollination of these trees by surrounding neighbors. The study provided evidence of contamination at least 40% of the seeds with pollen outside the stands, which indicates a high out-crossing rate for pine (Krzakowa, 1980).

In the seed orchard in Slovakia, the mating system of trees was analysed on the basis of five enzymes studies inheritance (Mrazikowa & Paule, 1990). The study was conducted simultaneously for macrogametophytes of embryos and seeds. The degree of foreign seed pollination with pollen from outside the plantation was shown.

4. Discussion

The isoenzyme markers are important tool in when we assay gene variability of forest trees. The complicity of genome of this organism makes it impossible in most cases to obtain information about particular genes. Nowadays, molecular markers are powerful tools, which enables to study genetic variation of the organism, but we are usually able to work with specific fragments DNA, not to assay the genes. The methods biochemical and molecular should be taken into account in case of genetic variability analyses of forest trees as complementary tools. It can be revealed by studies of migration paths of European beech using both types of markers (Magri et. al., 2006), where existence of one common refugium of the species was proved for most part of Central and West Europe.

Single genes are responsible to control proteins of particular features – genetic traits (Bergmann, 1991). They are not responsible to control the complex of morphological traits, physiological and adaptability of individuals as reaction of needs of environment, that are crucial for surviving of rooted in one site plants over their whole life.

Ecotype variation of forest tree species can be classified as relation to their geographic range (Müller-Starck et al., 1992):

- Species characterised large geographic range like Scots pine, Norway spruce and European beech and little genetic differentiation among populations within regions derived from the same refugia, but greater comparing various refugia
- Species with large geographic ranges, but with many subspecies like Pine species – *Pinus nigra, Pinus halepensis* with small interpopulational variation within subspecies, but differentiation among subspecies

- Species characterised small geographically ranges like fir but with great interpopulational differentiation and medium level of intrapopulational genetic variation – endemic species
- Species with extremely small geographic range like Siberian dwarf pine characterised relatively high interpopulational differentiation – relic species.

Using of isoenzyme studies as one of DNA responsible marker to solve the problems of reproductive and economic value of seed orchard stands is very crucial (Wheeler & Jech, 1992, Krzakowa, 1980).

A especially interesting seems to be using of isoenzyme analysis in estimation of gene flow in natural and artificial populations of forest trees, when the genetic values of artificial management stands is taking into account (Savolainen & Yazdani, 1991, Skrøpa, 1994). The reported by authors genetic diversities as well the level of outcrossing rates estimated on the basis of allozymes differentiation were comparable in natural and artificial stands. However, that was underlined the importance of possible changes in the important quantitative traits not revealed by neutral enzyme markers. It was undertaken (Skrøpa, 1994) the meaning of many aspects important for proper quality production of forest reproductive material e.g.: seed collection procedure, seedlings sylviculture management, progeny testing.

Low costs of the analyses are the reason why isoenzyme markers are good tools in pilot studies of gene pool, as well as conservation biology activities e.g. gene bank – enables choosing of proper sample for long time conservation. Using of isoenzyme analysis was the step to assess the genetic variation for *Sorbus torminalis* L. Krantz. natural populations in Poland (Bednorz et al. 2006), what was the basis to establish in the progeny stands in next step, as ex-situ measures for the species.

The present selection processes of forest stands should to maintain richness of natural diversity and do not allowed to use the trees with high economic as in of case natural populations of *Pinus wallichiana* A.B. Jacks (Blue Pine) in India (Bakshi & Könnert 2011).

5. Conclusions

The isoenzyme molecules are proteins, which are defined as first product of the first products coding region of DNA activity. Their heredity is known, when it is explained by Mendelian character of segregation it is possible to utilise as genetic markers.

Application of isoenzyme electrophoresis method is useful tool in forest trees genetic diversity assessment, in spite of their long history of their utilization, elaboration of DNA analysis markers as well as known limits of their possibilities to apply.

Especially the possibilities to use them on wild scale, because of low costs of analysis makes them important tools in:

- Genetic characteristics of different forest reproductive material of natural and artificial stands e.g.

- Mother stands and progeny stands gene flow
- Seed orchards gene flow
- Gene banks representation of populations assessment
- Solving of particular problems in case of:
  - clonal/pedigree identification/selection process
  - pollen contamination
  - mating system
  - patterns of gene flow
- Good tools to support conservation and management of forest genetic resources e.g. to support following activities:
  - identification of migration path of species from postglacial refugia
  - selection and protection of ecotypes
  - to asses initial gene pole for needs of effective gene conservation measures
  - to present selection processes of forest stands to maintain rich natural diversity.

Author details

Malgorzata K. Sulkowska
Forest Research Institute, Sekocin Stary, Raszyn, Poland

6. References

Bakshi, M., Könnert, M. (2011). Genetic diversity and differentiation through isozymes in natural populations of Pinus wallichiana A.B. Jacks (Blue Pine) in India. Annals Of Forest Research. 54(1): 23-37.

Bednorz L., Myczko Ł., Kosiński P. (2006). Genetic variability and structure of the wild service tree (Sorbus torminalis (L.) Crantz) in Poland. Silvae Genetica 55, 4-5: 197-201.

Bergmann, F. (1991). Isoenzyme gene markers. In: Müller- Stark, G., Ziehe, M. (Eds.), Genetic Variation in European Populations of Forest Trees. Sauerländer`s Verlag, Frankfurt am Main. p.: 67-78.

Brown, A.H.D., Moran, G. F. (1981). Isozymes and the Genetic Resources of Forest Trees. Proceedings of the Symposium on Isozymes of North American Forest Trees and Forest Insects. July 27, 1979, Berkeley, California. M. Thompson Conkle Publisher: Pacific Southwest Forest and Range Experiment Station Technical Report. Washington, DC: Department of Agriculture, US No. 48: 1-10.

Comps, B., Thiebaut, B., Paule, L., Merzeau, D., Letouzey, J. (1990). Allozymic variability in beechwoods (*Fagus sylvatica* L.) over central Europe: spatial differentiation among and within populations. Heredity 65: 407 -417.

Conkle, M.T. (1972). Analyzing genetic diversity in conifers - isozyme resolution by starch gel electrophoresis. USDA.Pacific Southwest Forest and Range Exp. Stn., Berkeley, Calif. Forest Serv. Res. Note PSW-264, 5 p.

Concle, M.T., Paul, D.H., Nunnnaly, L.B., Hunter, S.C. (1982). Starch gel Electrophoresis of Conifer Seeds: a laboratory manual. Pacific Southwest Forest and Range Experiment Station

Cuguen, J., Thiebaut, B., Ntsiba, F., Barriere, G. (1985). Enzymatic Variability of Beech stands (*Fagus sylvatica L.*) on three Scales in Europe: Evolutionary Mechanisms, w P. Jacquard (ed): Genetic differentiation and dispersal in plants. NATO ASI Series. Springer Verlag, Berlin - Heidelberg. Vol. G5: 17-39.

Crow, J.F., Kimura, M. (1970). Introduction to Population Genetics Theory. Harper and Row, New-York.

Dz.U. 04, nr 67, poz. 621 (2004).Rozporządzenia Ministra Środowiska z dnia 9 marca 2004 r. w sprawie wykazu obszarów i map regionów pochodzenia leśnego materiału podstawowego. The rules of Polish Ministry of Environment.

Dzwonko, Z. (1990). Ekologia. In: Buk zwyczajny *Fagus sylvatica*. Warszawa – Poznań: PWN. p. 237–328.

Giertych M. (1990). Genetyka. In: Buk zwyczajny *Fagus sylvatica*. Warszawa – Poznań: PWN. p. 193–237.

Gömöry, D., Paule, L., Schvadchak, M., Popescu, F., Sułkowska, M., Hynek, V., Longauer, R. (2003). Spatial patterns of the genetic differentiation in European beech (*Fagus sylvatica* L.) at allozyme loci in the Carpathians and adjacent regions. Silvae Genetica 52(2): 78–83.

Grotkopp, E., Rejmánek, M., Sanderson, MJ., Rost, TL., Soltis, P. (2004). Evolution of genome size in pines (*Pinus*) and its life-history correlates: supertree analyses. Evolution, 58:1705-1729.

Hazler, K., Comps, B., Šugar, I., Melovski, L., Tashev, A., Gračan, J. (1997). Genetic structure of *Fagus sylvatica* L. populations in Southeastern Europe. *Silvae Genetica* nr 46: 4: 229–236.

Hubby, J.L., Lewontin. R. C. (1966). A molecular approach to the study of genetic heterozygosity in natural populations. I. The number of alleles at different loci in Drosophila pseudoobscura. Genetics 54: 577-594.

Hunter, R.L., Merkert. C.L. (1957). Histochemical demonstration of enzymes separated by zone electrophoresis in starch gels. Science 125: 1294-1295

Huntley, B., Birks, H.J.B. (1983). An atlas of past and present pollen maps for Europe: 0–13000 years ago. Cambridge Univ. Press.

Gonczarenko, G.G., Padutow, B.E. (1988). Rukowodstwo po issliedowaniu drewiesnych widow metodom elektroforeticzeskowo analiza izofiermentow. Bieloruskij naucznoisledowatielnyj institut lesniewo hoziajstwa. Gomel.

Kim, Z.S. (1985). Viability selection at an allozyme locus during development in European beech (*Fagus sylvatica* L.). Silvae Genetica 34 (4–5): 181–186.

Krzakowa M. (1979). Enzymatyczna zmienność międzypopulacyjna sosny zwyczajnej (*Pinus sylvestris* L.) Wyd. UAM, ser. Biologia. Nr 17. p. 45.

Krzakowa, M. (1980). Variability of Glutamate-Oxalate-Transaminase (GOT-2.6.1.1) isoenzymes in open- pollinated progeny of homozygous Scots pine (*Pinus sylvestris* L.) trees. Acta Societatis Botanicorum Poloniae 49, 1/2: 143-147.

Liengsiri, C., Piewluang, C. and Boyle, T.J.B. (1990). Starch Gel Electrophoresis of Tropical Trees a Manual. ASEAN-Canada Forest Tree Seed Centre.

Magri, D., Vendramin, G. G., Comps, B., Dupanloup, I., Geburek, T., Gömöry, D., Latałowa, M., Litt, T., Paule, L., Roure, J. M., Tantau, I., van der Knaap, W. O., Petit, R., Beaulieu, J. L. (2006). A new scenario for the Quaternary history of European beech populations: paleobotanical evidence and genetic consequences. New Phytology 171(1): 199-221.

Matras, J. (2005). Ochrona leśnych zasobów genowych i ich wykorzystanie w selekcji drzew oraz nasiennictwie i szkółkarstwie leśnym. Międzynarodowa konferencja – naukowo techniczna – Ochrona leśnych zasobów genowych i hodowla selekcyjna drzew leśnych w Polsce – stan i perspektywy. Malinówka, czerwiec 2005. English summary. p.: 5-15.

Mejnartowicz, L. (1990). Enzymy nowym narzędziem w pracy genetyka leśnika. Las Polski 8: 9-10.

Müller-Starck, G. (1985). Genetic Differences between "Tolerant" and Sensitive" Beeches (*Fagus sylvatica* L.) in Environmentaly Stressed Adult Forest Stand. Silvae Genetica 34, 6: 241-248.

Müller-Starck, G., Baradat, P., Bergmann, F. (1992). Genetic variation within European tree species. *New Forests* 6: 23-47.

Nei, M. (1972). Genetic distance between populations. *American Nature* 106: 283–292.

Nei, M. (1973). Analysis of gene diversity in subdivided populations. Proceedings of the National Academy of Sciences, USA., 70 (12, Pt. 1), 3321-3323.

Nei, M. (1978). Estimation of average heterozygosity and genetic distance from a small number of individuals. *Genetics* 89: 583–590.

Paule, L., Gömöry, D., Vysny, J. (1995). Genetic diversity and differentiation of beech populations in Eastern Europe. Genetics and Sylviculture of Beech. Proceedings from 5th Beech Symposium of the IUFRO Project Group P1.10–00. 19–24 September 1994. Mogenstrup. Denmark. *Forskningsserien* 11:159–167.

Paule, L., Mrazikowa, M. (1990). Geneticka Analyza Potomstva Borowice Sosny (Pinus sylvestris L.) zo semenneho sadu. Lesnictvi, 36(LXIII), c.10: 843-854.

Prus-Głowacki, W., Stephan, B. (1994). Genetic variation of *Pinus sylvestris* L. from Spain in relation to other European populations. *Silvae Genetica* 43: 7–14.

Prus-Głowacki, W., Urbaniak, L., Żubrowska-Gil, M. (1993). Allozyme differentiation in some European populations of Scots pine (*Pinus sylvestris* L.). *Genetica Polonica* 34: 159–176.

Ralska-Jasiewiczowa, M. (1983). Isopollen maps for Poland: 0–11,000 years B.P. New Phytology 94: 133–175.

Rider, C.C., Taylor, C.B. (1980) Outline Studies in Biology London and New York Capman and Hall 150th Anniversary.

Rudin, D., and Lindgren, D. (1977). Isozyme studies in seed orchards. Studia Forestalia Suecica Nr 139, 23 p. Swedish Coll. Forestry, Stockholm.

Savolainen, O., Yazdani, R. (1991). Genetic comparisons of natural and artificial populations of *Pinus sylvestris* L. In: Müller- Stark, G., Ziehe, M. (Eds.), Genetic Variation in European Populations of Forest Trees. Sauerländer`s Verlag, Frankfurt am Main. p.: 228-234.

Skrøpa, T. (1994). Impacts of tree improvement on genetic structure and diversity of planted forets. Silva Fenica 28(4): 265-274.

Sulkowska, M. (2010). Genetic and ecotype characterization of European beech (*Fagus sylvatica* L.) in Poland. Acta Silv. Lign. Hung., Vol. 6: 115-122.

Sułkowska, M. (1995). Ocena zmienności genetycznej drzew leśnych na podstawie analiz izoenzymatycznych. Sylwan 139: (6): 23-35.

Sułkowska, M., Kowalczyk, J., Przybylski, P. (2008). Zmienność genetyczna i ekotypowa buka zwyczajnego (*Fagus sylvatica* L.) w Polsce. Leśne Prace Badawcze 69 (2): 133-142.

Sułkowska, M., Gömöry, D., Paule, L. (2012). Genetic diversity of European beech in Poland estimated on the basis of isoenzyme analysis. *Folia Forestalia Polonica* – series A Forestry. Vol. 54(1): 48-55.

Starke, R., Ziehe, M., Müller-Starck, G. (1996). Viability selection in juvenile populations of European beech (*Fagus sylvatica* L.). *Forest Genetics* 3(4): 217–255.

Swofford, DL., Selander, R.B. (1981). Biosys–1. User Manual. University of Illinois, 65 pp.

Szafer, W. (1935). The significance of isopollen lines for the investigation of the geografical distribution of trees in the Post-Glacial period. Bulletin de l'Academie Polonaise des Sciences et des Lettres, Serie B, 1: 235-239.

Tanskley, S.D., Orton, T.J. (1983). Isozymes in Plant Genetics and Breeding, Part A. Elsevier Science Publishers B.V.

Wang Xiao-Ru, Szmidt A.E. (1989). Staining Recipes for Starch Gel Electrophoresis to use with Pinus tabulaeformis Carr. Department of Forest Genetics & plant Physiology, Swedish University of Agricultural Sciences.

Wheeler, N. C. and Jech, K. S. (1992). The use of electrophoretic markers in seed orchard research. New Forests 6: 31 l-328.

Wright, S. (1969). Evolution and the Genetics of Populations, Vol. II. The Theory of Gene Frequencies. University of Chicago Press, Chicago.

Wright, S. (1978). Evolution and the Genetics of Populations, Vol. IV. Variability Within and Among Natural Populations. University of Chicago Press, Chicago.

Yazdani, R., Muona, O., Rudin, D., Szmidt, A.E. (1985). Genetic structure of *Pinus sylvestris* L. seed-tree stand and naturally regenerated understory. Forest Science, 31, 430–436.

Yeh, F.C., Boyle, T.B.J. (1997). Population genetic analysis of co-dominant and dominant markers and quantitative traits Belgian Journal of Botany 129: 157.

Chapter 10

Seminal Plasma Proteins as Potential Markers of Relative Fertility in Zebu Bulls (*Bos taurus indicus*)

Marcelo G.M. Chacur

Additional information is available at the end of the chapter

http://dx.doi.org/10.5772/45758

1. Introduction

Progress has been made in developing reliable indicators of ejaculate quality that allow exclusion of low quality ejaculates for use in artificial insemination (AI). Physical semen characteristics and sperm morphology measurements are not always indicative of fertility and reproductive performance in animals, and accurate and predictive genetic and protein markers are still needed (Foxcroft et al., 2008). Two-dimensional polyacrylamide gel electrophoresis (2-D PAGE) represents a valuable tool for the separation and characterization of proteins from complex biological samples (Killian, 1992; Killian et al., 1993).

In the case of dairy sires, data sets are avaliable to assess fertility of individual males based on insemination of large numbers of cows using semen that has been frozen by standardizing procedures. This information has been used to demonstrate that fertility associated proteins exist in bull seminal plasma (Killian et al., 1993). Seminal plasma is composed of secretions from the male accessory sex glands and epididymis, which contains many organic and inorganic components that have effects on sperm quality (Foxcroft et al., 2008). The proteins secreted into seminal plasma may play an important role during sperm capacitation and fertilization (Rodriguez-Martinez et al., 1998), and may also serve to protect sperm from damage or to maintain their longevity.

Specific proteins in seminal plasma have been identified as potential markers of male fertility or infertility in the human (Martinez-Heredia et al., 2008; Yamakawa et al., 2007). Comprehensive proteomic analyses have been conducted in the bull (Moura et al., 2006a; Moura et al., 2006b; Chacur et al., 2010a; 2011a). The literature on the effects of seasons on the semen characteristics and upon seminal plasma proteins in Nellore (*Bos taurus indicus*) cattle under natural conditions in Brazil has already been recently studied in Brazil (Chacur et al., 2006b; 2007; 2010b; 2011a).

There is evidence revealing that seminal plasma prevents premature capacitation of sperm (Eng & Oliphant, 1978) and protects sperm from peroxidative damage (Jones et al., 1979; Schönech et al., 1996). Its well known that low temperatures alter the function of spermatozoa (Watson, 1995). Cold shock results in the destabilization of sperm membranes and impairment of sperm function, and it is also well known that animal spermatozoa are sensitive to cold-shock stress as the bull (Schönech et al., 1996).

Seminal plasma has also been shown to have deleterious effect on bovine sperm during semen storage at ambient temperatures and a damaging effect during semen cooling and freezing. Recent studies have shown that proteins from bovine seminal plasma (BSP) may modulate sperm properties. Two proteins (26kDa, pI 6.2; 55kDa, pI 4.5) predominate in higher-fertility bulls and two proteins (16kDa, pI 4.1; 16 kDa, pI 6.7) predominate in lower-fertility bulls. The major protein fraction of bovine seminal plasma is represented by three acidic proteins, designated as BSP-A1/-A2, BSP-A3 and BSP-30kDa (collectively called BSP proteins). These proteins are secretory products of the seminal vesicles and ampullae, and their biochemical characteristics have been well-described (Moura et al., 2006a; 2006b).

Although many autors believe that the seminal plasma proteins may function to stabilize the sperm against premature capacitation and spontaneous acrosome reaction. Significantly, these proteins may also protect the sperm from cooling-induced damage, such as cryocapacitation. The establishment of 2-D PAGE reference map could represent a usefull tool for the study of the still poorly understood nature and functions of the seminal plasma proteins, for the identification of previously unknown proteins, and for the comparison of seminal plasma protein composition between males of differing fertility. Many autors believe that the bull seminal plasma contains fertility associated protein markers. Comparison of individual 2-D PAGE maps with the reference map could provide a useful key to relate protein pattern changes to some physiopathological events influencing the reproductive sphere (Mortarino et al., 1998).

The present chapter describes the use of electrophoresis in animal reproduction: Electrophoresis and fertility in animals; Approaches to use of electrophoresis; Effect of seasonal changes in seminal plasma proteins of zebu bulls (*Bos taurus indicus*).

2. Electrophoresis and fertility in animals

The development of genetic markers to identify bulls of high breeding value represents one of the ways of genetic gain achievement in dairy farming. Several studies have been performed in an attempt to uncover the relationship between semen quality and fertility (Linford et al., 1976; Saacke et al., 1994). Thus, sperm morphology and motility, the number of sperm of sperm per insemination, percentage of acrosome-reacted sperm, and *in vitro* fertilization have been extensively evaluated as an indication of sperm ability to fertilize an egg. Evidence suggests that seminal plasma, which is a complex mixture of secretions originating from the testis, epididymis, and accessory glands, contains factors that modulate the fertilizing ability of sperm (Amann & Griel, 1974; Henault et al., 1995).

Recent studies have shown that proteins from bovine seminal plasma (BSP) may modulate sperm properties (Killian, 1992; Bellin et al., 1994). Two proteins (26kDa, pI 6.2; 55kDa, pI 4.5) predominate in higher-fertility bulls and two proteins (16kDa, pI 4.1; 16 kDa, pI 6.7) predominate in lower-fertility bulls (Killian et al., 1993).

2.1. Electrophoresis and seminal plasma

Secretions from the accessory sex glans are mixed with sperm and ejaculation and contribute to the majority of semen volume and components. Some accessory sex glands proteins are known to bind to the spermatozoa membrane and affect its function and properties (Yanagimachi, 1994).

Mammalian seminal plasma is constituted by secretions of the male accessory glans in which the spermatozoa are suspended in semen. Seminal plasma is an extremely complex fluid containing a wide variety of both organic and inorganic chemical constituents, among which only proteins present high molecular masses. The protein composition varies from species to species but in all the cases investigated so far these components have important effects on sperm function (Killian et al., 1993). Two-dimensional polyacrylamide gel electrophoresis (2-D PAGE) represents a valuable tool for the separation and characterization of proteins from complex biological samples. Seminal plasma, a physiological secretion from multiple glands of the male reproductive tract, is the natural medium for maturation of the spermatozoa through hormonal, enzymatic, and surface-modifiyng events (Killian et al., 1993).

The establishment of a 2-D PAGE reference map could represent a useful tool for the study of the still poorly understood nature and function of the seminal plasma proteins, for the identification of previously unknown proteins, and for the comparison of seminal plasma protein composition between males of differing fertility. In particular, it seems likely that bull seminal plasma contains fertility-associated protein markers (Killian et al., 1993).

3. The use of the SDS-PAGE and 2-D PAGE

3.1. Proteins of the male accessory glands and fertility

In the case of dairy sires, data sets are avaliable to assess fertility of individual males based on insemination of large numbers of cows using semen that has been frozen by standardizing procedures. This information has been used to demonstrate that fertility associated proteins exist in bull seminal plasma (Killian et al., 1993).

The ability to analyze the components secreted exclusively from the accessory sex glands has provided unique information about proyeins they secret and that are correlated with fertility indexes. Relating expression levels of specific proteins to fertility phenotype should serve as a sound foundation to evaluate their role in sperm function and fertilization (Moura et al., 2006b).

3.1.1. *Heparin binding proteins (HBPs)*

Heparin, a commercially available, sulfated glycosaminoglycan, induces capacitation/ acrosome reactions in sperm from bulls (Lenz et al., 1983). Sperm from high-fertility bulls have a greater frequency of acrosome reactions in response to heparin-like material (Lenz et al., 1988). Heparin-binding proteins (HBPs) are produced by the male accessory glands, secreted into seminal fluid (Nass et al., 1990). Fertility for Group 1 (HBP-B5) in sperm membranes but undetectable HBP-B5 in seminal fluid) was 82% of 1,692 cows. Group 2 (HBP-B5 detectable in seminal fluid as well as in sperm membranes) had 67% of 919 cows pregnant. Group 3 (detectable HBP-B5 in the seminal fluid and undetectable HBP-B5 in the sperm membranes) and Group 4 (undetectable HBP-B5 in the seminal fluid and sperm membrane) had 63% of 747 and 1,208 cows pregnant, respectively (Bellin et al., 1994). These trials indicated that grouping bulls according to the presence or absence of the greatest affinity herarin-binding protein (HBP-B5) on sperm membranes and in seminal fluid was an effective means of identifying fertility potential. Thus, understanding the protein composition of sperm membranes can be directly applicable to field situations and result in a more efficient production of calves (Bellin et al., 1994).

Other studies have shown that accessory glands produced and secreted HBP into seminal fluid , and HBP bound to sperm at ejaculation (Miller et al., 1990; Nass et al., 1990). Miller et al. (1990) reported that HBP constituted 28% of the protein component of seminal fluid or approximately 19.6 mg HBP/mL of ejaculate, and HBP-B5 constituted 6% of the total HBP in fluid from vasectomized bulls. The average concentration of total HBP represents 19.2 mg and 0.14 mg per mL of ejaculate in bovine seminal plasma and sperm membrane, respectively (Bellin et al., 1994). Although all HBPs may bind to sperm surfaces (Bellin et al., 1996). The concentration of the 30-kDa HBP, namely fertility associated antigen (FAA), on bovine sperm has been paired with a greater fertility potential (Bellin et al., 1998). FAA represents 0.8% of total HBP and has been identified as a deoxyribonuclease I-like protein (McCauley et al., 1999).

Studies have shown that these BSP proteins facilitate capacitation by promoting cholesterol efflux from sperm membranes (Thérien et al., 1998). Because sperm membrane cholesterol has an important role in modulating membrane bilayer fluidity and stability (Bloch, 1985; Yeagle, 1985), its efflux may perturb membrane structure and thereby lead to capacitation (Davis, 1981; Ehrenwald et al., 1988). In the context of sperm cryopreservation may lead to a decrease in sperm resistance to cold shock (Darin-Bennett & White, 1977; White, 1993). Thus, changes induced by the BSP proteins in the sperm membrane may have influence on sperm fertilizing ability and the success of the cryopreservation process (Nauc & Manjunath, 2000).

The characterization of seminal plasma proteins including osteopontin (OPN) 55 kDa and the study of their binding proteins (HBPs) will be the first step in understanding their role in the fertilization and identification of HBPs would provide information that could improve the knowledge of this aspect of reproductive physiology in Nellore bulls (Fernandes et al., 2008). In Nellore bulls (HBPs) bands with molecular weights ranging from 15 to 63 kDa were observed: 15 kDa, 17 kDa, 22 kDa, 25 kDa, 39 kDa, 53 kDa, 58 kDa and 63 kDa (Fernandes et al., 2008).

3.1.2. Spermadesin

In dairy cattle the average intensity of spermadesin (14 kDa) was higher in bulls of low fertility. The most basic isoform of accessory sex gland fluid is equivalent to the one originally found as an antifertility factor in the seminal plasma (Moura et al., 2006b). Z13 is a seminal plasma protein made up of two disulfide-linked 13-kDa subunits. The data indicate that the protein is a dimer of 26 kDa in native conditions and a monomer of 13 kDa in the presence of reductants. Therefore it antifertility peptide reported by can be suggested that Z13 presents at least one intermolecular S-S bridge (Tedeschi et al., 2000).

The intensity of apermadhesin Z13 in AGF showed an inverse relationship with fertility (Moura et al., 2006b). The low-molecular-weight antifertility peptide reported by Killian et al. (1993) in the seminal plasma of Holstein bulls was also identified as spermadhesin Z13. The isoform originally described in the seminal plasma by those authors appears to be more basic (pI 6.7) than the ones found in the AGF as antifertility factors pI 6.5 and 6.3 (Moura et al., 2006b).

The 2-D PAGE reference map of bull seminal plasma proteins provides information about the presence, in this particular fluid, of polypeptides of specifical biological significance (Mortarino et al., 1998). The PDC-109 represents more than 30% of the total protein contained in bull seminal plasma (Manjunath & Sairam, 1987). Comparison of individual 2-D PAGE maps with the reference map could provide a useful key to relate protein pattern changes to some physiopatological events influencing the reproductive sphere (Mortarino et al., 1998).

Spermadhesin Z13 is a peptide yhat displays 50% and 43% homology with the acidic seminal fluis protein and seminal plasma motility inhibitor (SPMI), respectively (Tedeschi et al., 2000). The former has positive effects on bovine sperm *in vitro* when as average concentratons, but it can inhibit both sperm motility and mitochondrial activity when at high levels (Schoneck et al., 1996).

3.1.3. Family of proteins (BSP proteins)

Bovine seminal plasma contains a family of proteins designated BSP-A1/-A2, BSP-A3, and BSP-30-kDa (collectively called BSP proteins). These proteins are secretory products of seminal vesicles that are acquired by sperm at ejaculation, modifying the sperm membrane by inducing cholesterol efflux. Because cholesterol efflux is time and concentration dependent, continuous exposure to seminal plasma that contains BSP proteins may be detrimental to the sperm membrane, which may adversely affect the ability of sperm to be preserved (Manjunath et al., 2002). These proteins coat the surface of the spermatozoa after ejaculation and are believed to play an important role in membrane modifications occurring during capacitation. Isoforms of each BSP protein were found when purified iodinated proteins analysed by 2D-PAGE. BSP-A1was found at a M(r) of 16.5 kDa and in the range of pI of 4.7-5.0; BSP-A2 at 16 kDa and at a pI of 4.9-5.2; BSP-A3 at 14 kDa and a pI of 4.8-5.2, and BSP-30-kDa at 28 kDa and at a pI of 3.9-4.6 (Desnoyers et al., 1994). BSP protein are acidic and have several isoforms. Furthermore, they heterogeneity of BSP-30-kDa is mainly due to its sialic acid content (Desnoyers et al., 1994).

The concentration of BSP-A1/-A2 was much higher compared with other seminal plasma proteins, and this protein alone represented a average of 38% of the total protein fraction, whereas BSP-A3 and BSP-30 kDa represented 3% to 4% of the total protein fraction (Nauc et al., 2000). The determination of BSP protein content on sperm surface may be an index of individual bull fertilizing ability or post-thaw status of sperm membranes (Nauc et al., 2000).

3.1.4. Osteopontin (OPN)

Osteopontin (OPN) is an acidic glycoprotein of about 41.5 kDa that has been isolated from rat, human and bovine bone. It is rich in aspartic acid, glutamic acid and serine and contains about 30 monosaccharides, including 10 sialic acids (Butler, 1989). The 55 kDa protein, shown to be more prevalent in higher-fertility males, was determined to be osteopontin (OPN) (Cancel et al., 1997). Results of immunofluorescence analyses of the male reproductive tract paralleled results for tissue extracts and fluids, indicating that bovine OPN is secreted by the ampulla and seminal vesicle. Tissue sections of the testis, epididymis (caput, corpus, cauda), vas deferens, prostate and bulbourethral gland were negative when reacted with antibodies against bovine seminal plasma OPN (Cancel et al., 1999).

Brown et al. (1992) studied the expression of OPN in normal adult human tissues. They showed that OPN was present on the luminal surfaces of epithelial cells of the gastrointestinal tract, gall bladder, pancreas, urinary and reproductive tracts, lung, bronchi, mammary and salivary glands, and sweat ducts. In general Brown et al. (1992) found that OPN accumulated on surfaces of epithelia bordering the luminal compartment.

Bulls with the highest fertility scores had 2.3 times more of a 55 kDa osteopontin than bulls with above-average fertility and at least 4 times more than bulls with below average (Moura et al., 2006b). A secreted form of phospholipase A_2 (PLA_2) 58 kDa present in the accessory gland fluid was more prevalent in bulls of high fertility (Moura et al., 2006b).

3.2. Proteins of the cauda epididymal fluid and fertility

Factors isolated from epididymal fluid or epithelial cells in culture have been linked to sperm motility and protection of membranes against damage caused by cryopreservation (Reyes-Moreno et al., 2002), anticapacitation effects (Roberts et al., 2003), or sperm number (Gatti et al., 2004), but evidence linking epididymal proteins to male fertility indexes is limited (Moura et al., 2006a).

Because most proteins from the rete testis are not present in the milieu of the epididymis lumen (Olson & Hinton, 1985; Dacheux et al., 1989), there is a general assumption that proteins of the epididymal fluid are mainly the product of the epididymis itself. Numerous proteins have been detected in the epididymal milieu of mammalian species (Cornwall et al., 2002; Dacheux & Dacheux, 2002) but the exact roles of most of them in sperm maturation are yet to be determined (Gatti et al., 2004).

Immature spermatozoa newly formed in the seminiferous tubules have a period of transit through the epididymis where they become motile and undergo a series of events that include changes in the composition of membrane lipids and proteins (Sullivan et al., 2005). The epididymal epithelium secretes proteins that potentially affect not only sperm maturation (Dacheux and Dacheux, 2002) but also other aspects of sperm physiology while these cells are stored in the cauda compartment (Hinton et al., 1995).

Fluid produced by the epididymis is diluted about 8- to 10- fold when mixed with accessory sex glands secretions at ejaculation (Gerena et al., 1998). This makes it difficult to accurately identify epididymal proteins in the seminal plasma milieu, particularly those secreted in low abundance or if they are also secreted by other organs, such as the accessory sex glands (Moura et al., 2006a).

An average of 118 spots was detected in the 2-D maps of the cauda epididymal fluid (CEF) in Holstein bulls. The intensity of alfa-L-fucosidase and cathepsin D was 2.3- 2.4-fold greater in high-fertility bulls than in low-fertility bulls (Moura et al., 2006a). The intensity of 3 isoforms (24-27 kDa, pI 6.3-5.8) of prostaglandin D-synthase (PGDS) were from 3.2 to 2.2 fold greater in low-fertility sires. The findings suggest that molecular markers of male fertility are associated with both epididymal sperm physiology and postejaculation eventus regulated by accessory sex gland components (Moura et al., 2006a). PGDS could influence male fertiling by mediating the action of hydrophobic molecules on sperm during epididymal transit or cauda epididymal storage (Moura et al., 2006a).

P25b, a protein with predictive properties for bull fertility, is transferred from prostasome-like particles present in the cauda epididymal fluid (PLPCd) to the sperm surface. The pattern of distribution of the PLPCd transferred varied from one sperm cell to the other, with a bias toward the acrosomal cap (Frenette et al., 2002).

4. Seasonal variation of zebu bull seminal plasma proteins

4.1. Heat-shock proteins and seminal quality in (Bos taurus indicus)

There were a number of suggestions in the earli er literature exposed to heat can produce sperm which do not produce normal offspring in unexposed females (Setchell, 1998). Bulls were subjected to scrotal insulation for 48 hours and semen collected and cryopreserved 2 or 3 weeks later, Following *in vitro* fertilization with swin-up sperm from these samples, there were decreased rates of spermpenetration, pronuclear formation (Walters et al., 2006) embryo cleavage, development and blastocyst formation (Walters et al., 2005) with semen collected from two of the bullsthree weeks after the insulation, but not which semen from two others bulls, or with semen collected after two weeks.

In bulls the heat is lost from the testis and scrotum to the environment through the scrotal skin, which is well endowed with sweat glands (Setchell, 2006). The temperature on the surface of the scrotum is lower at its base than near the neck, but the temperature inside the testis is almost uniform, even slightly warmer at the base (Kastelic et al., 1996).

There is considerable variation between individual animals in their response to heat exposure. Of the six bulls subjected to scrotal insulation by Vogler et al. (1993), two showed a large increase in abnormal spermatozoa (to more than 60%) whereas others had as few as 23% abnormal cells. Likewise, 4 bulls used for semen collection for in *in vitro* fertilization showed widely variable effects of 48h scrotal insulation on pronuclear formation, embryo development and apoptosis, with two bulls classed as severe responders, one a moderate responder and one showing no response to scrotal insulation (Walters et al., 2006).

Progress has been made in developing reliable indicators of ejaculate quality that allow exclusion of low quality ejaculates for use in artificial insemination (AI). Physical semen characteristics and sperm morphology measurements are not always indicative of fertility and reproductive performance in animals, and accurate and predictive genetic and protein markers are still needed (Foxcroft et al., 2008).

There is evidence revealing that seminal plasma prevents premature capacitation of sperm (Eng & Oliphant, 1978) and protects sperm from peroxidative damage (Jones et al., 1979; Schönech et al., 1996). It`s well known that low temperatures alter the function of spermatozoa (Watson, 1995). Cold shock results in the destabilization of sperm membranes and impairment of sperm function, and it is also well known that animal spermatozoa are sensitive to cold-shock stress as the bull, rabbit and man (Schönech et al., 1996).

Specific proteins in seminal plasma have been identified as potential markers of male fertility or infertility in the human (Martinez-Heredia et al., 2008; Yamakawa et al., 2007). Comprehensive proteomic analyses have been conducted in the bull *Bos taurus taurus* (Moura et al., 2006b). The literature on the effects of seasons on the semen characteristics and upon seminal plasma proteins in Nellore and Tabapua (*Bos taurus indicus*) and Limousin, Brown-Swiss and Brangus cattle under natural conditions in Brazil has already been recently studied (Chacur et al., 2003; 2004; 2006a; 2006b; 2007; 2008; 2009; 2010a; 2010b; 2011a; 2011b; 2012).

4.2. Potential markers of relative fertility in *Bos taurus indicus*

Bos indicus bulls are less sensitive to the effects of high temperatures than *Bos taurus* or crossbred bulls, but as they are actually more sensitive to the effects of scrotal insulation (Brito et al., 2003). This would appear to be due to the greater ability of *indicus* animals to keep their testes cool (Brito et al., 2002). *Bos indicus* bulls have greater testicular artery length to testicular volume ratios, and smaller testicular artery wall thickness and arterial to venous distances, which may be responsible for greater cooling of the arterial blood in the spermatic cord (Brito et al., 2004).

In Brazil sixty-eight Nellore (*Bos indicus)* bulls were used, with twenty of the padron variety and forty-eight of the mocho variety with mean of 4 years old. There was no difference (P>0.05) for the spermatic morphology between padron and mocho variety, respectively with 5.06±8.20% and 5.32±6.40% of major defects; 9.91±6.7% and 8.36± 6.06% for minor defects; and 14.76±13.20% and 13.82±12.61% for the total defects. The electrophoresis of the seminal plasma showed protein bands with weights between 5- and 105-kDa. In 100% of the bulls with good semen the 13kDa protein were present, the same happens with the 18- and 20-kDa bands. The varieties padron and mocho revealed similar reproductive adaptation in front of the handling conditions and weather and looking very efficient (Chacur et al., 2006b).

Disruptions in sperm production include decreased sperm motility and increased of abnormal sperm. Seminal plasma appears to exert important effects on sperm function. The objective was to evaluate the dry and rainy season influence on the seminal characteristics and semen plasma proteins. Eleven bulls (*Bos taurus indicus*) with ages ranging from 34 to 38 months were submitted each one to 12 semen collect with eletroejaculation 6 on dry season and 6 on rainy season with 14 days interval, totalizing 144 samples. Qualitative and quantitative semen characteristics were evaluated. Samples of semen were centrifuged (1.500 g / 15 minutes) and conditioned and stored (–20ºC) until further processing. The proteins were extracted and quantified to electrophoresis performed. Variance analysis and Tukey test 5% was used. The semen vigor ($P<0.01$), minor defects and total defects ($P<0.05$) showed statistical difference between seasons, while the volume, motility and minor defects did not ($P>0.05$). The number of bands occurred between 6- and 125-kDa, see Table 1. The molecular band of 26 kDa was present in 100% of bulls in rainy season. The molecular bands of 6-, 9- and 125-kDa showed a high frequency in dry and rainy season. In conclusion, these results showed a band distribution variation throughout the season and the year seasons changed the semen quality with increase sperm vigor and reduction of abnormal sperm on dry season (Chacur et al., 2011b).

bulls	Proteins (kDa)	Dry season (n=144)	Rainy season (n=144)
a, b, c, d, e, f, h, i, j, k,	6	9/11 (81.81%)	10/11(90.90%)
a, b, c, d, e, f, g, h, j, k	9	10/11(90.90%)	8/11(72.72%)
d, e, f, h, i, j, k	12	7/11(63.63%)	5/11(45.45%)
a, b, c, d, e	13	5/11(45.45%)	1/11(9.09%)
a, b, c, e, f, g, j	17	3/11(27.27%)	7/11(63.63%)
a, d, e, f, j	20	3/11(27.27%)	3/11(27.27%)
a, b, c, d, e, f, g, h, i, j, k	26	6/11(54.54%)	11/11(100%)
a, b, c, e, f, h, i, j, k	35	4/11(36.36%)	7/11 (63.63%)
a, b, e, f, g, h, j, k	44	2/11(18.18%)	8/11(72.72%)
a, c, d, e, f, h, i, j	55	7/11(63.63%)	3/11(27.27%)
b, d, h, i, k	66	2/11(18.18%)	4/11(36.36%)
a, c, d, e, f, i, j, k	75	6/11(54.54%)	6/11(54.54%)
a, b, c, i	80	1/11(9.09%)	4/11(36.36%)

bulls	Proteins (kDa)	Dry season (n=144)	Rainy season (n=144)
a, c, f, i, j	105	4/11(36.36%)	3/11(27.27%)
a, b, c, e, f, g, h, i, j, k	125	9/11 (81.81%)	10/11(90.90%)

Chacur et al. (2011).

Table 1. Frequency of proteins bands in dry season (may-july) and rainy season (october-december).

Seminal plasma is a complex of secretions of the male accessory reproductive organs and appears to exert important effects on sperm function (Shivaji et al., 1990). The protein quality of the seminal plasma may affect positively the bulls' fertility (Killian *et al.*, 1993). Peptides of 55- and 66-kDa were present in bulls with excellent spermatic conditions for example motility and vigor. On the other hand, 16- and 36-kDa peptides were observed with unfavorable spermatic conditions (Chacur *et al.*, 2009). The objective was to determine the influence of season on seminal plasma proteins in Brown Swiss bulls. Semen from 33 Brown Swiss bulls 24 months of age were collected by electroejaculation during winter (from June to August) and summer (from December to February) in the southern hemisphere in 2008. Semen samples were collected with 14-day intervals totalizing 196 ejaculates. Samples of semen were centrifuged (1500*g*/15 min) and the seminal plasma was conditioned in cryotubes and stored at –20°C until further processing. Proteins were extracted from 200 µL of each sample in 2 mL of extraction buffer composed of 0.625 M Tris-HCl, at pH 6.8, in 2% SDS, 5% β-mercaptoethanol, and 20% of glycerol. Percentages of different plasma proteins by season were statistically compared by the chi-square test with significance level ($P < 0.05$). Proteins were quantified according to Bradford (1976) and electrophoresis was performed according to Laemmili (1970). Gels were fixed with isopropanol:acetic acid:water (4:1:5 v/v) for 30 minutes and stained in the same solution with 2% of Coomassie Blue R250. In 26 bulls, the absence of high molecular weight (HMW; 55 kDa, 66 kDa, and 80 kDa) proteins was found in the summer. There was a significant increase ($P < 0.05$) in total spermatic defects, acrosome defects, and distal cytoplasmatic droplets in these bulls. The 40-kDa protein that reflected low fertility was observed in 10 bulls in the summer with semen quality decreases. The 11 bulls showed presence of HMW (55 kDa) in the winter. In 11 bulls, HMW (55 kDa, 66 kDa, or 80 kDa) proteins were present with a satisfactory semen condition according to Killiam *et al.* (1993). In conclusion, the seasons of the year may influence the presence of proteins in seminal plasma. There was a direct relationship of the season with seminal plasma proteins. The presence of the proteins of 20 kDa, 55 kDa, 66 kDa, and 80 kDa suggested an increase of the semen quality during the winter (Chacur et al., 2010a; 2011a).

In Brazil semen from eleven Tabapua bulls, 30 months old, were collected by electroejaculation during winter (from June to August) and summer (from December to February) of 2007. From each bull a total of 132 semen samples were collected in an interval of 14 days. Samples of seminal plasma were centrifuged (1500g/15min) and conditioned in criotubes and stored at –20ºC until further processing. Proteins were extracted from 200 µL of each sample in 2 mL of extraction buffer composed by 0.625 M Tris-HCl, pH 6.8, 2% SDS, 5% β-mercaptoethanol and 20% of glycerol. Proteins were quantified according to Bradford (1976) and electrophoresis was performed according to Laemmli (1970). Gels were fixed with isopropanol: acetic acid: water (4:1:5 v/v) for 30 minutes, and stained in the same solution with 2% of Comassie Blue R250. Percentage of different seasons including plasma proteins were statistically compared by the Chi-square test with significance level at $P<0.05$. In two bulls, the absence of high molecular weight (HMW 55kDa, 66kDa and 80kDa) proteins was verified in the summer. There was a significant increase ($P<0.05$) in total spermatic defects in these two bulls. The protein of 40kDa which suppose to be of low fertility was observed in eight bulls in the summer with semen quality decrease. The eight bulls showed presence of HMW (55kDa) in the winter. In nine bulls HMW (55kDa, 66kDa or 80kDa) proteins were present with a satisfactory semen condition in accordance with Chacur et al. (2006a). The two bulls showed presence of HMW proteins (66kDa and 80kDa) in the summer. The results suggest that different seasons of the year may influence the presence of a variety of proteins in seminal plasma. There was a direct relationship of the season upon seminal plasma proteins. The presence of the proteins of 20kDa, 55kDa, 66kDa and 80kDa suggests an increase of the semen quality during the winter (Chacur et al., 2008).

Peptides of 55- and 66-kDa were present in bulls with excellent spermatic conditions for example motility and vigor (Chacur et al., 2009a). On the other hand, 13- and 33-kDa peptides were observed in association with unfavourable spermatic conditions (Chacur et al., 2009b). The objective of this study was to determine the profile SDS-PAGE of seminal plasma and evaluate the semen characteristics in Brangus and Brown-Swiss bulls. Semen from 14 Brangus, 36 months old, was collected by electroejaculation during summer of 2009-2010. A total of 84 semen samples were collected in an interval of 14 days. Semen volume, motility, vigor, major defects and minor defects were evaluated according to Brazilian College of Animal Reproduction (Manual..., 1998). Animals were divided in two groups: poor semen (motility <50% and major defects >10%) and good semen, and subsequently compared regarding the composition of seminal plasma proteins. Samples of seminal plasma were centrifuged (1500g/15min) and conditioned in criotubes and stored at –20ºC until further processing. Proteins were extracted from 200 μL of each sample in 2 mL of extraction buffer composed by 0.625 M Tris-HCl, pH 6.8, 2% SDS, 5% β-mercaptoethanol and 20% of glycerol. Proteins were quantified according to Bradford (1976) and electrophoresis was performed according to Laemmli (1970). Gels were fixed with isopropanol: acetic acid: water (4:1:5 v/v) for 30 minutes, and stained in the same solution containing 2% of Comassie Blue R250. Each semen collection was used in duplicate. The concentration of proteins was measured using a spectrophotometer PF-901(Chemistry Analyser Labsystems). Gels were submitted to a photodocumentation system (Bio Doc-IT and Visidoc-IT Gel Documentation systems, UVP) and analysed by Doc-IT-LS 6.0 software. GLM from SAS, version 6, was used in order to evaluate possible variations of seminal variables and protein molecular mass. Statistical significance was accepted from $P<0.05\%$. The means of semen variables were: volume (5±1 mL), motility (75±5%), vigor (4), major defects (7±2%) and minor defects (12±4%). The results of analyses of gels revealed a variety of proteins in each animal and among bulls. There were 28 different major polypeptides, ranging from 15 to 24 bands in each individual bull. In six Brangus bulls the presence of low molecular weight (LMW 13kDa and 33kDa) proteins was associated with low motility (35-40%) in accordance with Chacur et al. (2009a). There was a significant increase ($P<0.05$) in major spermatic defects in these six bulls (20.3±3.7%) associated with presence of proteins that had molecular weights of (23, 35 and 72KDa). In eight Brangus bulls, 55KDa, 66KDa or 80KDa proteins were present and associated with a satisfactory semen condition (motility 77±6% and major defects 5±2%) in accordance with Chacur et al. (2009b). In cattle, the 55-, 66- and 80-kDa proteins are associated positively with camp-dependent progressive motility (Shivaji et al., 1990). Consistently, in the present experiment, there was a positive relationship of presence of seminal plasma proteins 55kDa, 66kDa and 80kDa and semen quality (motility and major defects). The presence of these proteins suggests an increase in semen quality (Chacur et al., 2010b).

4.3. Interation between year seasons on the semen and hormones in *Bos indicus* and *Bos taurus* in Brazil

In Brazil the influence of four year seasons was study on semen characteristics and levels of testosterone and cortisol in Nelore and Simmental bulls. Five Nelore and five Simmental bulls with 48-72 months old, extensively managed were evaluated for sexual soundness using physical and morphological characteristics of semen and serum levels of testosterone and cortisol. There was decreased motility and vigor semen ($P<0.05$) during winter in Simmental bulls (Table 2). There was correlation ($P<0.01$) between testosterone x motility (0.69) and testosterone x vigor (0.57) in Simmental breed (Table 4) and cortisol x motility (0.68) and cortisol x vigor (0.65) in Nelore breed (Table 3). The effect of year seasons changed the semen quality with increase sperm motility and vigor on springer-summer in Simmental bulls. The cortisol level decreased on autumn in Nelore bulls (Chacur et al., 2012).

characteristics	breed	spring	summer	autumn	winter
Volume (mL)	S	8.80±0.65 Aab	9.85±0.65 Aa	8.76±0.58 Aab	7.35±0.65 Ab
	N	7.10±0.65 Aa	7.55±0.65 Ba	6.26±0.58 Ba	6.05±0.65 Aa
Motility (%)	S	70.00±5.83Aa	70.00±5.83Aa	60.80±8.21Aab	48.00±5.83Ab
	N	63.50±5.83Aa	60.00±5.83Aa	54.40±5.21Aa	53.50±5.83Aa
Vigor (1-5)	S	3.35±0.28 Aa	3.55±0.28 Aa	3.00±0.25 Aab	2.20±0.28 Ab
	N	3.05±0.28 Aa	2.95±0.28 Aa	2.32±0.25 Aa	2.50±0.28 Aa
Major defects (%)	S	11.33±1.31 Aab	8.00±1.07 Ab	10.75±0.98 Aab	12.18±1.20 Aa
	N	6.30±1.07 Ba	6.31±1.10 Aa	9.92±0.96 Aa	9.42±1.10 Aa
Minor defects (%)	S	7.88±0.83 Aa	7.25±0.79 Aa	8.25±0.72 Aa	10.25±0.88 Aa
	N	7.60±0.79 Aa	5.63±0.81 Aa	7.16±0.70 Aa	6.84±0.81 Ba
Total defects (%)	S	19.22±1.63 Aab	15.25±1.54 Ab	19.00±1.41 Aab	22.43±1.73 Aa
	N	13.90±1.53 Ba	11.97±1.59 Aa	17.08±1.38 Aa	16.36±1.59 Ba

characteristics	breed	spring	summer	autumn	winter
Concentration ($x10^9$/mL)	S	1.35±0.13 Aa	1.38±0.13 Aa	1.32±0.12 Aa	1.00±0.13 Aa
	N	0.88±0.13 Ba	1.14±0.13 Aa	0.95±0.12 Ba	0.91±0.13 Aa

Significance level 5% (P< 0,05); A, B – distinct letters in column (P<0,05); a, b – distinct letters in line (P<0,05).

Table 2. Semen characteristics in spring, summer, autumn and winter for Simmental (S) and Nelore (N) bulls.

	breed	spring	summer	autumn	winter	correlations
cortisol	S	0.6	0.6	0.5	1.1	
Volume (mL)	S	8.800	9.444	10.800	7.500	-0.87
color	S	1.700	2.111	1.400	1.200	-0.60
aspect	S	1.800	1.889	1.400	1.900	0.55
Mass moviment (1-5)	S	1.800	4.000	2.800	1.800	-0.52
Motility (%)	S	67.000	80.000	64.000	46.000	-0.83
Vigor (%)	S	3.000	4.111	3.200	2.100	-0.79
Concentration ($x10^9$/mL)	S	1.093	1.697	1.360	0.960	-0.66
Major defects (%)	S	13.889	6.222	10.800	10.875	0.10
Minor defects (%)	S	7.889	6.222	7.400	8.250	0.60
Total defects (%)	S	21.778	12.444	18.200	19.125	0.22
cortisol	N	1.68	3.10	1,36	3.06	correlations
Volume (mL)	N	8.278	8.286	5.700	5.800	0.14
color	N	1.444	3.000	1.600	2.200	0.87
aspect	N	1.889	2.429	1.500	2.000	0.85
Mass moviment (1-5)	N	2.222	3.714	1.300	2.200	0.75
Motility (%)	N	63.333	80.000	33.000	56.000	0.75
Vigor (1-5)	N	3.111	4.000	1.500	2.600	0.65
concentration ($x10^9$/mL)	N	0.913	1.329	0.701	1.080	0.91
Major defects (%)	N	5.889	5.857	10.700	9.200	-0.31
Minor defects (%)	N	7.000	4.571	8.000	7.500	-0.60
Total defects (%)	N	12.889	10.429	18.700	16.900	-0.43

color: 1 – white, 2 – White-Milk and 3 – White-yellow; aspect: 1 – aquous, 2 – viscous and 3 – cremous.

Table 3. Correlations between cortisol (μg/dL) and semen characteristics on year season in Simmental (S) and Nelore (N) bulls.

	breed	spring	summer	autumn	winter	correlations
testosterone	S	879.5	901.1	584.0	648.2	
Volume (mL)	S	8.800	9.444	10.800	7.500	-0.16
color	S	1.700	2.111	1.400	1.200	0.86
aspect	S	1.800	1.889	1.400	1.900	0.62
Mass moviment (1-5)	S	1.800	4.000	2.800	1.800	0.31
Motility (%)	S	67.000	80.000	64.000	46.000	0.69
Vigor (1-5)	S	3.000	4.111	3.200	2.100	0.57
Concentration ($x10^9$/mL)	S	1.093	1.697	1.360	0.960	0.37
Major defects (%)	S	13.889	6.222	10.800	10.875	-0.19
Minor defects (%)	S	7.889	6.222	7.400	8.250	-0.47
Total defects (%)	S	21.778	12.444	18.200	19.125	-0.26
testosterone	N	430.41	234.71	420.31	329.15	correlations
Volume (mL)	N	8.278	8.286	5.700	5.800	-0,28
color	N	1.444	3.000	1.600	2.200	-1.00
aspect	N	1.889	2.429	1.500	2.000	-0.89
Mass moviment (1-5)	N	2.222	3.714	1.300	2.200	-0.87
Motility (%)	N	63.333	80.000	33.000	56.000	-0.71
Vigor (1-5)	N	3.111	4.000	1.500	2.600	-0.70
Concentration ($x10^9$/mL)	N	0.913	1.329	0.701	1.080	-0.93
Major defects (%)	N	5.889	5.857	10.700	9.200	0.36
Minor defects (%)	N	7.000	4.571	8.000	7.500	0.82
Total defects (%)	N	12.889	10.429	18.700	16.900	0.56

color: 1 – white, 2 – White-Milk and 3 – White-yellow; aspect: 1 – aquous, 2 – viscous and 3 – cremous.

Table 4. Correlations between testosterone (ng/dL) and semen characteristics on year season in Simmental (S) and Nelore (N) bulls.

5. Summary and conclusions

Seminal plasma is composed of secretions from the male accessory sex glands and epididymis, which contains many organic and inorganic components that have effects on sperm quality (Foxcroft et al., 2008). The proteins secreted into seminal plasma may play an important role during sperm capacitation and fertilization (Rodriguez-Martinez et al., 1998), and may also serve to protect sperm from damage or to maintain their longevity.

Specific proteins in seminal plasma have been identified as potential markers of male fertility or infertility in the human (Martinez-Heredia et al., 2008; Yamakawa et al., 2007). Comprehensive proteomic analyses have been conducted in the bull (Moura et al., 2006a; Moura et al., 2006b; Chacur et al., 2010a; Chacur et al., 2010b; Chacur et al., 2011a; Chacur et al., 2011b). The literature on the effects of seasons on the semen characteristics and upon seminal plasma proteins in Nellore (*Bos taurus indicus*) cattle under natural conditions in Brazil has already been recently studied in Brazil (Chacur et al., 2003; 2004; 2006a; 2007; 2009a; 2010a; 2010b; 2011a; 2011b; 2012).

Its well known that low temperatures alter the function of spermatozoa. Cold shock results in the destabilization of sperm membranes and impairment of sperm function, and it is also well known that animal spermatozoa are sensitive to cold-shock stress as the bull (Watson, 1995).

Seminal plasma has also been shown to have deleterious effect on bovine sperm during semen storage at ambient temperatures and a damaging effect during semen cooling and freezing. Recent studies have shown that proteins from bovine seminal plasma (BSP) may modulate sperm properties. Many autors believe that the bull seminal plasma contains fertility associated protein markers. Comparison of individual 2-D PAGE maps with the reference map could provide a useful key to relate protein pattern changes to some physiopathological events influencing the reproductive sphere.

Factors isolated from epididymal fluid or epithelial cells in culture have been linked to sperm motility and protection of membranes against damage caused by cryopreservation (Reyes-Moreno et al., 2002), anticapacitation effects (Roberts et al., 2003), or sperm number (Gatti et al., 2004), but evidence linking epididymal proteins to male fertility indexes is limited (Moura et al., 2006a).

6. Future prospects

The determination of protein content on sperm surface and seminal plasma may be an index of individual bull fertilizing ability or post-thaw status of sperm membranes (Nauc & Manjunath, 2000). A reference map of seminal plasma proteins could be useful in relating protein pattern changes to physiopathological events influencing the reproductive sphere (Mortarino et al., 1998). PGDS could influence male fertility by mediating the action of hydrophobic molecules on sperm during epididymal transit or cauda epididymal storage (Moura et al., 2006a; 2006b). Although known functional attributes of these proteins provide some understanding of how they may influence male reproductive performance (Moura et al., 2006b; Chacur et al., 2010a; 2011a).

Author details

Marcelo G.M. Chacur
Animal Reproduction Department, University of Oeste Paulista (UNOESTE), Pres. Prudente-SP, Brazil

7. References

Amann & Griel (1974). Fertility of bovine spermatozoa from rete testis, cauda epididymis, and ejaculated sperm. *J. Dairy Sci.*, Vol.57, pp.212-219.

Bellin et al. (1994). Fertility of range beef bulls grouped according to presence or absence of heparin-binding proteins in sperm membranes and seminal fluid. *J. Anim. Sci.*, Vol.72, pp.2441-2448.

Bellin et al. (1996). Monoclonal antibody detection of heparin-binding proteins on sperm corresponds to increased fertility of bulls. *J. Anim. Sci.*, Vol.74, pp.173-182.

Bellin et al. (1998). Fertility-associated antigen on bull sperm indicates fertility potential. *J. Anim. Sci.*, Vol.76, pp.2032-2039.

Bloch (1985). Cholesterol evolution of structure and function. In: Vance, D.E., Vance, J.E. eds. Biochemistry of lipids and membranes. Amsterdam: Academic Press, pp.1-24.

Bradford (1976). A rapid and sensitive method for the quantitation of microgram quantities of protein utilizing the principle of protein-dye binding. *Analytical Biochemistry*, Vol.72, pp.248-254.

Brito et al. (2002). Effect of age and genetic group on characteristics of the scrotum, testes and testicular vascular cones, and on sperm production and semen quality in AI bulls in Brazil. *Theriogenology*,Vol.58, pp.1175-1186.

Brito et al. (2003). Effects on scrotal insulation on sperm production, semen quality and testicular echotexture in *Bos indicus* and *Bos indicus x Bos taurus* bulls. *Anim. Reprod. Sci.*, Vol.79, pp.1-15.

Brito et al., (2004). Testicular thermoregulation in *Bos indicus*, crossbred and *Bos taurus* bulls: relationship with scrotal, testicular vascular cone and testicular morphology, and effects on semen quality and sperm production. *Theriogenology*, Vol.61, pp.511-528.

Brown et al. (1992). Expression and distribution of osteopontin in human tissues: widespread association with luminal epithelial surfaces. *Mol. Biol. Cell.*, Vol.3, pp.1169-1180.

Butler (1989). The nature and significance of osteopontin. *Connect Tissue Res.*, Vol.23, No.2-3, pp.123-136.

Cancel et al. (1997). Osteopontin is the 55-kilodalton fertility associated protein in Holstein bull seminal plasma. *Biol. Reprod.*, Vol.57, pp.1293-1301.

Cancel et al. (1999). Osteopontin localization in the Holstein bull reproductive tract. *Biol. of Reprod.*, Vol.60, No.2, pp.454-460.

Chacur et al. (2003) Fertility selection in bulls and seminal plasma proteins, spermatic evaluation correlation. *Revista Brasileira de Reprodução Animal.* Vol.27, No.2, pp.185-186.

Chacur et al. (2004). Season influence upon seminal plasma proteins in bulls. Abstract...15th International Congress on Animal Reproduction, Porto Seguro, Brazil, Vol.1, pp.236.

Chacur et al. (2006a). SDS-PAGE seminal plasma proteins pattern and its relationship with the quality of Nelore bull (*Bos taurus indicus*) semen. *Veterinária Notícias*, Vol.12, pp.87-93.

Chacur et al. (2006b). Winter-springer and summer influence upon seminal plasma proteins in bulls. *Anim. Reprod.*, Vol.3, No.2, p.251.

Chacur et al. (2007). Influência da estação do ano sobre as proteínas do plasma seminal de touros Limousin. *Veterinária Notícias*, Vol.13, pp.47-53.

Chacur et al. (2008). Season influence upon seminal plasma proteins in Tabapua breed *Bos taurus indicus*. *Anim. Reprod.*, Vol.5, pp.522.

Chacur et al. (2009a). Season influence on the seminal characteristics and seminal plasma proteins in Tabapua breed *Bos taurus indicus*. *Arch. Zootec.*, Vol.60, No.230, pp.301-304.

Chacur et al. (2009b). Season influence upon seminal plasma proteins in Tabapua breed *Bos taurus indicus*. *Anim. Reprod.*, Vol.6, No.1, pp.339.

Chacur et al. (2010a). Season influence upon seminal plasma proteins in Brown-Swiss bulls. *Reprod. Fertil. Development*, Vol.22, No.1, pp.310.

Chacur et al. (2010b). Profile SDS-PAGE of seminal plasma in Brangus and Brown-Swiss bulls. *Anim. Reprod.*, Vol.7, pp.236.

Chacur et al. (2011a). Season influence upon seminal plasma proteins in Brown-Swiss bulls. *Arch. Zootec.*, Vol.60, pp.301-304.

Chacur et al. (2011b). Influence of the dry and rainy seasons upon plasma seminal (SDS-PAGE) and characteristics of the ejaculate from bulls (*Bos taurus indicus*). *Semina*, Vol.32, No.4, pp.1565-1574.

Chacur et al. (2012) Influence of year season on semen characteristics and hormonal levels in Nelore and Simmental bulls. *Arq. Bras. Med. Vet. Zootec.* (in press).

Cornwall et al. (2002). Gene expression and epididymal function. In: Robaire, B.; Hinton, B.T. eds. *The epididymis: From molecules to clinical Practice: A Comprehensive Survey of the Efferent Ducts, Epididymis and the Vas Deferens.* New York, NY: Kluver Academic/Plenum Publishers, pp.169-199.

Dacheux & Dacheux (2002). Protein secretion in the epididymis. In: Robaire, B.; Hinton, B.T. eds. *The epididymis: From molecules to clinical Practice: A Comprehensive Survey of the Efferent Ducts, Epididymis and the Vas Deferens.* New York, NY: Kluver Academic/Plenum Publishers, pp.151-168.

Dacheux et al. (1989). Changes in sperm surface membrane and luminal proteinfluid content during epididymal transit in the boar. *Biol. Reprod.*, Vol.40, pp.635-651.

Darin-Bennett & White (1977). Influence of cholesterol content of mammalian spermatozoa on susceptibility to cold-shock. *Cryobiology*, Vol.14, pp.466-477.

Davis (1981). Timing of fertilization in mammals: sperm cholesterol/phospholipid ratio as a determinant of the capacitation interval. *Proc. Natl. Acad. Sci. USA*, Vol.78, pp.7560-7564.

Desnoyers et al. (1994). Characterization of the major proteins of bovine seminal fluid by two-dimensional polyacrylamide gel electrophoresis. *Mol. Reprod. Dev.*, Vol.37, No.4, pp.425-435.

Ehrenwald et al. (1988). Cholesterol efflux from bovine sperm: II. Effect of reducing sperm cholesterol on penetration of zona-free hamster and *in vitro* matured bovine ova. *Gamete Res.*, Vol.20, pp.413-420.

Eng & Oliphant (1978). Rabbit sperm reversible decapacitation by membrane stabilization with highly purified glycoprotein from seminal plasma. *Biol. Reprod.*, Vol.19, pp.1083-1094.

Fernandes et al. (2008). Heparin-binding proteins of seminal plasma in Nellore bulls. *Ciência Rural*, Vol.39, No.1, pp.20-26.

Foxcroft et al. (2008). Identifying uneable semen. *Theriogenology*, Vol.70, pp.1324-1336.

Franzen & Heinegard (1985). Isolation and characterization of two sialoproteins present only in bovine calcified matrix. *Biochem. J.*, Vol.235, pp.715-724.

Frenette et al. (2002). Selected proteins of "prostasome-like particles" from epididymal cauda fluid are transferred to epididymal caput spermatozoa in bull. *Biol. of Reprod.*, Vol.67, No.1, pp.308-313.

Gatti et al. (2004). Post-testicular sperm environment and fertility. *Anim. Reprod. Sci.*, Vol.82-83, pp.321-339.

Gerena et al. (1998). Identification of fertility-associated protein in bull seminal plasma as lipocalin-typeprostaglandin D-synthase. *Biol. Reprod.*, Vol.58, pp.826-833.

Henault et al. (1995). Effect of accessory sex gland fluid from bulls of different fertilities on the ability of cauda epididymal sperm to penetrate zona-free ovine oocytes. *Biol. Reprod.*, Vol.52, pp.390-397.

Hinton et al. (1995). The epididymis as protector of maturing spermatozoa. *Reprod. Fertil. Dev.*, Vol.7, pp.731-745.

Jones et al. (1979). Peroxidative breakdown of phospholipids in human spermatozoa: spermicidal effects of fatty acid peroxides and protective action of seminal plasma. *Fertil. Steril.*, Vol.31, pp.531-537.

Kastelic et al. (1996). Contribution of the scrotum and testes to scrotal and testicular thermoregulation in bulls and rams. *J. Reprod. Fertil.*, Vol.108, pp.81-85.

Killian (1992). Fertility factors in seminal plasma. In: Proceedings of the 14th Technical Conference on Artificial Insemination and Reproduction, Milwaukee, pp.33-38.

Killian et al. (1993). Fertility-associated proteins in Holstein bull seminal plasma. *Biol. of. Reprod.*, Vol.49, No.6, pp.1202-1207.

Laemilli (1970). Cleavage of structural proteins during assembly of the head of the bacteriophage T. *Nature*, Vol.277, pp.680-685.

Lenz et al. (1983). Rabbit spermatozoa undergo an acrosome reaction in the presence of glycosaminoglycans. *Gamete Res.*, Vol.8, pp.11.

Lenz et al. (1988). Predicting fertility of dairy bulls by inducing acrosome reactions in sperm with chondroitin sulfates. *J. Dairy Sci.*, Vol.71, pp.1073.

Linford et al. (1976). The relationship between semen evaluation methods and fertility in the bull. *J. Reprod. Fertil.*, Vol.47, pp.283-291.

Manjunath & Sairam (1987). Electrophoresis, *Biochem. J.*, Vol.241, pp.685-692.

Manjunath et al. (1994). Major proteins of bovine seminal vesicles bind to spermatozoa. *Biol. of Reprod.*, Vol.50, No.1, pp.27-37.

Manjunath et al. (2002). Major proteins of bovine seminal plasma bind to the low-density lipoprotein fraction of hen`s egg yolk. *Biol. of Reprod.*, Vol.67, No.4, pp.1250-1258.

Manual para exame andrológico e avaliação de sêmen animal-CBRA (1998). Colégio Brasileiro de Reprodução Animal. 2ed. Belo Horizonte, Brasil, 49pp.

Martinez-Heredia et al. (2008). Identification of proteomic differences in asthenozoospermic sperm samples. *Hum. Reprod.*, Vol.23, pp.783-791.

McCauley et al. (1999). Purification and characterization of fertility-associated antigen (FAA) in bovine seminal fluid. *Mol. Reprod. Dev.*, Vol.54, pp.145-153.

Miller et al. (1990). Heparin-binding proteins from seminal plasma binding to bovine spermatozoa and modulate capacitation by heparin. *Biol. Reprod.*, Vol.42, No.899.

Mortarino et al. (1998). Two-dimensional polyacrylamide gel electrophoresis map of bull seminal plasma proteins. *Electrophoresis*, Vol.70, pp.797-801.

Moura et al. (2006a). Proteins of the cauda epididymal fluid associated with fertility of mature dairy bulls. *J. of Andrology*, Vol.27, No.4, pp.534-541.

Moura et al. (2006b). Identification of proteins in the accessory sex gland fluid associated with fertility indexes of dairy bulls: a proteomic approach. *J. of Andrology*, Vol.27, No.2, pp.201-211.

Nass et al. (1990). Male accessory sex glands produce heparin-binding proteins that bind to caudal epididymal spermatozoa and are testosterone dependent. *Mol. Reprod. Dev.*, Vol.25, No.237.

Nauc & Manjunath (2000). Radioimmunoassays for bull seminal plasma (BSP-A1/-A2, BSP-A3, and BSP-30-Kilodaltons), and their quantification in seminal plasma and sperm. *Biol. of Reprod.*, Vol.63, No.4. pp.1058-1066.

Ollero et al. (2000). Variation of docosahexaenoic acid content in subsets of human spermatozoa as different stages of maturation: implications for sperm lipoperoxidative damage. *Med. Reprod. Dev.*, Vol.55, pp.326-334.

Olson & Hinton (1985). Regional differences in luminal fluid polypeptides of the rete testis and epididymis revealed by two-dimensional gel electrophoresis. *J. Androl.*, Vol.6, pp.20-34.

Papa et al. (2008). Effect of seminal plasma removal on the viability of sperm from high and poor semen freezability bulls of artificial insemination center. *Acta Sci. Vet.*, Vol.36, pp.568.

Reyes-Moreno et al. (2002). Characterization of secretory proteins from culture cauda epididymal cells that significantly sustain bovine sperm motility *in vitro*. *Mol. Reprod. Dev.*, Vol.63, pp.500-509.

Roberts et al. (2003). Inhibition of capacitation-associated tyrosine phosphorylation signaling in rat sperm by epididymal protein Crisp-1. *Biol. Reprod.* Vol.6, pp.572-581.

Rodriguez-Martinez et al. (1998). Immunoelectroscopic imaging of spermadhesin AWN epitopes on boar spermatozoa *in vitro* to the zona pellucid. *Reprod. Fertil. Dev.*, Vol.10, pp.491-497.

Rutherfurd et al. (1992). Purification and characterization of PSP-I and PSP-II, two major proteins from porcine seminal plasma. *Arch Biochem. Biophys.*, Vol.295, pp.352-359.

Saacke et al. (1994). Relationship of semen quality to sperm quality, fertilization and embryo quality in ruminants. *Theriogenology*. Vol.40, pp.1207-1214.

Schönech et al. (1996). Sperm viability is influenced *in vitro* by the bovine seminal protein aSPF: effects on motility, mitochondrial activity and lipid peroxidation. *Theriogenology*, Vol.45, pp.633-642.

Schoneck et al. (1996). Sperm viability is influenced *in vitro* by the bovine seminal protein aSFP : effects on motility, mitochondrial activity and lipid peroxidation. *Theriogenology,* Vol.45, pp.633-642.

Setchell (1998). The Parkes lecture Heat and the testes. *J. Reprod. Fertil.*, Vol.114, pp.179-194.

Setchell (2006). The effects of heat on the testes of mammals. *Anim. Reprod.*, Vol.3, No.2, pp.81-91.

Shivaji et al. (1990). Proteins of Seminal Plasma, Wiley, New York, NY, USA, 526pp.

Sullivan et al. (2005). Role of exosomes in sperm maturation during the transit along the male reproductive tract. *Blood Cells Mol. Dis.*, Vol.35, pp.1-10.

Tedeschi et al. (2000). Purification and primary structure of a new bovine apermadhesin. *Eur. J. Biochem.*, Vol.267, No.20, pp.6175-6179.

Thérien et al. (1997). Major proteins of bovine seminal plasma modulate sperm capacitation by high-density lipoprotein. *Biol. Reprod.*, Vol.57, pp.1080-1088.

Thérien et al. (1998). Major proteins of bovine seminal plasma and high-density lipoprotein induce cholesterol efflux from epididymal sperm. *Biol. Reprod.*, Vol.59, pp.768-776.

Vogler et al. (1993). Effects of elevated testicular temperature on morphology characteristics of ejaculated spermatozoa in the bovine. *Theriogenology,* Vol.40, pp.1207-1219.

Walters et al. (2005). The incidence of apoptosis after IVF with morphologically abnormal bovine spermatozoa. *Theriogenology,* Vol.64, pp.1404-1421.

Walters et al. (2006). Assessment of pronuclear formation following *in vitro* fertilization obtained after thermal insulation of the testis. *Theriogenology,* Vol.65, pp.1016-1028.

Watson (1995). Recent development and concepts in the cryopreservation of spermatozoa and the assessment of their post-thawing function. *Reprod. Fertil. Dev.*, Vol.7, pp.871-891.

White (1993). Lipids and Ca2+ uptake of sperm in relation to cold shock and preservation: a review. *Reprod. Fertil. Dev.*, Vol.5, pp.639-658.

Yamakawa et al. (2007). Comparative analysis of fertile men with identification of potential markers for azoospermia in infertile patients. *J. Androl.*, Vol.28, pp.858-865.

Yanaginachi (1994). Mammalian fertilization. In: Knobil, E.; Nell, J.D., eds. *The Physiology of Reproduction,* New York, NY; Raven Press; pp.189-317.

Yeagle (1985). Cholesterol and the cell membrane. *Biochim. Biophys. Acta,* Vol.822, pp.267-287.

Chapter 11

Identification of Polymorphism in the Keratin Genes (KAP3.2, KAP6.1, KAP7, KAP8) and Microsatellite BfMS in Merino Sheep Using Polymerase Chain Reaction-Single Strand Conformational Polymorphism (PCR-SSCP) Analysis

Theopoline Omagano Itenge

Additional information is available at the end of the chapter

http://dx.doi.org/10.5772/45732

1. Introduction

Wool production is a major agricultural industry world-wide, the most important wool-growing countries include Australia, China, New Zealand, South Africa and countries within South America. In Australia for example, the world's largest producer of wool accounting for ~ 30% of the world production, wool industry is among the top industries in export revenue. While Australia has long been associated with the production of high-quality wool, the importance of this industry and the value of wool exports have been steadily declining.

1.1. Challenges facing the wool industry

The wool industry is faced with many challenges that require innovative solutions. The major competitors to the wool industry, cotton and synthetics, have developed new fibres that meet consumer needs such as being lightweight, soft and easy to care. These competitors have also made better productivity gains than wool, which has resulted in lower prices for all textile products. Today, there is much instability in wool prices, with a major problem facing the industry in faulty wool production. It has been observed that considerable variation exists both within and between fleeces across sheep breeds, as well as within inbred lines of sheep. Since the efficiency of wool processing is dependent on the consistency of wool fibre, it is of prime importance to wool producers that this variation is controlled. The wool characteristics that are of economic importance include fibre diameter (or fineness), grease and clean fleece weight, fleece strength and length, colour, yield, crimp and bulk. For Merino and halfbred wools, fibre diameter is the major factor that contributes to price variation as it significantly influences both fibre processing properties and ultimate product quality. The colour of wool is also important because superior colour (bright and white) can be dyed to the maximum range of shades and consequently is worth more than poorer coloured wool. Furthermore, the quantity of wool is important in overall wool production and in the efficiency of the production system.

1.2. Classical selective breeding – Not a simple solution

For many years, farmers have been using classical selective breeding, where by selection of breeding animals was traditionally based on the phenotype (that is appearance) of the individual animal, a rather slow method of selection. Each animal is assigned a breeding value (BV), which describes the future genetic potential of an animal. The BV is calculated by adjusting phenotype to exclude factors such as birth rank, lambing status and sex in order to give an estimate of the genetic merit. The desired goal of this strategy is the accumulation of "good" forms of genes for that particular trait in the population, over time. This has resulted in many breeds that are commercially important today. The domestic sheep *Ovis aries* today comprises over 500 different domestic breeds. However, wool characteristics, like many production traits (such as milk yield, growth rate, meat tenderness), do not exhibit simple Mendelian inheritance patterns (recessive or dominant). Instead, they are controlled by not only many genes, but also the interaction of these genes, each having small additive effects on the phenotype observed. Environmental and management factors also play a role. Thus, wool traits are quantitative and show continuous variation in phenotype, a fact that makes it difficult to deduce the genotype of an animal from its phenotype, and to relate genetic variation to differences in the phenotype. In other words, genetic improvement breeding programme select for "phenotypic superior" animals, without the knowledge of the actual genes that are being selected – which I will term as "**blind selection**" in this paper. Furthermore, other strategies to control environmental factors such as nutrition, time of shearing or mineral supplementation tend to be costly. In addition, wool production traits tend to only be fully expressed when an animal is mature, at least three years old, and therefore genetic progress using phenotypic selection and pedigree information is relatively slow.

1.3. Identification of gene markers: A possible solution

The answer to sidestepping this "**blind selection**", inaccuracy in describing the genetic potential of an animal and slow progress may lie in identifying specific genetic markers that are associated with wool production traits. Some sheep consistently produce quality or faulty wool, suggesting that genetic factors are an important key in determining wool characteristics. In addition, estimates for the heritability (h^2) of most wool traits are generally high ($h^2 = 0.3 - 0.6$), indicating that wool traits are under genetic control and that they can be selected for. A gene is a segment of DNA that provides the genetic information necessary to produce a protein. For almost all of the genes, there are two copies (alleles), one inherited from the mother and the other from the father. In any population of animals, there can be many different alleles. This is termed polymorphism or genetic variation. Polymorphism results from DNA mutation. It is this polymorphism that is taken advantage of, in order to identify genetic markers. A genetic marker for a particular characteristic can be defined as a piece of DNA that directly affects a phenotype and shows polymorphism. It can also be a piece of DNA that is closely linked to another piece of DNA that affects a phenotype. Genetic markers can either be genes or non-functional DNA segments such as microsatellites or minisatellites.

A number of different types of genetic markers are commonly used, including restriction fragment length polymorphisms (RFLPs), microsatellite and minisatellite DNA, and polymerase chain reaction-single strand conformational polymorphism (PCR-SSCP) variants. Restriction fragment length polymorphism results from the alteration of the restriction site(s) recognised by a specific restriction endonuclease or by the insertion or deletion of sequence between two restriction sites. The variation in fragment lengths is detected using gel electrophoresis. Although RFLPs were the first genetic markers developed, they are losing popularity as a screening method to identify genetic markers because they have the disadvantages of not identifying all of the polymorphism with a length of DNA, are time-consuming and restriction enzymes and consumables tend to be expensive. Simpler marker systems have subsequently been developed, many of these systems are now based on satellite DNA sequences.

Throughout the genome of higher eukaryotic organisms, there are a variety of different short DNA sequence repeats known as satellite DNA. These sequences do not code for protein and are highly variable from individual to individual in both the number and type of repeats (Groth *et al.*, 1987). Microsatellites are composed of DNA repeats in tandem at each locus. The tandem repeats are usually simple, and consist of either a single nucleotide or dinucleotide such as (CA)n, with each dinucleotide repeated about ten times. Minisatellites have longer repeated sequences than microsatellites, such as (ACTG)n. Since microsatellites and minisatellites show a substantial amount of polymorphism, they can serve as useful markers for the identification of genetic variation of value to animal breeding. Although the variation in the number of repeats can sometimes be detected using RFLP, PCR is generally used to amplify the polymorphic region and the amplimer analysed for length variation (a technique referred to amplified fragment length polymorphism – AFLP).

1.4. Polymerase chain reaction-single strand conformational polymorphism (PCR-SSCP) as a preferred type of genetic marker

PCR is also used in conjunction with SSCP. The PCR-SSCP technique offers a rapid, sensitive and relatively inexpensive way to screen for sequence variation with minimal sequencing. First described by Orita *et al.* (1989), this technique has become one of the preferred methods for screening samples to detect polymorphism because it is both simple and sensitive. In this techniques, regions of the gene of interest are amplified using PCR and the products denatured and then cooled rapidly to promote the formation of secondary structures due to internal base-pairing, which are in turn sequence dependent (Orita *et al.*, 1989). The folded single-stranded DNA molecules are separated by polyacrylamide gel electrophoresis under non-denaturing conditions. The folded secondary structures are affected by physical conditions such as temperature, percentage of polyacrylamide, ionic strength of the electrophoretic buffer, glycerol concentration (Spinardi *et al.*, 1991), ratio of acrylamide to bis-acrylamide, run length and run voltage. This can be exploited when optimising an SSCP protocol so that maximum variation can be detected in a given section of DNA. Molecules that differ by even a single nucleotide may form different conformers under a given set of conditions and, upon electrophoresis in a non-denaturing polyacrylamide gel, migrate differently. Many methods for viewing the folded DNA conformers have been described. These include the radioactive labelling of primers followed by autoradiography (Orita *et al.*, 1989), silver staining (Sanguinetti *et al.*, 1994), ethidium bromide staining (Yap and McGee, 1993) and more recently the use of fluorescently labelled primers and fluorescent dyes.

1.5. Methods used to identify genetic markers

There are several ways to identify genetic markers, but the two approaches most commonly used are the genome scanning or linkage analysis and the candidate gene approach. In the genome scan approach, the whole genome is searched to identify Quantitative Trait Loci (QTL) that affect any given trait. These are not necessarily the genes that are responsible for trait variation, but give an indication of where such genes may lie. Linkage analysis is an involved process. A map of the chromosomes, laying out the location, phase and order of genes and markers, and the distance between them, is required before linkage analysis can be performed. Firstly, a selection of about 200 markers distributed throughout the genome are genotyped, in the sire of the animals. Only the informative markers are genotyped in the progeny and each marker tested for suggestive linkage. Regions showing suggestive linkage are then studied by saturating the region with markers to identify those that are tightly linked. Phenotypic variation is then linked to the segregation of DNA markers within a population. Once the gene locus is identified by the tightly linked markers, the DNA can be sequenced. Linkage analysis can be an expensive and lengthy process requiring access to full chromosome libraries and arrays of markers.

In the candidate gene approach, known genes or gene markers that are thought to be responsible for the phenotypic variance of a trait are targeted for investigation. In this case, knowledge of the understanding of the genes that are likely to affect wool quality. The method requires a good knowledge of the physiological and biochemical processes of the gene product and can be a more direct method than the gene mapping approach, provided the right initial assumptions are made. One of the limitations of this approach is its "hit and miss" nature. A targeted gene may not be polymorphic in a population or genetic variation within the targeted gene may not affect the trait (Goddard, 2002). For the candidate gene approach to be useful, a quick and relatively inexpensive way to screen the target gene for polymorphism is essential.

The wool fibre is a complex structure composed primarily of proteins from the keratin family, which are the keratin intermediate-filament proteins (KRTs) and the keratin intermediate-filament associated proteins (KAPs). The KRTs form the skeletal structure of the wool fibre (microfibrils) and are embedded in a matrix of KAPs (Powell and Rogers, 1986), the different proteins being connected through disulphide cross-linkages (Powell, 1996). Therefore, genes that code for the KAPs and KRTs proteins are potential candidate genes in the identification of genetic markers associated with wool quality traits.

1.6. Half sib analysis

Half-sib analysis is a tool that allows genetic effects to be ascertained from field trials while controlling for environmental and management effects. Firstly, the gene being targeted must be polymorphic, with at least two alleles. A good sire is selected and mated to many ewes (at least 200 in number), that are selected at random from a range of environments, in order to maximize phenotypic variation in wool traits. The sire must be informative at each locus that is being investigated (i.e., the genotype of the sire must be heterozygous). If not, then the progeny does not get genotyped for those loci that the selected sire is homozygous. For those loci that the sire is heterozygous, the progeny born are genotyped soon after birth, and allowed to grow until their wool measurements can be taken at (12, 24 and 36 months of age). Suppose a sire has the genotype AB at the K33 locus, then all the progeny that have inherited the A allele from the sire are put in one group, and those that have inherited the B allele from the sire are put in another group. The means of the wool measurements from both groups are then compared. If the group of progeny that inherited the B allele from their sire are found to for example have a significantly stronger staple strength than those progeny that inherited the A allele from their sire, then this would give an indication that the K33 B allele might be associated with stronger staple strength.

1.7. Previously published association of genetic markers with wool traits

Numerous studies have described variation within both the KAP and KRT genes, including the work of Rogers *et al.* (1994a); Parsons *et al.* (1994a; 1996); McLaren *et al.* (1997); Beh *et al.* (2001); Itenge-Mweza *et al.* (2007). There are some reports associating variation in the KRT and KAP genes with variation in wool traits. Parsons *et al.* (1994b) and Beh *et al.* (2001) reported associations between variation in KAPs and mean fibre diameter in Merino sheep, while Rogers *et al.* (1994b) reported association between staple strength in Romney sheep and the region spanning the KAP1.1/KAP1.3/K33 loci on ovine chromosome 11. Itenge *et al.* (2009; 2010) reported association between variation in the KAP1.1 gene with variation in yield. In one of the half-sib families studied, variation in the K33 gene was associated with variation in staple strength. Markers, other than the KRT and KAP genes associated with wool traits have also been reported and these, together with reported keratin gene markers are summarised in Table 1.1.

1.8. Gel electrophoresis

Gel electrophoresis is the process in which an electrical current is applied to a gel to separate large molecules such as nucleic acids, from a mixture of similar molecules, based on differences or how they react to the electrical current. The technique relies on the fact that

Trait[1]	Analysis	Breed	Chromosome	Marker(s)	Gene(s)	Reference
Yield	Candidate gene Candidate gene	Merino Merino	11 11	- -	KAP1.1 KAP1.3	Itenge *et al.* (2009; 2010) Itenge *et al.* (2009)
FDSD	Candidate gene	Merino	11	-	K33	Itenge *et al.* (2009)
MFD	Candidate gene Genome Scan Genome Scan	Peppin Merino Merino Not specified	1 1 Not specified	- *Oar*DB6/RM65 Not specified	KAP6 and KAP8 - Not specified	Parsons *et al.* (1994a) Beh *et al.* (2001) Henry *et al.* (1998)
Staple strength	Candidate gene Candidate gene Candidate gene	Merino Romney Merino	11 11 11	- - -	K33 K33, KAP1.1 & KAP1.3	Itenge *et al.* (2009; 2010) Rogers *et al.* (1994a) Itenge *et al.* (2009)
Staple length	Candidate gene Candidate gene Candidate gene Segment mapping	Merino Merino Merino Synthetic breed INRA 401[2]	11 11 11 3, 7 & 25	- - - BMC1009, ILST005 & IDGVA8-IDGVA88	KAP1.1 KAP1.3 K33 - -	Itenge *et al.* (2009; 2010) Itenge *et al.* (2010) Itenge *et al.* (2009) Ponz *et al.* (2001)
CVD	Segment mapping	Synthetic breed INRA 401[2]	4, 7 & 25	McM218, ILST005 & IDGVA8-IDGVA88		Ponz *et al.* (2001)
Yellowness (Y-Z)	Candidate gene Genome Scan	Merino MRM[3] backcross	11 3	- TGLA77	KAP1.3 -	Itenge *et al.* (2009) McKenzie *et al.* (2001)
Brightness (Y)	Genome Scan	MRM[3] backcross	11	*Oar*FCB193	-	McKenzie *et al.* (2001)
Challenge colour	Genome Scan	Corriendale	11	*Oar*FCB193	-	McKenzie *et al.* (2001)
White colour	Candidate gene Candidate gene	Merino Merino	11 Not specified	- Not specified	KAP1.3 Not specified	Itenge *et al.* (2010) Benavides and Maher (2000)
CFW	Selected markers	Merino	Not specified	BfMS and DRB1	MHC region	Bot *et al.* (2003)

[1]MFD = mean fibre diameter; CVD = coefficient of variation of fibre diameter.
[2]A composite Romanov (prolific breed) and Berrichon du Cher (meat breed)
[3]MRM = Merino X Romney X Merino backcross

Table 1. Potential genetic markers for wool quality traits reported by various researchers.

nucleic acids are negatively charged because of the phosphate groups on the phosphodiester backbone of the nucleic acid strands (Nicholl, 1994). Nucleic acid molecules will migrate from the negative (black) terminal to the positive (red) terminal if put in solution and an electric field is applied, due to the net negative charge in solution. The gel matrix adds a sieving effect so that particles can be characterized by both charge and size.

Agarose is a macromolecular substance that is derived from seaweed. It can be purified to a whitish granular powder which, when mixed with water and heated, can be left to set like a jelly. This is called a gel and it acts like a sieve for the DNA molecules. To separate DNA molecules that are different lengths, agarose is used to produce a molecular sieve. The speed that the DNA travels through the gel is inversely proportional to the size of the DNA. In other words, small DNA particles migrate faster than large DNA molecules, as they are less physically restrained by the gel matrix. The length of a piece of DNA can be determined by comparing it to a molecular weight ladder. Agarose gel electrophoresis can be affected by:

1. The percentage of agarose, which affects the sieving of the DNA molecules.
2. The voltage applied during the electrophoresis, which cause the DNA molecules to move.

Typically, 1000 – 50,000 bp can be separated by 0.3% agarose, and 300 – 6000 bp can be separated by 1.4% agarose, while base pairs less than 500 are better separated using polyacrylamide gel, with gel percentage between 10-20. The polyacrylamide gel electrophoresis works under non-denaturing conditions.

After the electrophoresis is complete, the molecules in the gel can be stained to make them visible. Ethidium bromide, silver, or coomassie blue dye may be used for this process. Other methods may also be used to visualize the separation of the mixture's components on the gel. If the analyte molecules fluoresce under ultraviolet light, a photograph can be taken of the gel under ultraviolet lighting conditions, often using a Gel Doc. A molecular weight marker (MM) is often included on the gel to give an indication of the fragment size.

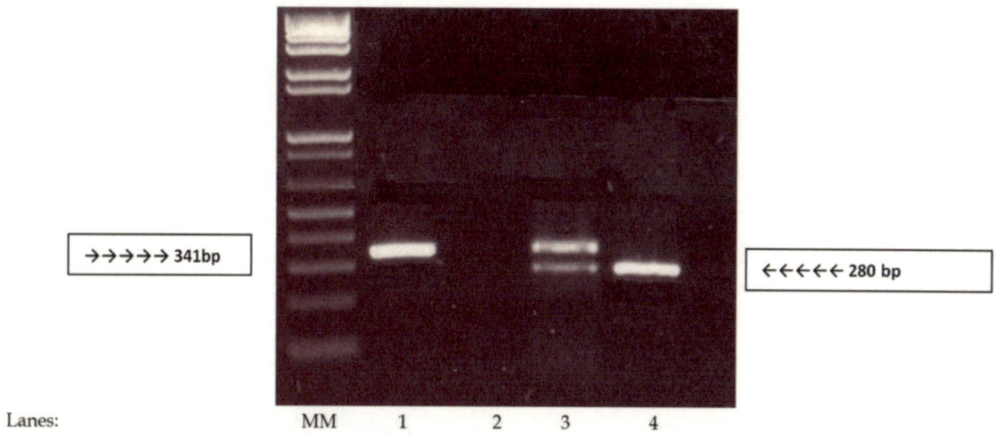

Figure 1. An example of a gel photo. MM is the molecular marker. Lane 1 has 341 bp DNA, while lane 4 has 280bp DNA. Lane 2 is blank.

1.9. Aim and objective of this paper

This paper discusses the identification of genetic variation in the KAP3.2, KAP6.1, KAP7, KAP8, KRT2.10 and BfMS loci in Merino sheep using polymerase chain reaction-single strand conformational polymorphism (PCR-SSCP) analysis. Polymorphism within these loci is likely to be in part responsible for the observed variation in wool characteristics and could result in the identification of gene markers to be used in gene marker-assisted selection programmes within the wool industry.

2. Materials and methods

2.1. Sheep used in the study

This study used two half-sib families referred to as Sire Line 1 (SL1) and Sire Line 2 (SL2). The SL1 half-sib was produced by mating a fine wool producer Merino ram to 150 Merino ewes, selected at random from a range of New Zealand environments, in order to maximise phenotypic variation in wool traits. In year one, the SL1 consisted of 131 pure New Zealand Merino lambs, with 128 of these surviving to the second shearing at 24 months. Following the second shearing the wether lambs and some of the ewe lambs were culled and only the remaining ewe lambs (n = 37) were shorn at 36 months of age. The SL2 half –sib consisted of

35 lambs (Merino x Romney ram x Merino ewes). Half-sib groups were kept as single flocks to minimise environmental variation between individual progeny and provide control. All lambs were tagged at birth to their dam and their gender and birth rank were recorded.

2.2. Wool shearing and sampling

Mid-side wool samples were collected at 12, 24 and 36 months of age for SL1 and at 12 months of age for SL2. Except for greasy fleece weight (GFW) which was determined at shearing, wool measurements were performed by the New Zealand Wool Testing Authority Ltd (NZWTA), Napier, New Zealand according to International Wool Textile Organisation (IWTO) standards. Measurements included comfort factor or the percentage of fibres of diameter greater than 30 μm (F<30), mean fibre diameter (MFD, IWTO-12-03), fibre diameter standard deviation (FDSD, IWTO-12-03), coefficient of variation of fibre diameter (CVD, IWTO-12-03) and curvature, were all measured using a Sirolan™ Laserscan Fibre Diameter analyser while the mean staple length (MSL, IWTO-30) and mean staple strength (MSS, IWTO-30) of each sample was determined using Automatic tester for Length and Strength (ATLAS). The colour (MY-Z) and brightness (MB) of the wool was measured using a reflectance spectrophotometer, where the tristimulus values Y-Z indicate the yellowness of the wool and the tristiulus value Y represents the brightness of the wool. The yield of wool, the weight of clean wool after impurities such as vegetable matter have been removed, expressed as a percentage of greasy wool weight was mathematically derived for the wool base (IWTO-19) measurements. Once yield measurements were obtained from the NZWTA, clean fleece weight (CFW) was calculated as the product of GFW and yield.

2.3. Blood sampling on FTA™ cards and DNA isolation

Blood samples (containing DNA) were collected from the progeny and sires onto FTA™ cards (Whatman, Middlesex, UK). These were stored at room temperature (See Figure 2.1). A small punch (1.2 mm in diameter) was taken from the blood on the FTATM cards using a Harris Micro Punch (Whatman International Ltd, UK) and put into a 200 μL tube. The DNA on the punches was isolated following a modified manufacturer's protocol. 200 μL of FTA™ reagent was added to each tube containing a 1.2 mm punch of FTA™ paper, containing the sample DNA. The tubes were incubated at room temperature for 60 minutes. Each tube was vortexed three times for about five seconds at the start of the incubation, half-way through the incubation, and after the incubation period. The FTA™ reagent was aspirated, and the cards were washed with 200 μL of TE buffer (1 M Tris and 0.5 M Na_2EDTA) for two minutes. The TE buffer was aspirated and the tubes were left open, but covered with a tube holder and stored at 4 °C and used for the subsequent PCR reaction.

2.4. Amplification of the loci using PCR

The PCR conditions for the loci that are described in the literature were initially used. However, re-optimisation was necessary for amplification in an i-Cycler PCR machine (Bio-Rad Laboratories Inc., Hercules, CA, USA). The PCR protocols were optimised by using a temperature gradient (to determine annealing temperature) coupled with a magnesium titration.

All the primer sequences used in the study were obtained from the literature (Table 2.1), and were synthesized by Invitrogen New Zealand Limited, Penrose, Auckland, New Zealand. PCR amplifications were performed in a reaction mixture containing ~ 50 ng of genomic DNA on a washed 1.2 mm punch of FTA™ paper, 1× PCR reaction buffer with 1U Taq polymerase (Qiagen, GmBH, Hilden, Germany). Table 2.2 lists the total reaction volume used along with the specific dNTP, primer, magnesium, and Q concentrations for each locus.

Amplification consisted of 1 minute denaturation at 95 °C, followed by 30 cycles of denaturation at 95 °C for 1 minute, annealing at temperatures specified in Table 2.3 for 1 minute and extension at 72 °C for 1 minute, with a final extension of 72 °C for 7 minute. All the primer sequences used in the study were obtained from the literature, and were synthesised by Invitrogen New Zealand Limited, Penrose, Auckland, New Zealand.

Figure 2. FTA™ cards of blood samples collected from the progeny of sire line 1

Locus	Primer Sequence	Source
Keratin genes		
KAP3.2	5'- CGAGACACCAAGACTTCTCTCATC-3' 5'- AGTGAGTGTTGAAGGCCAGATCAC-3'	McLaren *et al.* (1997)
KAP6.1	5'- CCAATGGCATGAAGGTGT-3' 5'- AAAAAGGGAAGGGTTGGTG-3'	McLaren *et al.* (1997)
KAP7	5'- CATCGGACAGCTTGAGGTAT-3' 5'- ACAGAGAATTGAGGGCGG-3'	McLaren *et al.* (1997)
KAP8	5'- GGACGCAACTGAGGACGCAACTG-3' 5'- ACACTTGGAATTCAATAAATATGTGTTGG-3'	Wood *et al.* (1992)
KRT2.10	5'- ATGGCCTGCCTGCTCAAGGAGTAC-3' 5'- CTTAGGACTGAGACTAGGATGAGG-3'	Rogers (1994)
Microsatellites		
BfMS	5'- CAACGGTCTGCAACCGAATTACC-3' 5'- CAATCCGTGGGTTGGAACACAA-3'	Bot *et al.* (2003)

Table 2. Primer sequences and source references for each locus investigated.

Locus	Total volume (μL)	Primer concentration (nM)	dNTP concentration (μM)	Mg^{2+} concentration (mM)	Q Conce-tration (×)
KAP3.2	25	350	175	1.0	1×
KAP6.1	25	400	200	1.0	1×
KAP7	25	350	175	1.0	1×
KAP8	25	350	175	1.0	1×
KRT2.10	20	400	200	1.0	-
BfMS	25	350	175	1.0	1×

Table 3. Optimised PCR conditions for each locus investigated.

Locus	Annealing temperature (°C)	Amplimer size (bp)
KAP3.2	58	424
KAP6.1	62	528
KAP7	63	413
KAP8	62	124*
KRT2.10	65	191
BfMS	58	200*

* = Variable length as amplification of a microsatellite region
# = Amplified with primers by Rogers *et al.* (1994b)

Table 4. Optimised annealing temperatures and predicted amplimer sizes for each locus investigated.

2.5. Agarose gel electrophoresis

Amplimers were analysed in 1.0% w/v SeaKem® LE agarose (FMC Bioproducts, Rockland, Maine, USA) gels prepared with 1× TBE buffer (89 mMTris, 89 mM orthoboric acid, 2 mM Na_2EDTA; pH 8) containing 0.1 mg/L ethidium bromide. Five μL of PCR product was added to 2.5 μL of loading dye (0.2% bromophenol blue, 0.2% xylene cyanol, 40% (w/v) sucrose) and the gels were electrophoresed at a constant 10 Vcm^{-1} for 30 minutes. A molecular weight marker (Invitrogen Life Technologies) was included on the gel to give an indication of the fragment size. DNA bands were viewed on a UV transilluminator (254 nm) and a photograph taken for records.

2.6. Optimisation of SSCP gels

PCR-SSCP conditions were available in the literature for KAP3.2 (McLaren *et al.*, 1997), however these were deemed to be insufficiently stringent. For this reason, the PCR-SSCP protocols used in this study were established empirically using template DNA from two small half-sib families (to observe inheritance of allele-specific banding pattern) and DNA samples of other unrelated Merino sheep (for increased genotypic variation). Many different gel conditions (gel percentage, voltage, time of running, temperature, addition of glycerol) were assessed to determine the optimum combination of conditions to resolve allele specific banding patterns in a reproducible manner. Amplimers from sires of the SL1 and SL2 and their selected progeny were also included on the optimising gels in order to ascertain allele banding patterns by following inheritance, and to determine whether the sires were heterozygous, and therefore informative, for the locus genotyped. Alleles were named in the order they were identified using letters of the alphabet.

2.7. Detection of sequence variation using PCR-SSCP

Each locus used specific SSCP gel conditions, and these are summarised in Table 2.4. Polyacrylamide (37.5:1 acrylamide / bis-acrylamide, Bio-Rad Laboratories, Hercules, Ca, USA) vertical gels (Protean II 16 x 16 cm, 1.0 mm thick spacers, 28 well comb, Hoefer, Inc., San Francisco, Ca, USA) were prepared containing 0.5× TBE (44.5 mMTris, 44.5 mM orthoboric acid, 1 mM Na_2EDTA [pH 8.0]) and polymerised using 10% ammonium persulphate and TEMED. Gels were pre-electrophoresed at running temperatures and voltage for one hour. Amplimers were mixed with 50 μL loading dye (95% formamide, 10 mM Na_2EDTA, 0.025% bromophenol blue, 0.025% xylene cyanol), denatured by heating at 95 °C for five minutes and immediately placed on wet ice before loading 15 μL aliquots. The gels were then electrophoresed at the optimum gel conditions with 0.5× TBE running buffer, followed by silver-staining according to the method of Sanguinetti *et al.* (1994).

2.8. Cloning of allele standards

For KAP3.2, KAP7 and KRT2.10 loci, genomic DNA was obtained from the sire and this DNA was amplified using the PCR conditions described above and the amplimers were

Locus	Gel % (37.5:1)[1]	Run length (hours)	Temperature[2](°C)	Voltage (V)
KAP3.2	8	17	30	250
KAP7	10	4	20	200
KAP8	10	4	20	200
BfMS	12	7	15	300

[1] Acrylamide/Bis-acrylamide ratio

[2] The vertical gel electrophoresis tanks were connected to a circulating water chiller to maintain a constant gel temperature. This is the temperature listed in the above table.

Table 5. Optimised SSCP conditions for the loci investigated.

subsequently cloned using the Promega pGEM® - T Easy Vector System I (Promega Corporation, Madison, WI, USA). Since each plasmid can only accept one molecule of DNA and therefore only one allele. Ligation reactions were performed in a total reaction volume of 10 μL containing three units T4 DNA ligase, 50 ng of plasmid DNA and 1× ligation buffer, and incubated overnight at 4 °C. Constructs were transformed into competent *E. coli* cells (Invitrogen™, One Shot™, INVαF') using the manufacturer's protocol. Sixty μL and 150 μL of the transformation mix were spread on labelled LB (0.5 % casein hydrolysate, 0.25 % yeast extract, 85.6 mM NaCl; pH 7.0) agar plates containing 100 μg/mL ampicillin that had been spread with 40 μL of 40 mg/mL X-Gal (BDH Laboratory Suppliers, Poole, England). The plates were incubated overnight at 37 °C. Six colonies for each representative allele were selected and cultured overnight in terrific broth (Invitrogen Corporation, Paisley, Scotland, UK), supplemented with 50 μg/mL ampicillin, for plasmid isolation. Colonies were screened for the correct alleles using a rapid boiling-PCR method, where by Fifty μL aliquot of an overnight culture (bacterial cells with gene of interest cultured in terrific broth) was centrifuged at 13,000 rpm for 2 minutes, the supernatant was discarded and 30 μL TE (1 M Tris, 0.5 M Na_2EDTA) buffer added, boiled for 10 minutes, centrifuged at 13000 rpm for 2 minutes and then 1μL of the supernatant was used as the template for the appropriate PCR. Amplimers were run on 2% agarose gels next to the original genomic PCR amplimers for comparison. Plasmid DNA was then isolated from clones, which had banding patterns corresponding to the original banding pattern seen from amplimers of genomic DNA, using the FastPlasmid™ Mini Kits (Eppendorf, Hamburg, Germany) following the manufacturer's instructions. These amplified plasmid DNAs were subsequently sequenced and used as standards for scoring unknown genotypes.

2.9. DNA sequencing

Plasmid standards were sequenced in the forward and reverse directions using the M13 forward and reverse primers at the Waikato University DNA Sequencing Facility, University of Waikato, New Zealand or Lincoln University Sequencing Facility, Lincoln, New Zealand. The sequences were compiled using DNAMANTM version 4.0 (Lynnon Biosoft, Quebec, Canada) and the electropherograms. To minimise the likelihood of PCR and sequencing errors, sequence data was derived from four separate colonies, at least two of which were from independent PCR amplifications. When sequencing data was consistent, the sequences were submitted to NCBI GenBank (http://www.ncbi.nlm.hih.gov). These were *Ovis aries* keratin intermediate-filament Type II (KRT2.10) gene: Accession number AY437406; *Ovis aries* high-sulphur keratin IF-associated protein 3.2 (KAP3.2) gene: Accession number AY483216 and *Ovis aries* high-glycine/tyrosine Type II keratin protein 6.1 (KAP6.1) gene: Accession number AY483217.

2.10. Statistical analyses

In order for any of the loci to be informative, they have to be heterozygous in the chosen sires allowing the segregation of the sire alleles to be followed in the progeny and segregation analyses performed. Segregation of the sire alleles within SL2 was observed and a chi-square goodness of fit test performed to ascertain whether the sire alleles inherited by the progeny occurred in a 1:1 ratio within the population. Any progeny which had the same genotype as both its sire and dam was excluded from the association analysis since it was not possible to determine which of the alleles had been inherited from the sire. The association of alleles of KAP8 with all measured wool traits (MFD, FDSD, CVD, curvature, yield, yellowness, brightness, comfort factor, staple length, staple strength, GFW and CFW) was then analysed for each year of phenotypic data using an analysis of variance (ANOVA) tests using SPSS version 13 (SPSS Science Inc., Chicago, IL, USA). The ANOVA model included sire allele and gender as factors and a full factorial model was used. The analysis used assumed that the ewe's alleles effects were distributed randomly in progeny. The date of birth was not included in the ANOVA because the progeny were half-sibs born in a five weeks period, and it was assumed that variation in birth date was balanced across the half-sib in the segregation analyses, and that none of the genes analysed had a significant effect on gestation length.

3. Results

Six loci (KAP3.2, KAP6.1, KAP7, KAP8, KRT2.10 and BfMS) were included in the study. All of them were amplified successfully using PCR and polymorphism was detected in three loci (KAP3.2, KAP8 and BfMS). Of the loci which were polymorphic, only KAP8 was heterozygous for SL2 (Tables 3.1), and thus potentially informative as a genetic marker. The remaining loci appeared to be homozygous in the sires, and thus uninformative. Table 3.2 shows the genotype of SL2 progeny at KAP8 locus.

Locus	No. of alleles detected	SL1 genotype	SL2 genotype	Informative[1] (Yes / No)
KAP3.2	3	AA	AA	No
KAP6.1	1	AA	AA	No
KAP7	1	AA	AA	No
KAP8	4	AA	AB	Yes[2]
KRT2.10	1	AA	AA	No
BfMS	3	AA	CC	No

[1]Heterozygous = informative; homozygous = uninformative
[2]Informative for SL2 only.

Table 6. Genotype results for the loci investigated in the study, indicating whether the sire genotype was informative (heterozygous) or non-informative (homozygous).

Lamb identity	Ewe identity	Lamb genotype
1027	86	AB
1028	162	AA
1029	57	BB
1030	114	AA
1031	59	AB
1032	59	AA
1033	105	AA
1034	89	AA
1035	51	AB
1036	49	AD
1037	120	AA
1038	56	AB
1039	65	AC
1040	47	AB
1041	119	BB
1045	77	BB
1046	155	AA
1048	14	BB
1049	130	AA
1050	161	AA
1051	84	BB
1052	150	BB
1053	17	BB
1054	NT	AA
1055	113	AB
1056	140	BB
1057	21	AB
1058	87	AA
1059	135	AB
1060	137	AA
1061	117	AC
1062	38	AB
1063	148	AB
1064	68	AB
1065	.	?
1067	116	AA
1068	58	?
1069	64	AA

[?]The genotype of the sheep could not be ascertained

Table 7. Genotype of KAP8 SL2 progeny

3.1. KAP8 (Polymorphic and informative in SL2)

Four banding patterns were identified for the KAP8 microsatellite amplimer using PCR-SSCP typing methods, and these were named A, B, C and D (Figure 3.1). The alleles were not sequenced. Mendelian inheritance was observed in SL2 half-sib family for KAP8 (Table 3.2). A Chi-square goodness of fit analysis to test whether the segregation of the sire alleles differed from a 1:1 ratio confirmed normal Mendelian segregation (Table 3.3).

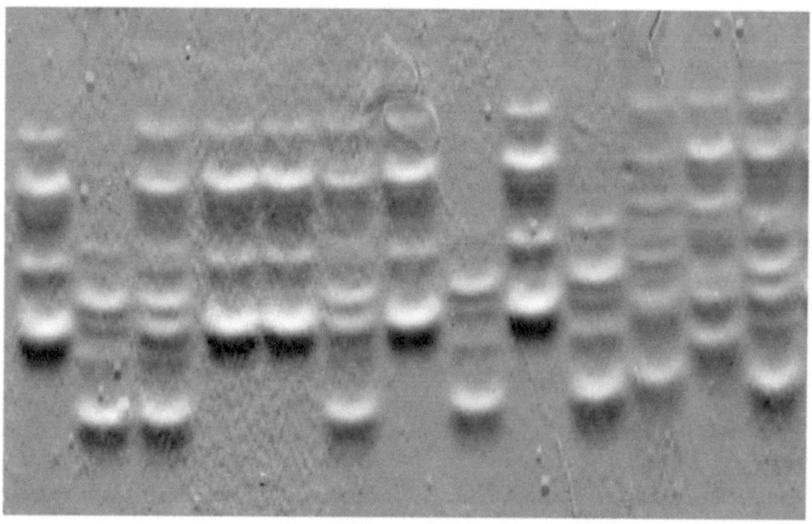

AA BB **AB** *AA AA AB AA BB AA BB AC AD AB*

Figure 3. PCR-SSCP of the 124 bp amplimer of the KAP8 microsatellite showing the four alleles identified. Amplimers were electrophoresed on a 10% non-denaturing acrylamide/bis-acrylamide gel for 4 hours, 200 V at room temperature (20 °C). Genotype of an individual animal is shown below each lane. SL2 genotype (AB) is bolded, and his randomly selected half-sib progeny are shown in italics.

SL2 Genotype	AB
Number of progeny inheriting allele A	17
Number of progeny inheriting allele B	12
Number of progeny genotyped same as the sire	5
Total number (n)	34
$\chi 2$	0.8621
P-value1	0.3532

Table 8. Segregation of sire alleles within SL2 sire-line. Chi-square goodness of fit was used for to ascertain whether the sire alleles inherited by the progeny occurred in a 1:1 ratio within the population. Probability values (*P* values) are given. [1]A *P*-value > 0.05 means that the allele segregation did not differ significantly from a 1:1 ratio.

3.1.1. Genotype

SL1 was homozygous at the KAP8 locus based on SSCP gel patterns, and hence uninformative. SL2 was heterozygous at the KAP8 locus, having the genotype AB. Eleven out of 36 progeny had the genotype AB (Table 3.2), which was the same as that of the sire. The genotype of the ewes for these lambs was subsequently determined. Five of the ewes genotyped as AB, and the progeny of these ewes were excluded from further statistical analysis as the allelic contribution from the sire could not be determined.

3.1.2. Association between segregating sire alleles and wool traits

The sire alleles at the KAP8 locus showed a Mendelian pattern of inheritance and segregated in a 1:1 ratio in the progeny of each half sib (Table 3.3). Statistical analyses within sire SL2 half-sib family showed that there were no association between the sire alleles (or gender) and variation of wool traits.

3.1.3. Power analyses

The number of differences between alleles within sire-lines which were not statistically significant suggested the possibility of Type II errors (failing to detect a difference when in fact there is one). To address this issue, a power analysis was conducted for each trait within each of the sire-lines to determine whether the sample sizes available were adequate to detect at least 10% differences between alleles, within each sire-line, at P<0.05 with 80% power, i.e. $n_{per\ allele} = (8 \times 2 \times \text{ERROR MEAN SQUARE}_{estimate}) / (0.1 \times \text{TRAIT AVERAGE}_{across\ sire\text{-}lines})^2$.

This equation was then rearranged to allowed the actual detectable difference to be calculated for each sire-line, i.e. % detectable difference = $[\sqrt{(8'2'\text{EMS}/n_{per\ allele})}\ /\text{TRAIT AVERAGE}_{across\ sire\text{-}lines}] \times 100$. A power analysis was performed for the KAP8 data. Wool trait measurements were only taken at 12 months for the SL2 half-sib family. There were inadequate SL2 progeny numbers (n=29) to detect a 10% difference between sire allele groups for yield, curvature, CVD, FDSD, staple length, brightness and yellowness (CFW and GFW were not measured). A comparison of the smallest detectable difference between sire-allele groups with the progeny numbers used with the observed difference between the sire-allele groups is shown in Table 3.4.

3.2. KAP3.2 and BfMS (Polymorphic, but uninformative)

KAP3.2 and BfMS were found to be polymorphic in the progeny used in this study, although they appeared to be homozygous for both sires used (Figures 3.2 and 3.3, respectively). This was confirmed with cloning and sequencing amplimers derived from sire SL1.

Sire-line	N_{lambs}	Trait[1]	Trait average[2]	EMS[3]	$N_{per\ allele}$ to detect at least a 10% difference	N_{lambs} required to detect a 10% difference
SL2	29	Prickle factor	1.61	1.75	1080	2159
	29	MFD	19.07	10.57	46	93
	29	FDSD	3.89	67.23	7103	14207
	29	CVD	20.46	91.34	349	698
	29	Curvature	94.58	0.45	0	0
	29	Yield	71.84	9.79	3	6
	29	Staple length	73.33	2.04	1	1
	29	Staple strength	31.30	12.58	21	41
	29	Brightness	69.92	0.27	0	0
	29	Yellowness	-2.83	135.71	27195	54389

[1]MFD: mean fibre diameter; FDSD: fibre diameter standard deviation; CVD: coefficient of variation of fibre diameter; GFW: greasy fleece weight; CFW: clean fleece weight. [2]Across all progeny measured. [3]Error Mean Square-taken from the ANOVA for each individual trait.

Table 9. Sample size required to detect at least a 10% difference between KAP8 sire allele groups in the wool traits list for each sire-line, at $P<0.05$ with 80% power.

Sire-line	N_{lambs}	Age (months)	Trait[1]	Trait average[2]	Smallest detectable difference (%)[3]	Difference observed between alleles (%)
SL2	29	12	Prickle factor	1.77	86.3	29.5
	29	12	MFD	19.21	17.9	3.2
	29	12	FDSD	3.91	221.3	1.9
	29	12	CVD	20.45	49.1	-0.4
	29	12	Curvature	97.04	0.7	6.4
	29	12	Yield	72.69	4.6	1.1
	29	12	Staple length	33.56	2.0	4.3
	29	12	Staple strength	69.06	11.9	22.6
	29	12	Brightness	-2.97	0.8	-2.1
	29	12	Yellowness	72.60	-433.1	7.1

[1]MFD: mean fibre diameter; FDSD: fibre diameter standard deviation; CVD: coefficient of variation of fibre diameter. [2]Across all progeny measured. [3]At 80%.

Table 10. A comparison of the smallest detectable difference between KAP8 sire-allele groups with the progeny numbers used and the observed difference between the sire-allele groups means for each wool trait measured.

3.3. KAP6.1, KAP7 and KRT2.10 (Non-polymorphic in SL1 and SL2)

Polymorphism could not be detected at the KAP6.1, KAP7 and KRT2.10 loci in any of the animals used in this study. KAP 7 was sequenced, and nucleotide sequences from SL1 KAP7 amplimer (GenBank accession number AY791846) was aligned with the published KAP7 gene by Kuczek and Rogers (1987); GenBank accession number X05638) which shows two unique sequences (Figure 3.4).

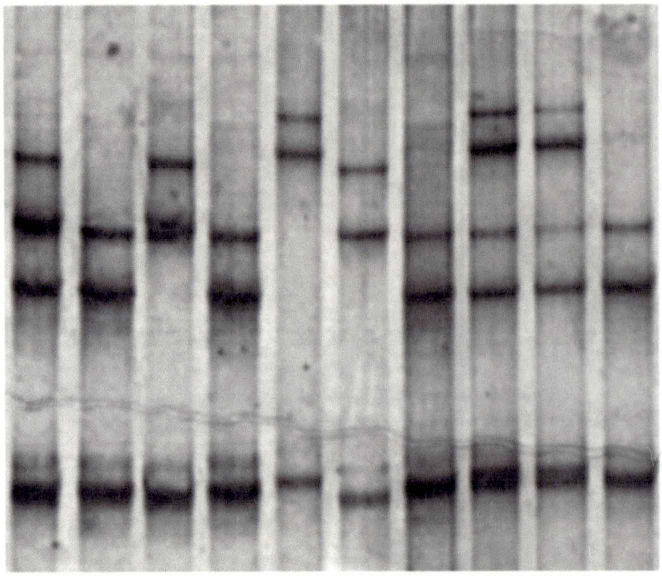

AB AA BB AA CC BB AA AC AC AA

Figure 4. PCR-SSCP analysis of the 424 bp amplimer of the KAP3.2 gene showing the three alleles identified (A, B and C). The genotype of an individual animal is shown below each lane.

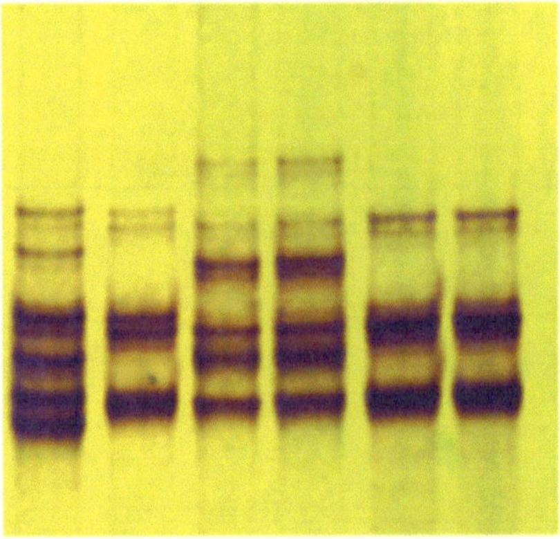

AC AA AB AB **AA** **AA**

Figure 5. PCR-SSCP analysis of the 200 bp amplimer of the BfMS microsatellite showing the three alleles identified (A, B and C) using a half-sib test family. The genotype of an individual animal is shown below each lane. Sires SL1 and SL2 genotypes are bolded.

```
KAP7 (AY791846)  CATCGGACAGCTTGAGGTATAAAAGG.TCCCGTGCAGGAC   39
KAP7 (X05638)    --------------------------t---------tt--   40

KAP7 (AY791846)  GAGAACTTCATTCCTTCTTGGGTAACTTGCTCTTCACATT   79
KAP7 (X05638)    ----------------------------------------   80

KAP7 (AY791846)  CTATCCAAATCCTTCCCACTCCTGCCACAATGACTCGTTT  119
KAP7 (X05638)    ----------------------------------------  120

KAP7 (AY791846)  CTTTTGCTGCGGAAGCTACTTCCCAGGCTATCCTTCCTAT  159
KAP7 (X05638)    ----------------------------------------  160

KAP7 (AY791846)  GGAACCAATTTCCACAGGACCTTCAGAGCCACCCCCCTGA  199
KAP7 (X05638)    ----------------------------------------  200

KAP7 (AY791846)  ACTGCGTTGTGCCCCTTGGCTCTCCCCTTGGTTATGGATG  239
KAP7 (X05638)    ----------------------------------------  240

KAP7 (AY791846)  CAATGGCTACAGCTCCCTGGGCTACGGTTTCGGTGGAAGC  279
KAP7 (X05638)    ----------------------------------------  280

KAP7 (AY791846)  AGCTTTAGCAACCTGGGCTGTGGCTATGGGGGCAGCTTTT  319
KAP7 (X05638)    ----------------------------------------  320

KAP7 (AY791846)  ATAGGCCATGGGGCTCTGGCTCTGGCTTTGGCTACAGCAC  359
KAP7 (X05638)    ----------------------------------------  360

KAP7 (AY791846)  CTACTGATGGACCATGGCTCCAGATGACTACGGG.ACCCG  398
KAP7 (X05638)    ----------------------------------a-----  400

KAP7 (AY791846)  CCCTCAATTCTCTGT                           413
KAP7 (X05638)    ---------------                           415
```

Figure 6. Alignment of the KAP7 gene sequence cloned from sire MV144-58-00 (Accession number AY791846) with Kuczek and Rogers (1987) published KAP7 gene (Accession number X05638). Upstream and downstream primers are underlined and the start and stop codons are bolded). Dashes represent same nucleotides to the nucleotide above and dots represent nucleotides missing in the other sequence.

Sex Average (cM)	Female (cM)	Male (cM)	Locus code	Marker	Marker description or associated gene
101.9	84.3	119.3	\BM4129	BM4129	Sequence – tagged site
104.1	86.6	122.0	\UCDO31	UCD031	RAPD Marker
107.1	89.0	125.2	\MCM58	MCM58	Microsatellite
111.3	91.7	130.5	\BL41	BL41	VANGL1
111.3	91.7	132.2	\BM723	BM723	STS
111.3	91.7	132.2	\BM723	BM723A	STS
113.8	92.9	133.9	\OARAE57	AE57	Microsatellite
123.4	105.2	142.2	\MCMA6	MCMA6AH	?
124.5	106.8	142.2	\MCMA6L	MCMA6AL	?
124.5	106.8	142.2	\BMS482	BMS482	Sequence – tagged site
124.5	106.8	142.2	\CSSM054	CSSM54	Phosphoglycerate dehydrogenase
126.0	107.9	143.2	PRPF3	BMS963	PRPF3 protein
126.0	108.9	143.2	ARNT	RME23	Aryl hydrocarbon receptor nuclear translocator
126.0	108.9	143.2	THH	TRHY	Trichohyalin
127.3	111.1	144.3	\RM065	RM65	Dinucleotide repeat
132.0	115.3	149.1	~CSAP033E	CSAP33E	Microsatellite
134.9	120.0	150.4	IGSF9	KIA1355	Immunoglobulin superfamily 9
135.7	121.5	150.4	ATP1A2	INRA6	ATPase
137.0	121.5	152.9	ADAMTS4	ADAMST4	ADAM metallopeptidase
139.8	122.7	156.8	\URB006	URB006	Sequence – tagged site
143.6	127.6	160.4	\BM6438	BM6438	Sequence – tagged site
143.6	127.6	160.4	OLIG2	OLIG2	Oligodendrocyte transcription factor 2
144.8	127.6	162.2	\SRCRS23H	SRCR23H	?
144.8	127.6	162.2	\TGLA49	TGLA49	Microsatellite
144.8	127.6	162.2	\DVEPC88	DVEPC88	Neu associated kinase
145.3	127.6	163.1	KRTAP7-1	KAP7HAP	Keratin associated protein 7.1
145.3	127.6	163.1	KRTAP7-1	KAP7_B	Keratin associated protein 7.1
145.3	127.6	163.1	KRTAP8-1	KAP8	Keratin associated protein 8.1
145.3	127.6	163.1	KRTAP7-1	KAP7_M	Keratin associated protein 7.1
145.3	127.6	163.1	KRTAP11-1	1-105	Keratin associated protein 11.1
145.4	127.6	163.2	KRTAP6-1	KAP6	Keratin associated protein 6.1
145.8	127.6	164.0	GRIK1	GRIK1	Glutamate receptor, ionotropic, kainite 1
149.6	131.5	168.1	APP	APPO10	Amyloid beta (A4) precursor protein
150.5	131.5	169.5	\BMS574	BMS574	Sequence – tagged site
150.5	131.5	170.2	\DVEPC117	DVEP117	Sequence – tagged site
150.5	131.5	170.2	\DVEPC117	DVEPC96	Sequence – tagged site
152.1	132.8	171.3	\BMS2321	BMS2321	Sequence – tagged site
153.2	132.8	173.1	\DVEPC128	DVEP128	Neural cell adhesion molecule 2
157.1	138.0	176.7	\RM095	RM095	Dinucleotide repeat
158.1	138.0	177.6	\MAF64	MAF64	Dinucleotide repeat
169.2	150.1	188.2	\ILSTS004	ILSTS04	Sequence – tagged site
171.1	152.6	188.2	\DVEPC54	DVEPC54	Microsatellite
174.4	154.8	194.2	\MCMA8	MCMA8	Sequence – tagged site
176.0	154.8	197.2	\MNS94	MNS94A	Microsatellite
193.1	169.9	216.0	\CSSM004	CSSM04	Microsatellite
195.3	171.1	219.6	\BMS4000	BMS4000	Sequence – tagged site
200.0	177.1	223.6	\UCDO46	UCD046	?

Figure 7. Linkage map for part of ovine chromosome 1 (modified from http://rubens.its.unimelb.edu.au/~jillm/jill.htm). The bolded genetic markers were investigated in this study.

Sex Average (cM)	Female (cM)	Male (cM)	Locus Code	Marker	Marker description or associated gene
149.6	151.4	148.8	\BMS695	BMS695	Sequence – tagged site
149.6	151.4	148.8	\BM827	BM827	Microsatellite
152.5	151.4	153.4	\MCM141	MCM141A	?
153.1	151.4	154.3	\OARSHP2	SHP2	Microsatellite
153.1	151.4	154.3	\ILSTS042	ILSTS42	Sequence – tagged site
154.1	151.4	156.3	\BMS424	BMS424	Sequence – tagged site
163.1	155.2	170.4	\BP1	BP1	Blood pressure QTL1
163.1	155.2	170.4	\DU469297	DU469297	?
165.6	160.5	170.4	\EPCDV025	EPCDV25	?
167.1	166.2	170.4	KITLG	SCF	KIT Ligand
168.7	166.6	172.4	\UCDO13	UCD013	?
170.7	166.6	174.3	KERA	KERA	Keratocan
170.7	166.6	174.3	LUM	CSAP19E	Lumican
177.8	172.1	182.4	\AGLA293	AGLA293	Microsatellite
179.4	174.9	183.5	~CSAP017E	CSAP17E	Microsatellite
179.4	174.9	183.5	\OARFCB5	FCB5	Dinucleotide repeat
179.4	174.9	183.5	GLYCAM1	GLYCAM1	Glycosylation dependant cell adhesion molecule
179.4	174.9	183.5	\OARHH38	HH38	Microsatellite
180.0	176.2	183.5	\ILSTS022	ILSTS22	Sequence – tagged site
180.0	176.2	183.5	RARG	RARG	Retinoic acid receptor 8
182.9	178.7	186.5	KRTHB*	KRT2.10	Keratin
183.9	181.2	186.5	KRTHB*	KRT2.13	Keratin
183.9	181.2	186.5	\BMC1009	BMC1009	Similar to intermediate filament type II keratin
186.3	181.2	190.9	\CABB011	CABB11	Genomic survey sequence
188.2	185.3	190.9	\CSSM034	CSSM34	Microsatellite
188.2	185.3	190.9	HDAC7A	KD103	Histone deacetylase 7A
188.2	185.3	190.9	\UCDO52	UCD052	?
195.5	188.6	201.0	\BL4	BL4	Bell-like homeodomain protein 4
197.0	190.4	202.8	LYZ	LYZ	Lysozyme
198.6	191.8	204.6	\CSRD2125	CSRD125	?
199.1	191.8	205.5	IFNG	KP6	Interferon gamma
199.1	191.8	205.5	IFNG	IFNG	Interferon gamma
199.1	191.8	205.5	IFNG	IFNGHAP	Interferon gamma
202.2	195.4	207.8	\BMS1617	BMS1617	STS
204.1	196.7	210.6	\OARVH34	VH34	Microsatellite
206.2	197.7	213.7	\BR2936	BR2936	Sequence – tagged site
207.0	197.7	215.5	\OARVH130	VH130	Microsatellite
207.0	197.7	215.5	\MAF23	MAF23	Microsatellite
209.0	199.0	218.1	\OARCP43	CP43	Microsatellite
214.7	208.6	219.8	\RM154	RM154	Tandem repeat region
218.5	211.3	223.7	IGF1	IGF1	Insulin like growth factor
218.5	211.3	223.7	IGF1	IGF1.B	Insulin like growth factor
218.5	211.3	223.7	IGF1	IGF1HAP	Insulin like growth factor
218.5	211.3	223.7	IGF1	CSAP40E	Insulin like growth factor
223.9	215.4	231.6	\CSRD2111	CSRD111	?
224.4	215.4	232.5	~CSAP009E	CSAP09E	?

Figure 8. Linkage map for part of ovine chromosome 3 (modified from http://rubens.its.unimelb.edu.au/~jillm/jill.htm). The bolded genetic markers were investigated in this study.

4. Discussion

Four alleles, designated A, B, C and D were identified at the KAP8 (CA)n repeat microsatellite locus using PCR-SSCP in this study. The microsatellite at the KAP8 locus was included in the study because this region is highly polymorphic, with 15 alleles previously reported (Wood *et al.*, 1992) using denaturing polyacrylamide gel electrophoresis, while Parsons *et al.* (1994a) detected four allelic fragments (123, 125, 133 and 139 bp) at the same locus using the methods by Wood *et al.* (1992) in a Merino half-sib family. Only SL2 was heterozygous at the KAP8 microsatellite in this study. SL1 was homozygous, despite the reported highly polymorphism in this locus (Wood *et al.*, 1992). The method used to detect polymorphism in this study differed to that of (Wood *et al.*, 1992), which used denaturing polyacrylamide gel electrophoresis. In this study, PCR-SSCP was used because this technique is simple, sensitive, relatively inexpensive and routinely used in the laboratory where the research was carried out. It is possible that if the original technique was employed, more alleles may have been observed at this locus.

Neither of the SL2 alleles were associated with variation in the wool traits that were measured (data not shown). The possibility of this locus having an affect on wool traits cannot be ruled out however, because only two alleles (that were the genotype of SL2) were analyzed, and that the sample numbers used in the study were relatively small (n = 29). Power analysis results (Table 9.5) showed that the observed differences between the sire allele groups were smaller than the smallest detectable difference for MFD, FDSD, CVD, curvature, yield, staple length, brightness and yellowness and therefore the possibility of making a Type II error (i.e. not detecting an association when there was one) is likely. Variation in MFD has previously been significantly associated with alleles at the KAP8 locus (Parson *et al.*, 1994a). The authors did not describe the alleles associated, and no sequence data was presented. Though alleles at the KAP8 microsatellite locus were not sequenced in this study, it is possible that SL2's alleles were different from those associated with differences in average MFD by Parsons *et al.*, (1994a).

Three alleles, designated A, B and C were identified at the KAP3.2 locus. However, both sire lines were homozygous, and thus uninformative. McLaren *et al.* (1997) identified two alleles at the KAP3.2 locus using PCR-SSCP methods. KAP3.2 (together with KAP1.1, KAP1.3 and K33) have been mapped to ovine chromosome 1 (Figure 3.5). Variations in all of the three genes (KAP1.1, KAP1.3 and K33) have been previously associated with variation in wool traits (Itenge *et al.*, 2009; Itenge *et al.*, 2010; Rogers *et al.*, 1994b). It is therefore suggested that sires that are heterozygous get investigated in further studies. Three alleles, designated A, B and C were identified at the BfMS microsatellite. Bot *et al.* (2003) reported eight alleles at the BfMS locus. Two of these alleles were significantly associated with CFW and GFW. However, both sire lines were homozygous, and thus uninformative at the BfMS locus.

Polymorphism could not be detected at the KAP6.1, KAP7 and KRT2.10 loci in this study, although all of these genes have been reported to be polymorphic in the literature (Parsons *et al.*, 1993; McLaren *et al.*, 1997). The reported polymorphism in KRT2.10 (two alleles) and KAP7 (four alleles) was identified using PCR-RFLP (McLaren *et al.*, 1997) whereas the polymorphism within KAP6.1 (two alleles) was revealed with PCR-SSCP of *Alu*I-digested PCR amplimers (McLaren *et al.*, 1997). Parsons *et al.* (1993) reported a diallelic polymorphism using *Bam*HI PCR-RFLP to give alleles designated A1 (24.5 kb) and A2 (14.1 kb). However, no sequence data was presented. Since only two KAP6.1 alleles have previously been reported, thus it was accepted that SL1 was homozygous at this locus without further sequencing although this locus could still be polymorphic which only sequencing would reveal. The KAP6.1 amplimers were also subjected to a variety of PCR-SSCP conditions in an effort to detect sequence variation. Digestion of the amplimer with *Alu*I or *Bam*HI as per McLaren *et al.* (1997) was not performed, however, and it is possible that this may have revealed variation at KAP6.1. KRT2.10 has been mapped to ovine chromosome three and two alleles have been reported at the KRT2.10 locus using a *Bsr*DI PCR-RFLP (McLaren *et al*, 1997). Genes coding for the KRT proteins are highly conserved during evolution (Powell, 1996; Marshall and Gillespie, 1982), and do not have much variation within them. Therefore, it was easy to accept that the KRT2.10 locus (with only two alleles) was likely to be homozygous. The fact that the KRT proteins are highly conserved during evolution (Powell, 1996; Marshall and Gillespie, 1982) suggests that genes coding for these proteins are intolerant to major changes and that they are very important to the integrity of the wool fibre.

5. Conclusion

Loci that were polymorphic, but uninformative in this study (KAP3.2, BFMS))need to be investigated further. Sires that are heterozygous at these loci need to be identified and used in half-sib analysis. Other loci that map to the same chromosome regions as the keratin genes investigated in this study are also worth of investigating in the future as potential gene markers for wool quality traits. On chromosome 1, future genes of interest include KAP11.1 and genes coding for trichohyalin (a very important wool follicle protein) (refer to Figure 3.5). On chromosome 3, loci of interest include KRT2.13, BMC1009 (Similar to intermediate filament type II keratin), RARG (Retinoic acid receptor 8) and IGF1 (insulin like growth factor) (refer to Figure 3.6). It is worth noting that previous studies by Damak *et al.* (1996) have shown positive effects of IGF1 on wool traits. Transgenic sheep produced by pronuclear microinjection with a mouse ultra-high-sulphur keratin promoter linked to an ovine IGF1 resulted in significant increase of CFW and bulk in transgenic sheep compared to non-transgenics, although MFD did not show significant differences (Damak *et al.*,1996).

There are other genes that have not been positioned on the linkage map that may be potential gene markers for wool quality traits. Some of these have already been associated with wool quality traits. These include the retinoic acid receptor (RARα) (Nadeau *et al.*, 1992), homeobox proteins (HOX2) (Nadeau *et al.*, 1992) and growth hormone (Hediger *et al.*,

1990). Retinoic acid induces expression of genes such as homeobox and KRTs and there is a possibility that retinoic acid is involved in the regulation of KAPs, given its genomic position on chromosome 11 (Parsons *et al*, 1994c). Growth hormone has been positioned on chromosome 11 through *in situ* hybridization (Hediger *et al.*, 1992). Furthermore, there have been numerous reports with variable effects of growth hormone on wool characteristics. For example, Ferguson (1954) and Johnson *et al.* (1985) observed significant increase in GFW during the injections of growth hormone. In contrast, no effect of recombinant growth hormone on CFW was found in a study by Zainur *et al.* (1989). Wheatley *et al.* (1966) found that growth hormone suppressed wool growth and that there was accelerated wool growth after withdrawal of growth hormone. Polymorphism at the genes encoding growth hormone have been reported (Valinsky *et al.*, 1990; Wallis *et al.*, 1998; Sami *et al.*, 1999), and different alleles of growth hormone may affect wool growth in different ways.

5.1. Genetic markers versus genetic engineering

The search for genetic markers affecting wool quality traits is very different to genetic engineering (GE) and transgenesis. While GE involves the manipulation or modification of genetic composition of an organism, and transgenesis requires the development and use of transgenic animals, the former detects changes within the genetic make-up of an organism, but does not alter it. Marker-assisted selection may therefore be better preferred within the wider "non-scientific" community, than the use of transgenic sheep to produce superior wool traits. Transgenesis in sheep is also still in its infancy, and successful transgenesis rates are very low (less than 13%) (Powell *et al.*, 1994). This makes marker-assisted-selection a more efficient, relatively cheaper and easier technique to improve wool quality traits than sheep transgenesis. The debate on GE will most likely continue and intensify especially where animals are involved. However, marker-assisted technology in livestock offers a powerful "green" alternative to gene manipulation.

5.2. Advantages of marker-assisted selection

Genetic markers are not affected by environmental noise and would allow sheep breeders to select animals with improved wool characteristics at an early age and cull the non-desirable lambs. This would speed up the process of genetic selection and decrease the generation interval. There is therefore a potential to select superior animals very early in life and not have to wait for an animal to reach its adult life to demonstrate that it has superior wool quality. This has the advantage of overcoming the limitation of "blind" selection, and increase the accuracy and efficiency of selection and result in a more profitable wool industry with direct benefits of cost to the consumer.

Author details

Theopoline Omagano Itenge
Department of Animal Science, Faculty of Agriculture and Natural Resources, University of Namibia, Namibia

Acknowledgement

I thank the Almighty God for everything in my life. I wish to thank my Honours supervisor, Prof. Jim Reynoldson from Murdoch University and my PhD supervisors (Prof. Jon G. H. Hickford from Lincoln University and Dr. Rachel Forrest from Eastern Institute of Technology). I am very grateful for the AUS-AID and NZAID scholarships that I received from the Australian and New Zealand governments, respectively. I am also very grateful for the Staff Development Fellowship award that I received from the University of Namibia.

6. References

Beh, K.J., Callaghan, M. J., Leish, Z., Hulme, D. J., Lenane, I. and Maddox, J. F. (2001) A genome scan for QTL affecting fleece and wool traits in Merino sheep. *Wool Technology and Sheep Breeding*. 49:88-9.

Benavides, M. V. and Maher, A. P. (2000) Quantitative genetic studies on wool yellowing in Corriedale sheep. II. Clean wool colour and wool production traits: Genetic parameter estimates and economic returns. *Australian Journal of Agricultural Research*. 51:191-196.

Bot, J., Karlsson, L. J. E., Greef, J. and Witt, C. (2003) Association of the MHC with production traits i 50:349-358.

Damak, S., Su, H., Jay, N. P., and Bullock, D. W. (1996) Improved Wool Production in Transgenic Sheep Expressing Insulin-like Growth Factor 1. *Biotechnology*. 14:185 – 188.

Ferguson, K. A. (1954) Prolonged stimulation of wool growth following infections of ox growth hormone. *Nature*. 174:411.

Goddard, M. E. (2002) Breeding Wool Sheep for the 21st Century. *Wool Technology and Sheep Breeding*. 50:349-358.

Groth, D. M., Taylor, L., Wetherall, J. D., Carrick, M. J. and Lee, I. R. (1987) Nature and applications of a DNA probe recognising biological individuality. In: *Merino Improvement Programs in Australia.* (Edited by McGuirk B. J.), Australian Wool Corporation, New South Wales, Australia.

Hediger, R., Johnson, S. E., Barendse, W., Drinkwater, R. D., Moore, S. S. and Hetzel, J. (1990) Assignment of the growth hormone gene locus to 19q26-qter in cattle and to 11q25-qter in sheep by in situ hybridization. *Genomics*. 8:171-174.

Henry, H. M., Dodds, K. G., Wuliji, J., Jenkins, Z. A., Beattie, A. E. and Montgomery, G. W. (1998) A genome screen for QTL for wool traits in a Merino X Romney backcross flock. *Wool Technology and Sheep Breeding*. 46:213-217.

http://www.ncbi.nlm.hih.gov

http://rubens.its.unimelb.edu.au/~jillm/jill.htm

Itenge, T. O., Hickford, J. G. H., Forrest, R. H. J., McKenzie, G. W. and Frampton, C. (2009) Improving the quality of wool through the use of gene markers. *South African Journal of Animal Science*. Vol. 39 no.1, pp. 219-223.

Itenge, T. O., Hickford, J. G. H., Forrest, R. H. J., McKenzie, G. W. and Frampton, C. (2010) Association of variation in the Ovine KAP1.1., KAP1.3 and K33 genes with wool traits. International Journal of sheep and wool science. 58:1-20.

Itenge-Mweza, T.O., Forrest, R.H.J., McKenzie, G.W., Hogan, A., Abbott, J., Amoafo, O. &Hickford, J.G.H. (2007) Polymorphism of the KAP1.1, KAP1.3 and K33 genes in Merino sheep. *M olecular and Cellular Probes*. 21: 338-342.

Johnson , I. D., Hart, I. C. and Butler-Hogg, B. W. (1985) The effects of exogenous bovine growth hormone and bromocriptins on growth, body development, fleece weight and plasma concentration of growth hormone, insulin and prolactin in female lambs. *Animal Production*. 41:207-217.

McKenzie, G., Maher, A. P., Dodds, K. G., Beattie, A. E., Henry, H. M., Hickford, J. G. H. and Montgomery, G. W. (2001) The search for quantitative trait loci affecting Wool Colour. *Proceedings of the New Zealand Society of Animal Production*. 61:104-108.

Kuczek, E. S. and Rogers, G. E. (1987) Sheep wool (glycine + tyrosine)-rich keratin genes. A family of low sequence homology. *European Journal of Biochemistry*. 166:79-85.

Marshall, R. C. and Gillespie, J. M. (1982) Comparison of samples of human hair by two-dimensional electrophoresis. *Journal of Forensic Science Society*. 22:377-385.

McLaren, R. J., Rogers, G. R., Davies, K. P., Maddox, J. F. and Montgomery, G. W. (1997) Linkage mapping of wool keratin and keratin-associated protein genes in sheep. *Mammalian Genome*.8:938-940.

Nadeau, J.H., Compton, J. G., Giguère, V., Rossant, J. and Varmuza, S. (1992) Close linkage of retinoic acid receptor genes with homeobox- and keratin-encoding genes on paralogous segments of mouse chromosomes 11 and 15. *Mammalian Genome*. 3:202-208.

Nicholl, D. S. T. (1994) An introduction to genetic engineering. Cambridge University Press, Cambridge.

Orita, M., Suzuki, Y., Sekiya, T. and Hayashi, K. (1989) Rapid and sensitive detection of point mutations and DNA polymorphisms using the polymerase chain reaction.*Genomics*. 5:874-879.

Parsons, Y. M., Cooper, D. W., Piper, L. R. and Powell, B. C. (1993) A *Bam*HI polymorphism in ovine glycine/tyrosine-rich type II (KRTAP) keratin sequences. *Animal Genetics*. 24:218.

Parsons, Y. M., Cooper, D. W. and Piper, L. R. (1994a) Evidence of linkage between high-glycine-tyrosine keratin gene loci and wool fibre diameter in a merino half-sib family. *Animal Genetics*. 25:105-108.

Parsons, Y. M., Piper, L. R., Powell, B. C. and Cooper, D. W. (1994b) Wool keratin gene polymorphisms and production characters in Australian Merino. *Animal Genetics*. 21:113-116.

Parsons, Y. M., Piper, L. R. and Cooper, D. W. (1994c) Linkage relationships between keratin associated protein (KRTAP) genes and growth hormone in sheep. *Genomics*. 20:500-502.

Parsons, Y. M., Cooper, D. W. and Piper, L. R. (1996) Genetic variation in Australian Merino sheep. *Animal Genetics*.27:223-228.

Ponz, R., Moreno, C., Allain, D., Elsen, J. M., Lantier, F., Lantier, I., Brunel, J. C. and Pérez-Enciso, M. (2001) Assessment of genetic variation explained by markers for wool traits in sheep via a segment mapping approach. *Mammalian Genome*. 12:569-572.

Powell, B. C. (1996) The keratin proteins and genes of wool and hair. *Wool Technology and Sheep Breeding*. 44:100-118.

Powell, B. C. and Rogers, G. E. (1986) Hair keratin: composition, structure and biogenesis, Volume 2. In: *Biology of the Integument.* (Edited by Bereiter-Hahn J., Matoltsy A. G. and Richards K. S.), pp. 695-721. Springer-Verlag, Berlin.

Powell, B. C., Walker, S. K, Bawden, C. S., Sivaprasad, A. V and Rogers, G. E. (1994) Transgenic sheep and wool growth: Possibilities and curent status. *Reproduction, Fetility and Development*. 6:615-623.

Rogers, G. R., Hickford, J. G. H. and Bickerstaffe, R. (1994a) A potential QTL for wool strength located on ovine chromosome 11. *Proceedings, 5th World Congress on Genetics Applied to Livestock Production, University of Guelph, Guelph, Ontario, Canada.* 21:291-294.

Rogers, G. R., Hickford, J. G. H. and Bickerstaffe, R. (1994b) Polymorphism in two genes for B2 high sulphur proteins of wool. *Animal Genetics*. 25:407-415.

Sami, A. J., Wallis, O. C. and Wallis, M. (1999) Production and characterisation of deletion mutants of ovine growth hormone. *Journal of Molecular Endocrinology*. 23:97-106.

Sanguinetti, C. J., Neto, E. M. and Simpson, A. J. G. (1994) Rapid silver-staining and recovery of PCR products separated on polyacrylamide gels. *Biotechniques*. 17:915-919.

Spinardi, L., Mazars, R. and Theillet, C. (1991) Protocols for an improved detection of point mutations by SSCP.*Nucleic Acids Research*. 19:4009.

Valinsky, A., Shani, M. and Gootwine, E. (1990) Restriction fragment length polymorphism in sheep at the growth hormone locus is the result of variation in gene number. *Animal Biotechnology*. 1:135-144.

Wallis, M., Lioupis, A. and Wallis, O. C. (1998) Duplicate growth hormone genes in sheep and goat. *Journal of Molecular Endocrinology*. 21:1-5.

Wheatley, I. S., Wallace, A. L, C. and Bassett, J. M. (1966) Metabolic effects of ovine growth hormone in sheep. *Journal of Endocrinology*.35:341-353.

Wood, N. J., Phua, S. H. and Crawford, A. M. (1992) A dinucleotide repeat polymorphism at the glycine- and tyrosine-rich keratin locus in sheep. *Animal Genetics*. 23:391.

Yap, E. P. H. and McGee, J. O. D. (1993) Nonisotopic discontinuous phase single strand conformation polymorphism (DP-SSCP): genetic profiling of D-loop of human mitochondrial (mt) DNA. *Nucleic Acids Research*.21:4155.

Zainur A. S., Tassell, R., Kellaway, R. C. and Dodemaide, W. R. (1989) Recombinant growth hormone in growing lambs: effects on growth, feed utilization, body and carcase characteristics and on wool growth. *Australian Journal of Agricultural Science.* 40:195-206.

Chapter 12

Temporal Expression of Isozymes, Alozymes and Metabolic Markers at the Early Ontogeny of *Prochilodus argenteus* (Characidae – Prochilodontidae) from São Francisco Basin, Três Marias, Minas Gerais, Brazil

Flavia Simone Munin, Maria Regina de Aquino-Silva,
Maria Luiza Barcellos Schwantes, Vera Maria Fonseca de Almeida-Val,
Arno Rudi Schwantes and Yoshimi Sato

Additional information is available at the end of the chapter

http://dx.doi.org/10.5772/45949

1. Introduction

It is well-known phenomenon that all changes in a population depend on reproduction, growth and mortality. Analysis of differential expression of genes which encode enzymes has made it possible to relate developmental changes at the molecular level to the general physiological changes whit accompany differentiation. According to [1], the specific protein expression during the different stages of development indicates the gene activity that can be started or turned off during embryogenesis. These properties make the multiple forms particularly interesting for the beginning ontogenetic gene regulation. Many enzymes exist as isozymes, and these isozymes are often differentially expressed during embryogenesis [2]. Thus, the enzymatic studies, including isozymes and allozymes must be informative about gene's activity and regulation during the early development. The electrophoretic and kinetics studies can be employed to investigate when genes are started during development, and how these enzymes are increased or reduced during ontogeny.

The Três Marias hydroelectric station was built in 1960s in the main canal of the São Francisco River in Minas Gerais state. According to [3] it has been observed that several migratory fish collected at downstream region closet to the dam, are smaller in size and have immature gonads during the spawning season. A distinct condition is observed 30 Km downstream from the dam, where these animals generally are normal-sized and have developed gonads. These facts reveal that conditions on this region are less favorable to they reproduction. There are two possible factors, among others caused by hydroelectric station, as a lower water temperature and oxygenation.

The Prochilodontidade family is composed by iliophagous-migratory fish, that swimming upstream to deposit their eggs every year in the end of dry season. *Prochilodus argenteus,* is an endemic fish specie, that migrate to the water spring for the spawning. It is called curimbatá, and today is endangered specie of fish from São Francisco basin.

2. Objective

The general objectives are to verify the correspondence among enzymatic levels, early ontogeny and physiological activities, as well as the correspondence between spatial and temporal enzymatic activities and their metabolic role, through electrophoretic and spectrophotometric studies, verifying the moment of gene activation, monomorphism or polymorphism of the enzymatic systems, synchrony or asynchrony of paternal and maternal genes, what kind of metabolism (glycolytic or aerobic) is predominant and changes in enzymes activities that occurring during the early ontogeny of *P. argenteus.*

3. Material and methods

- Five adults and mature couples of *P. argenteus* were induced to reproduce through a hypophysation process. After the gametes extrusion, fertilized eggs or larvae were collected at 0, 4, 8, 12, 16, 20, 36, 60, 87, and, 135 hours post fertilization (h.p.f.) and immediately iced. Temperature, solved oxygen and pH of water measured in each incubation tank, whit a Horiba U10 instrument. After the reproduction, the adults were sacrificed and, muscle, heart, and liver were collected and iced.
- Were performed electrophoretic analysis for alcoholic dehydrogenase (ADH, EC 1.1.1.1), glucose phosphate dehydrogenase (GPI, EC 5.3.1.9), glucose 6 phosphate dehydrogenase (G6PDH, EC 1.1.1.49), isocitrate dehydrogenase NADP-dependent (IDHP, EC 1.1.1.41), lactate dehydrogenase (LDH, EC 1.1.1.27); malate dehydrogenase (MDH, EC 1.1.1.37), malic enzyme NADP-dependent (ME, EC 1.1.1.38), 6-phosphogluconate dehydrogenase (6PGDH, EC. 1.1.1.44), and enzymes activities analysis for lactate dehydrogenase (LDH, EC1.1.1.27), malate dehydrogenase (MDH, EC1.1.1.37), piruvate kinase (PK, EC 2.7.40), and citrate synthase (CS, EC 4.1.3.7).
- A small piece of each tissue for adults or a few eggs or larvae was/were homogenized (w/v) in 50mM phosphate potassium buffer (pH 7.0), using a manual homogenizer and centrifuged at 27000g for 30 min at 4ºC in a Sorvall RC5B centrifuge. The resulting crude extracts were used for electrophoretic analysis. Electrophoreses were carried out employing a horizontal gels containing 13% (pH 8.7) (for GPI, G6PDH, LDH and, 6PGDH) or 14% (pH 6.9)(for ADH, IDHP, sMDH, mMDH, ME) corn starch prepared according to [4]. A voltage gradient of 5V/cm was applied for 12-14 h at 4ºC. After electrophoreses, the gels were sliced lengthwise and the lower halves incubated in a specific staining solution. The histochemical solutions used were described by [5] and modified by us, for G6PDH,

6PGDH, IDHP, LDH and ME. For ADH and MDH, we used the solution described by [6] with modifications and, for GPI [7]. Nomenclature for the gene loci and iso/allozymes was taken from [8]. Alleles were designated by number with **100* representing the most frequent allele. Subsequent numbers refer to their relative mobility.

- The activities enzymes analyses were performed using revised proceedings by [9]. Was used a GENESYS 2 spectrophotometer at 25°C for activities measures. After each assay the total protein was measured by Bradford method, using a wavelength 595 nm. The graphics were performed in Excel and Origin 6.0 by Windows program, were calculated median and stand deviation and the statistics Student's Test.

4. Results

Conditions of temperature, solved oxygen and pH of the incubation water are in table 1.

Offsprings	Temperature (°C)	Solved oxygen (mg/ l)	pH
1 (couple 1)	25.2	4.24	6.15
2 (couple 2)	25.1	4.27	6.12
3 (couple 3)	25.1	4.43	6.04

Table 1. Temperature, solved oxygen and pH measured of the incubation water in nursery tank.

5. Electrophoresis

Zimograms with phenotypes patterns detected from the couples of *P. argenteus* used for reproduction are in table 2 and figure 1.

	Couple 1		Couple 2		Couple 3	
	Male1	Female1	Male 2	Female 2	Male 3	Female 3
*GPI-A**	**100*	**100*	**100*	**100*	**100*	**100*
*GPI-B**	**-100/50*	**-100*	**-100/50*	**-150/50*	**-150*	**50*
*G6PDH**	**100*	**100*	**100*	**100*	**100*	**100*
*IDHP-A**	**100*	**100*	**100*	**100*	**100*	**100*
*IDHP-B**	**100*	**100*	**100*	**100*	**100*	**100*
*LDH-A**	**100*	**29/100*	**29/100*	**29/100*	**29/100*	**29/100*
*LDH-B**	**100*	**100*	**100*	**100*	**100*	**100*
*MMDH**	**100*	**100*	**100*	**100*	**100*	**100*
*SMDH-A**	**100*	**100*	**100*	**100*	**100*	**100*
*SMDH-B**	**100*	**100*	**100*	**100*	**100*	**100*
*ME-1**	**100*	**100*	**100*	**100*	**100*	**100*
*ME-2**	**100*	**100*	**100*	**100*	**100*	**124*
*6PGDH**	**100*	**88/100*	**100*	**100*	**88*	**100*

Table 2. Phenotypes detected in zimograms from the 3 couples of *P. argenteus* used for reproduction.

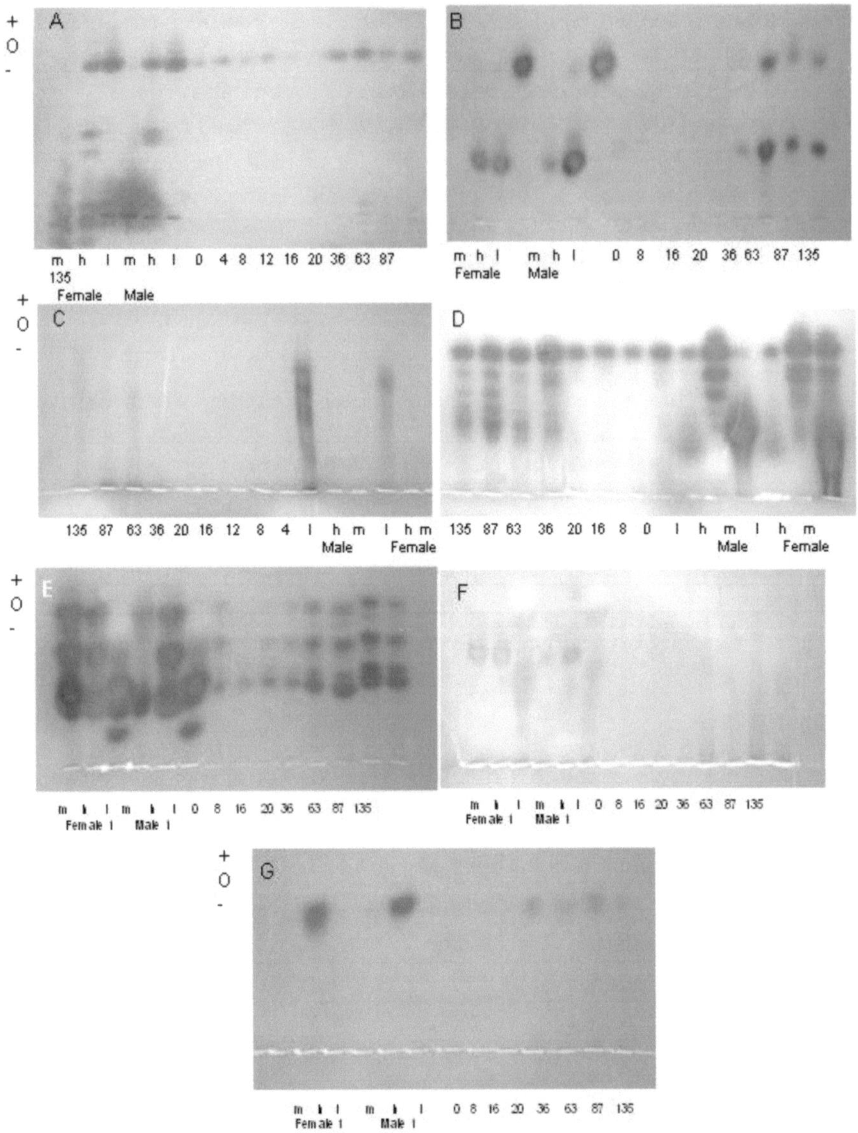

Figure 1. Electrophoretic patterns detected in couple 1 and offspring: A- GPI; B- G6PDH; C- IDHP; D- LDH; E- MDH; F- ME and G- 6PGDH. m - muscle; h - heart; l - liver.

The zimograms showed variant loci: *GPI-B** (**-100, *200* and, **-300*), 6PGDH* (**88* and **118*), *ME-1** (*124) and, *LDH-A** (**16* and **-16*). Another systems: G6PDH, IDHP and, MDH showed no variation. For the *G6PDH**, *IDHP-A**, *B**, *s-MDH-A**, *B**, *m-MDH** and *ME-1** and *ME-2** loci was not possible to detect the asynchrony or synchrony of gene expression, because the paternal and maternal phenotypes were identical or both had 1 common allele in heterozygotes (table 2).

6. Enzymes activities

The low high ratios L/H are presented in Figure 2. There were detected decreasing of these ratios during the development, what indicated decreasing of B subunits and/or A subunits synthesis. These subunits A synthesis were expected, because the muscle predominant subunit is type A, the product of *LDH-A** was detected late, 36 hours post fertilization, in heteropolymeric form.

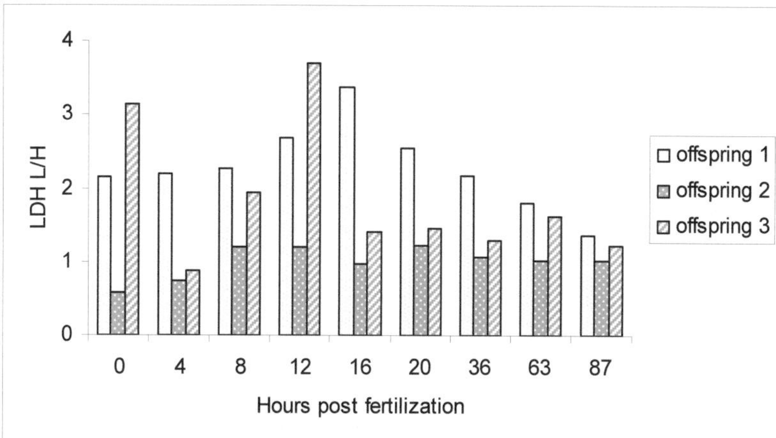

Figure 2. Low/high LDH ratios in different developing phases of offsprings of 3 couples of *P. argenteus*.

The obtained ratios between the glycolytic enzyme LDH and MDH from different phases of offspring developing from each pair, as well the medians calculated by the 3 couples together offspring showed the greater oxidative metabolism/malate aspartate shuttle until 87 hours post fertilization are in Figure 3.

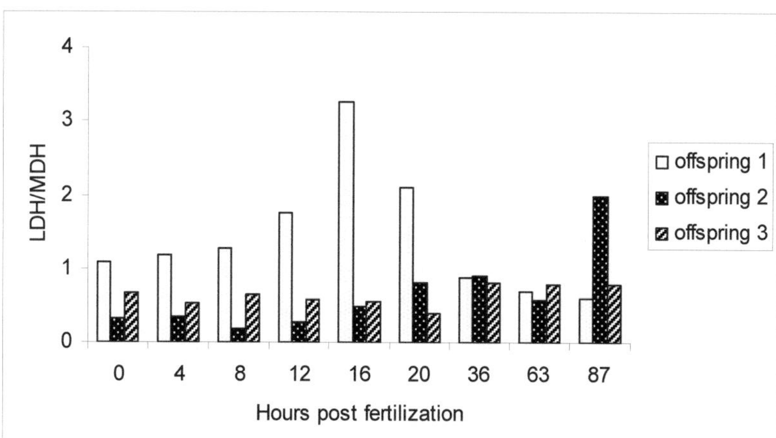

Figure 3. LDH/MDH ratios in different developing phases of offsprings of 3 couples of *P. argenteus*.

Differently of the observed for the LDH/MDH ratios, the LDH/CS observed ratios, for the 3 progenies, as well as the ratios obtained by the median of each enzyme for the 3 issues from the 3 couples presented all the values greater than one, and the greatest values in the stages between 12 and 16 hours post fertilization, decreasing after that (Figure 4). The PK/CS ratios and its medians observed were not great when compared with LDH/CS ratios, showing the maximum values at 16 hours post fertilization (Figure 5). The median value obtained for the 4 enzymes activities, showed (Figure 6).

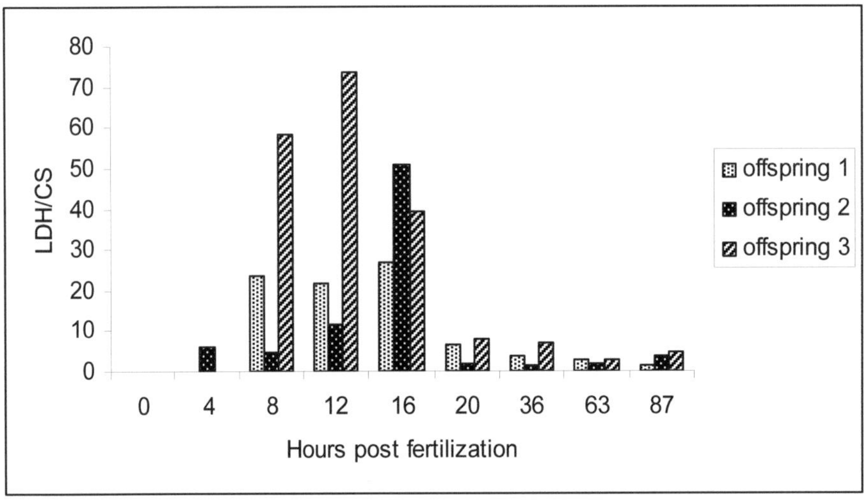

Figure 4. LDH/CS ratios in different developing phases of offsprings of 3 couples of *P. argenteus*.

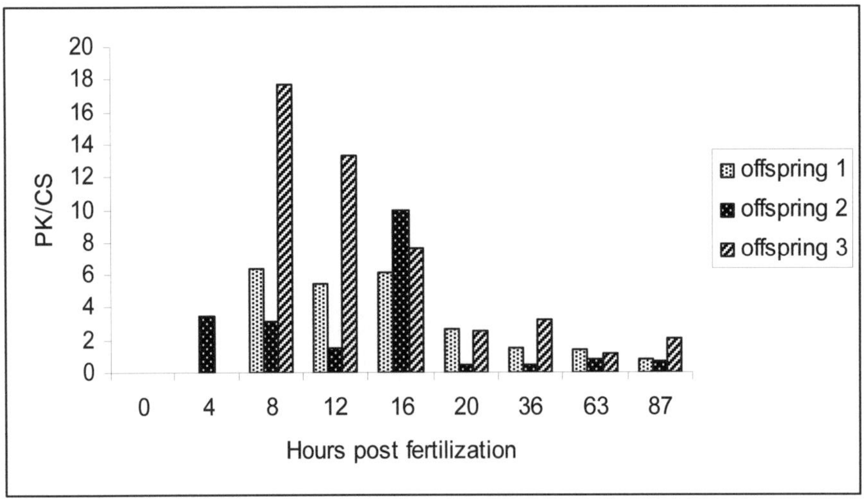

Figure 5. PK/CS ratios in different developing phases of offsprings of 3 couples of *P. argenteus*.

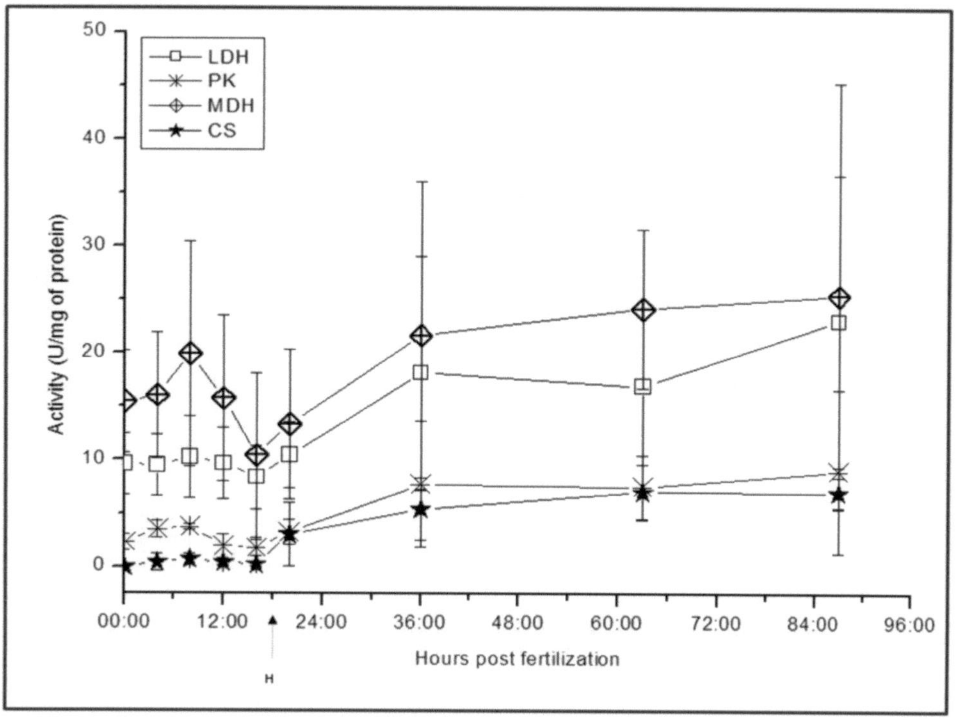

Figure 6. Quantitative glycolytics and e oxidatives average enzymes alterations in the early ontogeny of the offsprings of the 3 couples of *P.argenteus*, and H= hatching; 36 hours: starting *LDH-A**.

7. Discussion

Temporal iso/allozymes expression during early ontogeny of *P. argenteus.*

The investigation of gene expression in vertebrate's embryos indicated that developing program is determined in part by the transcripts of matter genes during the egg maturation and the other part by the expression of the embryos' genes after the fertilization [10].

The later *GPI** heteropolimeric detection at 16 hours post fertilization, indicates the both A and B subunits were simultaneously synthesized in many cells of embryos. These B homodimers were firstly detected, late at 36 hours post fertilization, like [11] also detected late expression for this locus in *L. cyanellus,* between 25 and 38 h.p.f.. According these authors, during the development of the *L. cyanellus,* the A homodimer levels were constant before the B subunits levels have been increased. According to [12] *P. scrofa,* detected the B homodimer activity after 12.5 hours after fertilization. According [13], A/B ratios change due the miotonic differentiation. This hypothesis can be accepted since the first skeletal muscle contractions were visible in *L. cyanellus, P. scrofa* and *P. argenteus* simultaneously at early subunits B expression.

The G6PDH from adults and their offspring presented one monomorphic locus liver restricted. In ontogenetic studies the single band was detected 36 hours post fertilization, revealing the late expression, like was observed by [11] in *L. cyanellus* at between 11 and 14 h.p.f.

The IDHP (IDH NADP^{+}- linked) catalyzing the carboxylation of 2-ketoglutarate, is not a strictly a Krebs cycle enzyme. Nevertheless, it is sometimes utilized by organisms to generate NADPH for fat biosynthesis. Under this conditions the carbon source for the reaction is usually glutamate, which is transaminated 2-ketoglutarate, the immediate substrate for the cell simultaneously generates isocitrate and thus helps to augment the pool of Krebs cycle intermediates (anaplerotic reactions) [14]. On the adult animals studied the expression of the IDHP loci was bidirectional divergent due the *IDHP-A** is skeletal and heart muscle predominant and the *IDHP-B** is liver restricted. The *IDHP-A** locus had an early expression while the *IDHP-B** had a late expression at 36 h.p.f.

On this study the LDH was product from two loci: the polymorphic *LDH-A** in skeletal muscle with two variant alleles **-16* and **29* and, the monomorphic *LDH-B** predominant in liver and heart. In the electrophoretic studies realized by the 3 offspring, the B subunit was the only one band observed in all stages of developing in the same level. This early expression of the B subunit was described for other teleosts as well as for other vertebrates [15; 16; 17]. The *LDH-A** loci product, the A subunit had been detected 36 h.p.f. in heteropolimeric form. Author [15] detected this locus activity at 34 h.p.f. it can be related to firstly muscle larvae contractions and has an important biochemical role, at anaerobic energetic modulation, that could be one requisite to muscle cells differentiation.

The sMDH-A* liver predominant, and muscle predominant sMDH-B* were non monomorphics and did not presented the typical temporal divergence, probably reflecting the heart loci expression, like detected by author [18] in P. scrofa. In L. cyanellus, the homodimer sMDH-A* and heterodimer sMDH-B were detected at the same time, early of development. B homodimer was visible at 28-31 h.p.f. [11].

There are two forms from vertebrate's malic enzyme (ME), the mitochondrial (less anodic mME) and citosolic (more anodic sME) [19]. In this work we called the ME-2 for mME and ME-1for sME . The ME-2 detected in muscle and heart appeared at 36 h.p.f. and the ME-1 liver-restricted was detected at 63 h.p.f. These enzymes provide the Krebs cycle with metabolic intermediates and play an important role in muscular gluconeogenesis *in situ* [20]. In the adult electrophoretic analysis the *ME-1** isoform product was detected only in liver, and the *ME-2** in heart and skeletal muscle, bidirectional divergent expression pattern. During the ontogeny both loci showed later expression: the *ME-2** at 36 hours post fertilization and the *ME-1** at 63 hours post fertilization.

The 6PGDH liver-restricted enzyme, showed a new variant allele, the **88*. And only one band for this locus was detected at 16 hours post fertilization. Thus there are two loci detected at 16 hours post fertilization: the *GPI-B** muscle predominant and *6PGDH** liver restricted, from glycolytic and 6-phosphogluconate pathways respectively. On this moment the larvae movements were stronger inside of the egg, two hours before the hatching. Were detected, at the early stage of development, enzymes from glycolysis represented by GPI and LDH, the Krebs cycle by IDHP, mMDH, sMDH, gluconeogenesis, and lipogenesis shuttle malate-aspartate in the mitochondria by sMDH and ME activities. The Krebs cycle, it is important to mention, not only provides the catabolism of energy compounds, but also serves in the formation of equivalent reducing required to other metabolic pathways in the α-ketoglutarate generation of anabolic ways precursors like as oxaloacetate, (precursors in the amino acids formation) and citrate (which can be diverted to the fatty acids synthesis). Only ME, whose principal function is the provision of equivalent to reducing the fatty acids synthesis did not take any of his two loci expressed since the beginning of development.

The ADH and SOD, enzymes related to detoxification, studied in adults, showed no embryonic activity.

With 16 (6PGDH) and 36 (G6PDH) h.p.f. has been detected by the activity of the 6-fosfogluconate, important NADPH generator which is used in the synthesis of fatty acids, and pentose for the synthesis of nucleic acids. Author [21] analyzing the initial ontogeny of the *Erymizon sucetta,* there ova and in the early stages of their development, high levels of G6PDH and 6-PGDH suggesting that much of the glucose available free at that time, be directed towards the pathway of 6-fosfogluconate.

In *P. argenteus,* the asynchrony from paternal alleles were detected within 2 loci only at crossings where the parents had not alleles present in mothers, *GPI-B** and *6PGDH** and maternal alleles in the *GPI-B** during *P. scrofa* ontogeny asynchrony during activation of paternal alleles for *GPI-B** was also detected [12]. Beyond this locus, [1, 15, 18] show asynchrony paternal in: *LDH-A**, *sMDH*-A* and *B** and *GPI- A** in *P. scrofa*. Author [22] found, in studies of early ontogenetic development of zebra fish, *Danio rerio,* that the expression of the mRNA of the way ADH3 of alcohol dehydrogenase is of maternal origin, there asynchrony of expression between maternal and paternal genes.

This some late expression loci could be a result of its regulation on molecular and cellular levels as proposed by [10] and [23]. The molecular events that lead to patterns of ontogenetic early enzyme and isozyme expression are under genetic control, direct or indirect, and represent the gene regulation in its broadest sense [24, 25]. Although this gene regulation occurs in the embryo, many genes appear to be cast from the first moments of the early ontogeny, ensuring that the basic enzymatic machinery is in operation in the most critical phases of the development process, in which the body just gets energy from the egg reserve and formed all the structures needed for their survival in the subsequent phases of life.

Apparently the asynchrony for father genes expression is more common that the mother because the egg or ovum brings a wealth of content cytoplasmic, and organelles. According

[26] cases of asynchrony are rare in offspring from couples of the same species, and can be caused by genetic changes in regulatory loci.

8. Enzymatic activities

According to [27], fat acids, glycogen and adenilic nucleotides (ATP, ADP e AMP) are the more important energetic substrates from mature egg.

The LDH low/high ratios calculated in the initial phases of the early ontogeny of *P. argenteus* (of 0 the 135 hours after the fertilization), average L/H values of each phase of the initial ontogeny of the 3 couples sample that in the interval of 0 to the 12 hours after fertilization occurred the biggest inhibition of the LDH and, therefore, the biggest amount of B subunits. After this period, occurs the reduction of this inhibition and reasons L/H, produced, probably, for the synthesis of subunits, whose detection occurs in electrophoresis 36 hours after the fertilization.

The analyses of the reasons between indicating enzymes of anaerobic metabolism and the MDH must be seen with exceptions when we try to measure the level of aerobic metabolism, a time that this enzyme is involved also in the malate-aspartate shuttle and, at least part of its activity, it must be related with this function. When we used laccolitic enzyme in the reason is the LDH, average values shows predominance of the oxidative metabolism / malate aspartate shuttle up to 87 hours after the fertilization, with its higher value with 8 hours after the fertilization when it is differentiated head extremity and the tail extremity. After this phase, considering the average reasons, an increase of the glycolytic metabolism occurred, and in the morphogenetic development, the movements if they intensify inside of the egg. When the glycolytic enzyme of the reason was the PK, in all phases and in the averages of these ratios we observed predominance of the oxidative metabolism and/ or malate-aspartate shuttle. The biggest reason was detected 36-87 hours after the fertilization when was detected an increase of the glycolytic metabolism.

The observed values in the offspring of the 3 couples of *P. argenteus,* for LDH/CS and average of these (even so not homogeneous, what it can have been consequence of the crude extract use), had disclosed predominance of the anaerobic metabolism in all the analyzed phases, being this, bigger between 12 and 16 hours after fertilization decaying after that. The values of PK/CS and average of these not so high how much of LDH/CS, they after show to its maximum (anaerobic metabolism) 16 hours post fertilization.

The water of the nursery used for the *P. argenteus* development, contained on average 4.3 mg oxygen/liter, more than what the water where the adult fish had been collected, therefore the water that comes of the barrage passes for aeration process before arriving at the nurseries. Thus, the higher enzymatic activity, in all the phases of the initial ontogeny of the 3 couples of *P. argenteus,* kept in these conditions, was of the LDH, characterizing the predominance of the anaerobic metabolism. The second more active enzyme was the

MDH what it would characterize high activity of the function malate-aspartate shuttle. The aerobic metabolism, characterized for the activity of the CS, lowest of the 4 analyzed, it was remained low. Soon after the hatching, the activities of 4 enzymes had suffered increase what it would be in accordance with the active movement of the larvae soon after hatching according to [28] with increase of the carbohydrates metabolism described for [29].

Studies carried through with the teleost *Misgurnus fossils* [30] show the dependence, during its initial development, of the use of the stored glycogen, as energy source. If it will be possible to surpass these results for other species of teleosts, this would be the only substratum of glycolysis in the embryos. According to [25], glycolysis and the phosphogluconate way are the more important energy sources for the biosynthetic activities and maintenance of the embryo morphology. The glucose is, doubtlessly, important in the period between the fertilization and after-hatching, as indicated for the high levels of B subunits from LDH and A subunits from GPI, as well as the high activity of the LDH, during this period and for the appearance of the delayed subunits, in *P.argenteus, P.scrofa* [1, 12, 15, 18] and in *E. sucetta* [21].

The subunits B predominance since the first phases of the initial ontogeny of *P.argenteus*, kinetically adjusted to the aerobic metabolism, in the lactate oxidase function, would supply, it would supply to private the shuttle function malate-aspartate of the MDH, enzyme with the highest activity detected here.

The adults of this species are migratory in time of piracema (November to February) and the São Francisco basin river rough relief and the fish have to use of essentially anaerobic metabolism (it has pulled out) to cross these obstacles.

Author details

Flavia Simone Munin
University of São Paulo, Pirassununga, São Paulo, Brazil

Maria Regina de Aquino-Silva
University of Paraíba Valley – UNIVAP, São José dos Campos, São Paulo, Brazil

Maria Luiza Barcellos Schwantes
Federal University of São Carlos, São Carlos, São Paulo, Brazil

Vera Maria Fonseca de Almeida-Val
National Institute for Amazon Research – INPA, Manaus, Amazonas, Brazil

Arno Rudi Schwantes
Federal University of São Carlos, São Carlos, São Paulo, Brazil

Yoshimi Sato
Hydrobiology and Pisciculture Station of CODEVASF, Três Marias, Minas Gerais, Brazil

9. References

[1] Fenerich-Verani, N. expressão gênica diferencial de sistemas isozímicos e alozímicos durante fases iniciais da ontogenia do Curimbatá, Prochilodus scrofa (Characiformes, Prochilodontidae). Tese de Doutorado. Departamento de Ciências Biológicas. Universidade Federal de São Carlos. São Carlos, SP, 1987.

[2] Markert, C.L. and Moller, F. (1959) Multiple forms of enzymes tissue, ontogenic and species specific patterns. Proc. Natl. Acad. Sci., 45: 753-763.

[3] Sato, Y., Miranda, M.O.T., Bazzoli, N. and Rizzo, E. (1995) Impacto do Reservatório de Três Marias sobre a Piracema à Jusante da Barragem. XI Encontro Brasileiro de Ictiologia, p.2.

[4] Val, A.L., Schwantes, A.R., Schwantes M.L.B. and De Luca, P.H. (1981) Amido hidrolisado de milho como suporte eletroforético. Ciênc Cult 33:992-996.

[5] Allendorf, F.M., Mitchel, N., Ryman, N. and Stahl, G. (1977). Isozyme Loci in Brown Trout (*Salmo trutta* L.): Detection and Interpretation from Population Data. Hereditas, 86: 179-190.

[6] Shaw, C.R. and Prasad, R. (1970) Starch Gel Electrophoresis of Enzymes: A Compilation of Recipes. Biochem. Genet., 4: 279-320.

[7] Coppes de Achaval, Z., Schwantes, M.L., Schwantes, A.R., De Luca, P.H. and Val, A,L. (1982) Adaptative Features of Ectothermic Enzymes. III. Studies on Phosphoglucose Isomerase (PGI) from Five Species of Tropical Fishes of the superorder Ostariophisy. Comp. Biochem. Physiol, 72 B: 201-214.

[8] Shaklee, J.B., Allendorf, F.W., Morizot, D.C. and Whitt, G.S. (1989) Genetic nomenclature for proteins-coding *loci* in fish: Proposed guidelines. Trans. Amer. Fish. Soc. 118: 218-27.

[9] Driedzic, W.R. and Almeida-Val, V.M.F. (1996). Enzymes of cardiac energy metabolism in Amazonian teleosts and fresh-water stingray (Potamotrygon hystrix). J Exp Zool 274:327-333.

[10] Davidson, E.H. (1976). Gene activity in early development. Academic Press, New York, 211 – 218; 223 – 230pp.

[11] Champion, M. J. and Whitt, G. S. (1976). Differential gene expression in multilocus isozyme systems of the developing sunfish. J. exp. Zool. 196, 263-282.

[12] Fenerich-Verani, N.; Schwantes, M.L.B. and Schwantes, A.R. (1990c) Patterns of gene expression during Prochilodus scrofa (Characiformes: Prochilodontidae) embriogenesis – III. Glucose-6-phosphate isomerase. Comp. Biochem. Physiol. 97B: 247-255.

[13] Dando, P. R. (1974). Distribution of multiple glucose phosphate isomerases in teleostean fishes. Comp. Biochem. Physiol. 47, 663-679.

[14] Peter W. Hochachka and George N. Somero (1984). Biochemical Adaptation. Princeton, University Press, New Jersey, 537 p.

[15] Fenerich-Verani, N.; Schwantes, M.L.B. and Schwantes, A.R. (1990a) Patterns of gene expression during Prochilodus scrofa (Characiformes: Prochilodontidae) embriogenesis – 1. Lactate dehydrogenase. Comp. Biochem. Physiol. 97B: 235-246.

[16] Schulte, P.M., Gómez-Chiarri, M., Powers, D.A. (1997). Structural and functional differences in the promoter and 5′ flaking region of LDH-B within and between populations of the teleost Fundulus heteroclitus, Genetics. 145(3): 759- 769.

[17] Ahmad, R. and A. Hasnain, (2005). Ontogenetic changes and developmental adjustments in lactate dehydrogenase isozymes of an obligate air-breathing fish Channa punctatus during deprivation of air access. Comp. Biochem. Physiol. Biochem. Mol. Biol., 140: 271-278.

[18] Fenerich-Verani, N.; Schwantes, M.L.B. and Schwantes, A.R. (1990b). Patterns of gene expression during Prochilodus scrofa (Characiformes: Prochilodontidae) embriogenesis – II. Soluble malate dehydrogenase. Comp. Biochem. Physiol. 97B: 247-255.

[19] Vuorinen, J. (1984). Duplicate loci for supernatant and mitochondrial malic enzymes in vendace, Coregonus albula (L.). Comp Biochem Physiol B., 78(1):63-6.

[20] Peter W. Hochachka (1980) Living without oxygen. Closed and open systems in hypoxia tolerance. Harvard University Press, New York.

[21] Shaklee, J.B., Champion, M.J., Whitt, G.S. (1974). Developmental genetics of teleosts: a biochemical analysis of lake chubsucker ontogeny. Dev Biol. Jun; 38(2): 356-382.

[22] Dasmahapatra, A.K., Doucet, H.L., Bhattacharyya, C. and Carvan, M.J. (2001). Developmental expression of alcohol dehydrogenase (ADH3) in zebrafish (Danio rerio). Biochemical and Biophysical Research Communications 286:1082-1086.

[23] Davidson, E. H. and Britten, R. J. Q. (1979) Regulation of gene expression: possible role of repetitive sequences. Science 204 1052-1059.

[24] Paigen, K. (1979). Gene factors in developmental regulation. pp. 1-61. In: Physiological Genetics, Edited by J. G. Scandalios, Academic Press, New York.

[25] John, G. Scandalios (1979) Control of gene expression and enzyme differentiation. In: Scandalios, J. G. Physiological genetics. New York: Academic Press, p. 63-107.

[26] Whitt, G. S. (1981) Developmental genetics of fishes: lsozymic analyses of differential gene expression. Am. Zool. 21, 549-572.

[27] Boulekbache, H. (1981). Energy metabolism in fish development. *Amer. Zool.* 21: 377 – 389

[28] Sato, Y.; N. Bazzoli; E. Rizzo; M.A. Boschi and Miranda, M.O.T. (2003). Impacto a jusante do reservatório de Três Marias sobre a reprodução do peixe reofilico curimatá-pacu (*Prochilodus argenteus*), p. 327-345. *In*: H.P. GODINHO and A.L. GODINHO (Eds). Águas, peixes e pescadores do São Francisco das Minas Gerais. Belo Horizonte, PUC-Minas, 458p.

[29] Yamauchi, T., Goldberg, E. (1974) Asynchronous expression of glucose-6-phosphate dehydrogenase in splake trout embryos. Dev Biol., Jul;39(1):63-8.

[30] Yurowitsky, Y.G.; Milman, L.S. (1973). Changes in enzyme activity of glycogen and hexose phosphate metabolism during oocyte maturation in teleost. Wihelm Roux. Arch. 177: 81-88.